The Materials Research Society Series

The Materials Research Society Series covers the multidisciplinary field of materials research and technology, publishing across chemistry, physics, biology, and engineering. The Series focuses on premium textbooks, professional books, monographs, references, and other works that serve the broad materials science and engineering community worldwide. Connecting the principles of structure, properties, processing, and performance and employing tools of characterization, computation, and fabrication the Series addresses established, novel, and emerging topics.

Atsushi Yoshizawa

Liquid-Crystalline Supermolecules

Hierarchical Structures, Symmetry Breaking, and Biological Activities

MATERIALS RESEARCH SOCIETY®

Springer

Atsushi Yoshizawa
Hirosaki University
Hirosaki, Japan

ISSN 2730-7360 ISSN 2730-7379 (electronic)
The Materials Research Society Series
ISBN 978-981-92-2523-1 ISBN 978-981-92-2524-8 (eBook)
https://doi.org/10.1007/978-981-92-2524-8

Cover credit: Marina Lohrbach_shutterstock

This Springer imprint is published by the registered company Springer Nature Singapore Pte Ltd.
The registered company address is: 152 Beach Road, #21-01/04 Gateway East, Singapore 189721, Singapore

Preface

Liquid crystals were discovered in 1888 and enabled the creation of large-screen flat-panel displays in the twenty-first century. This state of matter exhibits a combination of the softness of a liquid and the order of a crystal. These components can be readily aligned by an electric field, a property that renders them highly advantageous in displays employed in mobile phones, calculators, personal computers, and flat-screen televisions. Liquid crystals have become ubiquitous in our daily lives. This field has been the subject of extensive research in the context of soft matter, with new discoveries and applications continually being reported. The following are illustrative examples. The concept of chirality in the liquid crystal phase has undergone a transformation. The prevailing hypothesis posited that for liquid crystals to form helices or exhibit ferroelectricity, the molecules themselves must possess a chiral origin such as an asymmetric carbon. However, in 1996 it was reported that bent-core molecules lacking a chiral origin exhibit ferroelectricity, and subsequently spontaneous mirror symmetry breaking was repeatedly observed in achiral liquid crystal phases. In 2017, ferroelectric nematic liquid crystals were discovered. These discoveries stemmed from a molecular design that overturned the conventional wisdom that liquid crystal molecules are rod-shaped and lack a head-tail distinction. The existence of blue phases had been recognized since the advent of liquid crystals; however, their expression temperature range was exceedingly limited (approximately 1 K), thereby confining their utilization to the domain of fundamental research. However, in 2002, the expansion of the blue phase temperature range through polymer stabilization was reported, leading to rapid progress in applications for displays and photonic crystals. The blue phases are no longer regarded as a distinct liquid crystal phase; rather, they are considered a foundation for the development of soft functional materials. Since their discovery, liquid crystals have been noted for their connection to life. Liquid crystals have been identified in a variety of living systems. The parallels between liquid crystals and cell membranes have recently garnered significant interest, prompting research endeavors to explore the potential of medical applications of liquid crystals. This book delineates the progress in new worlds of liquid crystals. The focal point of this book lies in molecular design, a fundamental aspect that has led to the realization of these significant findings.

Chapter 1 of this book provides a foundation of knowledge regarding liquid crystals, which is essential for comprehending the subsequent material. Chapter 2 introduces liquid-crystalline supermolecules that exhibit hierarchical and/or dissipative structures. The molecular design process is informed by an understanding the dynamic behavior of molecules. In Chap. 3, the molecular behavior in the smectic phases was initially elucidated using ^{13}C nuclear magnetic resonance spectroscopy. In the following section, we delineate the characteristics of the ferroelectric smectic phase and the challenges inherent in its practical application. This is followed by a discussion of the discovery of ferroelectric nematic liquid crystals and the recent research trends in this field. Chapter 4 provides an overview of the structural characteristics, materials used, and display functions of the blue phases. The text provides comprehensive coverage of amorphous blue phase III, the structure of which has recently been elucidated. In Chap. 5, the reader is introduced to the process of constructing chiral higher-order structures in achiral systems. This mechanism facilitates the occurrence of homochirality in natural phenomena. In the final Chap. 6, after introducing the relationship between biological tissues and liquid crystals, we proceed to a discussion of liquid crystal sensors, drug delivery systems, and the antitumor effects of liquid crystals.

It is recommended that readers first read Chap. 1, and then proceed to any chapters that pique their interest. Extensive use has been made of figures to facilitate reader comprehension. The author would be delighted if readers could sense how these designed molecules open up new worlds.

This book comprises a series of studies conducted in the author's laboratory at Hirosaki University. I would like to express my profound gratitude to all of my students and collaborators, whose intellectual contributions and unwavering dedication have been instrumental in the success of this endeavor. Additionally, the author expresses gratitude to the Japan Science and Technology Agency, the Japan Society for the Promotion of Science, and Hirosaki University for their financial support. I would like to express my gratitude to Mr. Shinichi Koizumi and all the staff at Springer Nature for their patient support in the writing of this book.

Hirosaki, Japan Atsushi Yoshizawa

Competing Interests The author has no competing interests to declare that are relevant to the content of this manuscript.

About This Book

This book explores the world of liquid crystals created by strategically designed supermolecules. Addressing fundamental aspects of liquid crystal material science, it explores new applications and highlights the interdisciplinary nature of liquid crystals, which range from information displays to medical uses. These new worlds enable polar order in anisotropic fluids, alignment-free high-contrast liquid crystal displays, and chiral materials generated in an achiral system. They also enable selective antitumor activity. Additionally, the book presents significant basic scientific findings, such as dissipative structures through fluctuations and spontaneous mirror symmetry breaking. Topics include hierarchical and dissipative structures, ferroelectric liquid crystals, blue phases, supramolecular chirality transfer, and biological liquid crystals. An introduction to liquid crystals is included to facilitate understanding of the aforementioned cutting-edge fields. Distinguished by its unique perspective and insight on an emerging topic, this book serves as an authoritative guide for researchers and graduate students.

Contents

About the Author

Atsushi Yoshizawa was born in Niigata, Japan in 1956. He earned his Doctor of Engineering degree in synthetic chemistry from Kyoto University. In 1985, He joined the Central Research Laboratories at Nippon Mining Co., Ltd. In 2000, He joined Hirosaki University as a full professor. He served as Dean of the Graduate School of Science and Technology and then as Executive Director of Planning and Vice President before retiring from Hirosaki University in 2022. He is now a professor emeritus at Hirosaki University. He received the Best Paper Award at the 14th International Display Workshops in 2007, as well as the Japanese Liquid Crystal Society Awards for Outstanding Achievement in 2012 and Distinguished Achievement and Contributions in 2023. His research interests include liquid-crystalline supermolecules with hierarchical structures, blue phase liquid crystals, ferroelectric liquid crystals, chirality in liquid crystals, and liquid-crystalline medicine.

Chapter 1
Introduction to Liquid Crystals

Abstract Liquid crystals exhibit the anisotropy characteristic of crystals and the softness associated with liquids. This unique state of matter is referred to as the "fourth state," distinguishing it from the conventional phases of crystals, liquids, and gases. Molecules in the liquid crystal state exhibit a high degree of sensitivity to external stimuli, such as electric or magnetic fields, which can prompt a change in their orientation. This, in turn, results in the manifestation of distinct physical properties. The application of liquid crystals to display materials paved the way for the development of flat panels, which utilize liquid crystal displays. Conversely, liquid crystallinity has been linked to biological function. In this chapter, an overview of liquid crystals is given, followed by an introduction to their formation, characterization and electro-optical effects. These introductory descriptions are designed to facilitate the reader's comprehension of liquid crystals in general and to provide the necessary background knowledge for the subsequent chapters.

Keywords Phase transition · Lyotropic liquid crystals · Thermotropic liquid crystals · Characterization

1.1 Overview

Liquid crystals combine the optical anisotropy of solids with the fluidity of liquids. Therefore, liquid crystals change their orientation state due to external fields such as electric and magnetic fields, and exhibit different physical properties depending on their orientation. The liquid crystal state was discovered by the Austrian botanist Friedrich Reintzer in 1888. He synthesized cholesteryl benzoate by reacting cholesterol extracted from carrots with benzoic acid. It was observed that the substance underwent a transition from a crystalline state to an opaque liquid at 145.5 °C and subsequently to a transparent liquid at 178.5 °C. German physicist Otto Lehman demonstrated that this opaque state is physically unitary, fluid, and optically anisotropic. A comprehensive account of the discovery of liquid crystals can be found in the seminal work by Dunmur and Sluckin [1]. The dynamic scattering

A. Yoshizawa, *Liquid-Crystalline Supermolecules*, The Materials Research Society Series, https://doi.org/10.1007/978-981-92-2524-8_1

mode in the liquid crystal state was introduced in the 1960s, and this was followed by the practical application of liquid crystal displays in electronic desktop calculators. Subsequent to the year 2000, they have been extensively utilized as flat panel displays for personal computers and televisions. Furthermore, it was determined that high-strength fibers could be obtained by subjecting the fibers to spinning in a liquid crystal state. These fibers were subsequently utilized in the field of engineering plastics.

In the crystalline state of a material, long-range order exists in the position and orientation of the constituent molecules. External factors, such as heating, cause the loss of their two orders, resulting in an isotropic liquid. In the liquid crystal state, the order of position is partially or completely lost, while the order of orientation remains intact. Liquid crystal is a term employed to describe a state of matter analogous to crystal or liquid. However, it is also occasionally utilized to denote a material that exhibits a liquid crystal state. Conversely, a state in which positional order persists yet orientational order is absent is designated as a plastic crystal. The formation of the liquid crystalline phase is influenced by two factors. One such example is lyotropic liquid crystals, which are defined by the appearance of the liquid crystalline phase under the action of a solvent, typically an amphiphilic molecule such as phospholipid or surfactant. The second category comprises thermotropic liquid crystals, which are produced by means of temperature change. Depending on their molecular structure, they are broadly classified into three categories: rod-shaped molecules, disk-shaped molecules, and main-chain or side-chain polymers. Typical molecular structures showing the liquid crystalline phase are illustrated in Fig. 1.1.

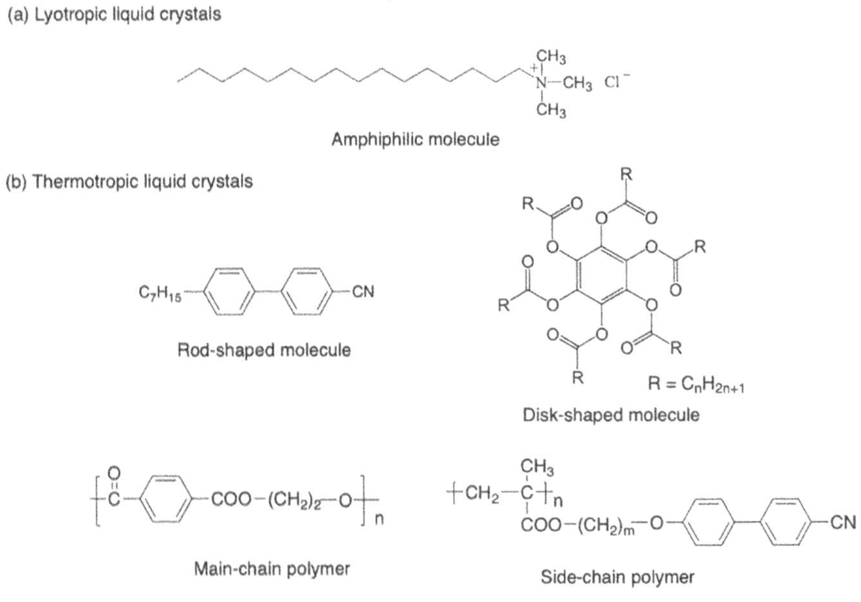

Fig. 1.1 Molecular structures showing **a** lyotropic and **b** thermotropic liquid crystals

As illustrated in Fig. 1.2, the phase transition of 4-cyano-4′-hepthylbiphenyl (**7CB**) exhibits a specific sequence of events [2]. In the solid state, there is long-range order in position and orientation. Upon heating, the solid undergoes a phase transition at 30 °C, transforming into a cloudy white liquid. This phenomenon is known as the crystal–nematic liquid crystal phase transition. Positional order is completely lost, but orientation order remains. Subsequent heating of the sample results in a phase transition to a transparent liquid phase. In the liquid state, both position and orientation orders are completely lost. The temperature at which the substance undergoes a phase transition from a crystalline state to a liquid crystal state is designated as the melting point. The temperature at which a liquid crystal transitions to a liquid state is designated as the clearing temperature.

As illustrated in Fig. 1.3, the relationship between Gibbs energy and temperature is demonstrated for three distinct states: crystal, liquid crystal, and liquid. The gradient of the *G*–*T* line exhibits state-dependent variation, with an increase in slope as the state transitions from crystal to liquid crystal and subsequently to liquid. It has been demonstrated that, under conditions of heating, the energy of the crystal *G*–*T* line is the lowest of the three *G*–*T* lines until T_1. At T_1, the energies of the crystal and liquid crystal are equal. A phase transition from a crystal to a liquid crystal occurs at a temperature designated T_1. Whitin the range from T_1 to T_2, the energy of the liquid crystal is minimal, and the substance is in its liquid crystal state. T_2 is the temperature at which the liquid crystal-isotropic liquid phase transition occurs.

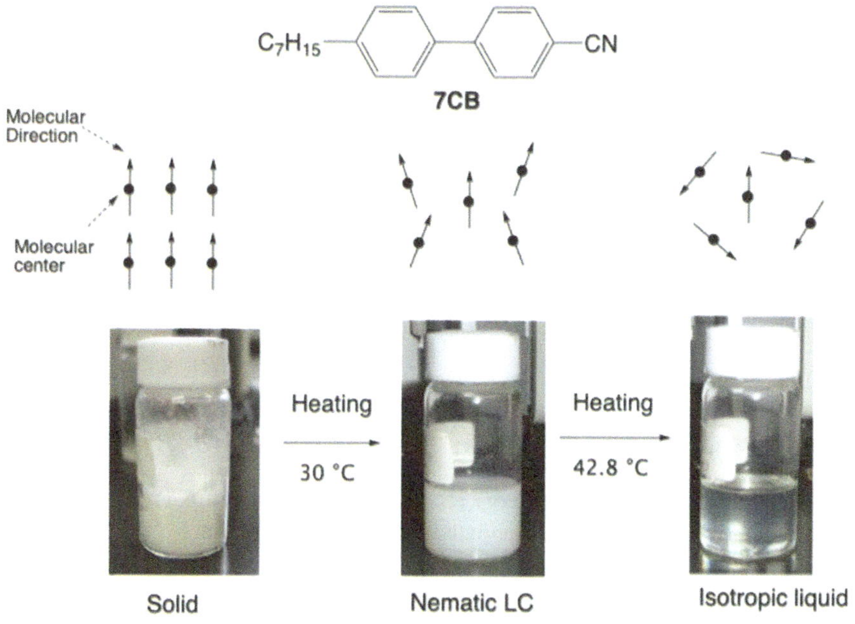

Fig. 1.2 Appearance and order in the phase transition of **7CB**. Reprinted with permission from [3]. Copyright [2022] Springer Nature

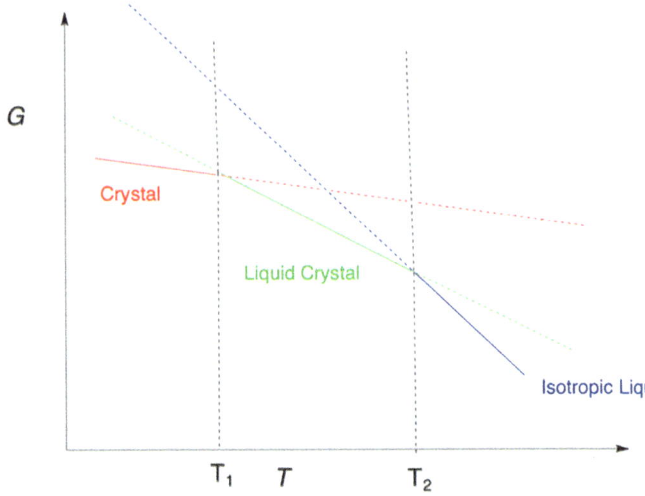

Fig. 1.3 Relationship between Gibbs energy and temperature for three states: crystal, liquid crystal, and liquid

It is imperative to ascertain the significance of the slope. The slope depicted in Fig. 1.3 signifies the temperature dependence of Gibbs energy under constant pressure, and can be written as $(\partial G/\partial T)_P$. The ensuing discussion will demonstrate the aforementioned correspondence with entropy. The Gibbs free energy can be expressed from enthalpy (H), temperature (T) and entropy (S) as indicated in Eq. 1.1.

$$G = H - TS \tag{1.1}$$

The total differentiation of Eq. 1.1 results in Eq. 1.2. Enthalpy is defined by Eq. 1.3, and its total differential is given by Eq. 1.4.

$$dG = dH - TdS - SdT \tag{1.2}$$

$$H = U + PV \tag{1.3}$$

$$dH = dU + PdV + VdP \tag{1.4}$$

The alternation in internal energy (U) is delineated by the quantity of incoming heat (q) and work received, as articulated in Eq. 1.5. Substituting Eq. 1.5 into Eq. 1.4 and subsequently substituting it into Eq. 1.2 yields Eq. 1.6.

$$dU = dq - PdV = TdS - PdV \tag{1.5}$$

$$dG = VdP - SdT \tag{1.6}$$

On the other hand, the total derivative of the Gibbs free energy can also be expressed in Eq. 1.7. A comparison of the coefficients of dT in Eqs. 1.6 and 1.7 results in Eq. 1.8.

$$dG = \left(\frac{\partial G}{\partial P}\right)_T dP + \left(\frac{\partial G}{\partial T}\right)_P dT \tag{1.7}$$

$$\left(\frac{\partial G}{\partial T}\right)_P = -S \tag{1.8}$$

In the field of statistical mechanics, entropy is defined as indicated in Eq. 1.9, where k_B denotes Boltzmann constant and $W(E)$ represents the microscopic number of states with energy E. Entropy serves as a metric for measuring the disorder in a system.

$$S = k_B \ln W(E) \tag{1.9}$$

The gradient of the linear segment in each state in Fig. 1.3 is indicative of the degree of disorder in that specific state. As the phase transition from crystal to liquid crystal and then to isotropic liquid progresses, an increase in disorder is observed.

Nematic liquid crystals exhibit liquid-like fluidity, rending them susceptible to facile modification of orientation in response to external stimuli. As illustrated in Fig. 1.4, the application of voltage in the nematic phase of **7CB** results in a change in the orientation of liquid crystal molecules. In rod-shaped liquid crystal molecules, the dielectric constants in the long-axis and short-axis directions of the molecules are different. The value obtained by subtracting the dielectric constant in the short axis direction from the dielectric constant in the long axis direction is referred to as dielectric anisotropy (Δε). **7CB** has a cyano group along the molecular long axis, thereby positive dielectric anisotropy. When this compound is sandwiched between two glass substrates with electrodes and voltage is applied in the vertical direction

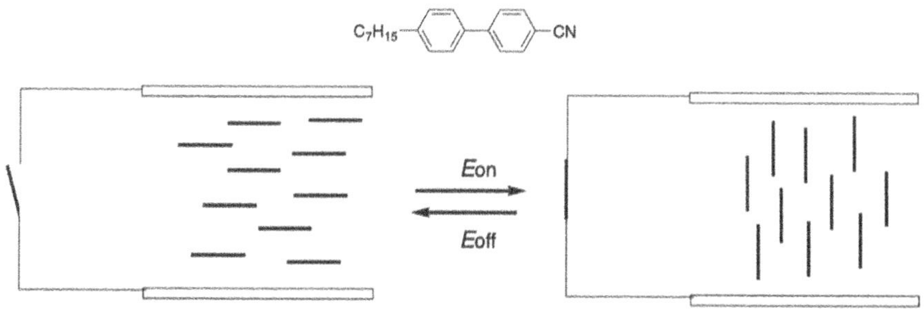

Fig. 1.4 Change in orientation of **7CB** in the N phase with and without an electric field [3]

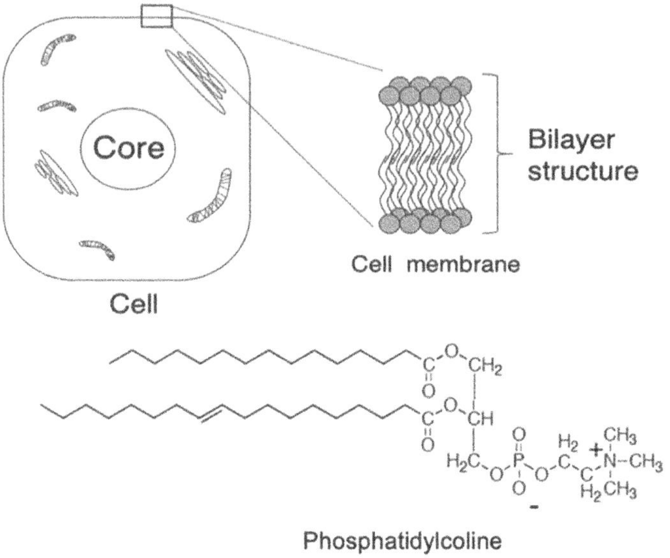

Fig. 1.5 Schematic diagram of cell structure

in the nematic phase, the molecules are oriented perpendicular to the glass substrate due to the interaction of the electric field and dielectric anisotropy.

The similarity between liquid crystals and biological systems was noted early on. Living organisms are composed cells. Each cell is surrounded by a cell membrane consisting of a lipid bilayer (Fig. 1.5). Cell membranes are composed of phospholipids and exhibit a liquid crystalline state. Phosphatidylcholine, a representative lipid molecule, is shown in the figure. The application of liquid crystals in biomedical contexts, including tumor markers and drug delivery systems, is a growing field of research. The potential of these compounds as pharmaceuticals, such as anti-cancer drugs, has also been documented [4]. The biological activities will be described in Chap. 6.

1.2 Formation of Liquid Crystal Phases

Lyotropic LCs emerge as a result of a mixture of an amphiphilic substance and a solvent. A typical example of such a mixture is a mixture of soap and water. As illustrated in Fig. 1.6, the formation of a lamellar phase is the result of the interaction between amphiphilic molecules and water. The amphiphilic molecules are oriented in such a manner that their hydrophilic groups interact with the water molecules and their hydrophobic groups interact with each other. The manifestation of lyotropic LC phases within a surfactant-solvent system is contingent upon the solvent concentration. To increase the hydrophilic surface area that interacts water molecules, the amphiphilic molecules self-organize to form supramolecular assemblies. In the context of an amphiphilic surfactant–water system, the phase sequence of lamellar, bicontinuous cubic (Cub_V), hexagonal columnar (Col_h), and micellar cubic (Cub_I) undergoes a transition as the volume fraction of a solvent is increased. This sequence is characteristic of the morphologies of mesophases within these systems, as illustrated in Fig. 1.7. In the context of an amphiphilic surfactant–oil system, the reversed phases manifest. It is noteworthy that the lyotropic LC phases are also present in the living systems, such as cell membranes and proteins.

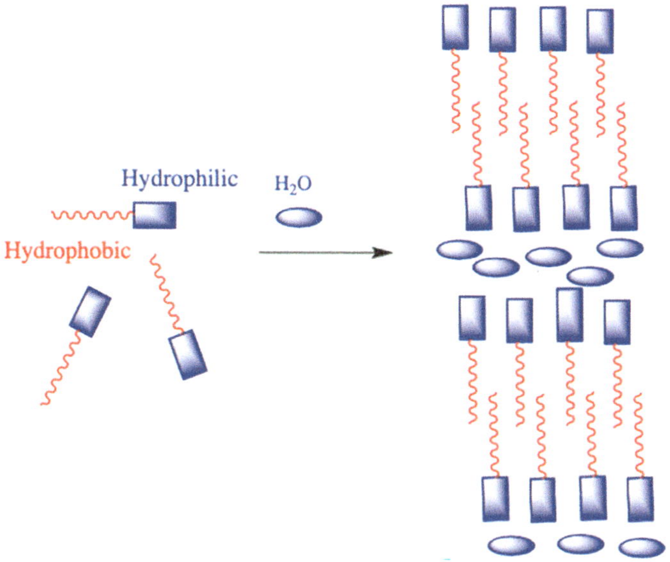

Fig. 1.6 Formation of a lyotropic lamellar phase in a mixture of an amphiphilic molecule and water. Reprinted with permission from [4]. Copyright [2022] Springer Nature

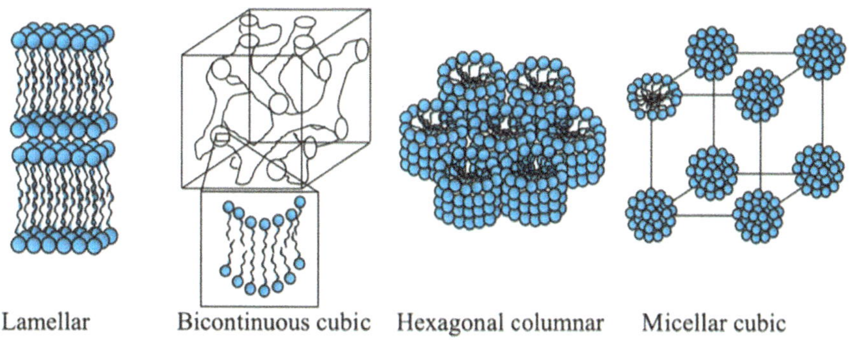

Fig. 1.7 Phase structures of lyotropic liquid crystals. Reprinted with permission from [4]. Copyright [2022] Springer Nature

Amphiphilic surfactants are composed of a hydrophilic moiety, designated as the head group, and a hydrophobic part [5, 6]. The head group is responsible for the water solubility of the compound. Examples of some common surfactants are listed in Fig. 1.8 [5]. When water-miscible surfactants are dissolved in water, they form aggregates termed micelles above the critical micelle concentration (CMC). As the concentration increases, these aggregates will be further assembled into the lyotropic liquid-crystalline phases, which possess different structures (see Fig. 1.7). The phenomenon of the structural transitions can be elucidated is by the critical packing parameter (CPP), which is predicated on the spatial stacking of amphiphilic molecules. This theoretical framework facilitates the prediction of the system's structure [7–9]. The CPP is defined as $CPP = V/(al)$, where V represents the hydrophobic chain volume, a represents the cross-section area of the hydrophilic group, and l represents the hydrophobic chain length in the molten state (Fig. 1.9). In the case of $CPP < 1$, normal phases, such as the micellar cubic phase, hexagonal phase, and bicontinous cubic phase, are observed. Conversely, their reversed phases are developed with $CPP > 1$. When CPP is approximately 1, the lamellar phase is formed. The CPP of amphiphilic molecules is influenced not only by their molecular structure but also by environmental factors, including pH and additives.

The formation of thermotropic liquid crystals can be classified according to intermolecular interactions. One such interaction is molecular packing, which is observed in monophilic compounds. The other interaction is the microphase separation, which is observed in amphiphilic compounds. The interactions exhibited by monophilic rod-shaped molecules are characterized by a duality: repulsion and attraction. First, we must address the repulsive interactions [10]. It is imperative to consider the same same rod-shaped molecules i and j (see Fig. 1.10). In the event of an approaching configuration in which molecules i and j are oriented in parallel alignment, the center

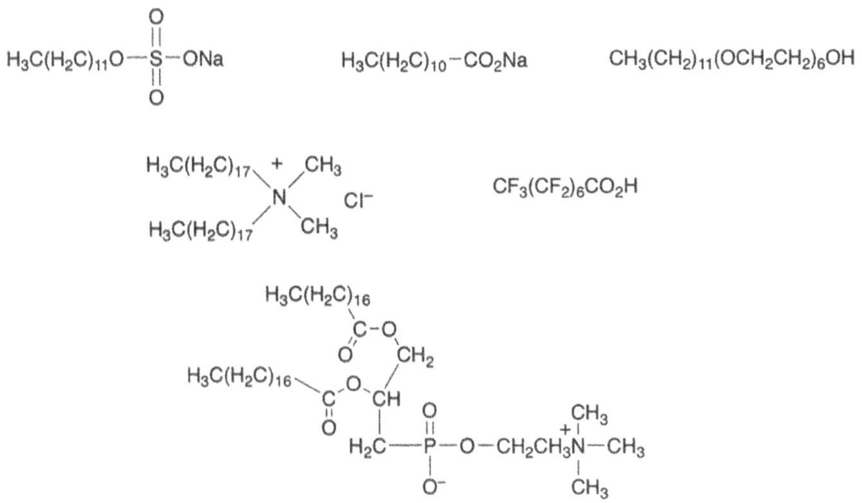

Fig. 1.8 Molecular structures of some common surfactants

Fig. 1.9 Illustration of CPP

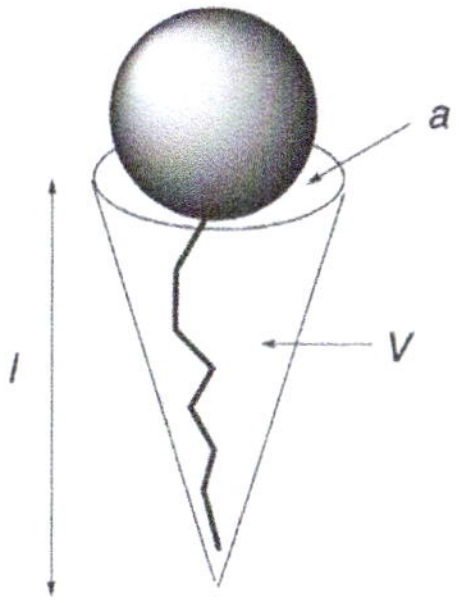

of molecule j is incapable of entering the dotted line delineation. This phenomenon can be attributed to the destabilization caused by the repulsive force acting between the molecules. The region *V*a, delineated by the dotted line A, is designated as the excluded volume of molecule i. On the other hand, when molecule j approaches molecule i perpendicularly, the excluded volume is defined as the region *V*b, as illustrated by the dotted line B. Iin the event that the molecule is elongated, as illustrated in Fig. 1.10, it can be deduced that *V*a will be smaller than *V*b. The efficient filling of molecules is facilitated by a smaller excluded volume. In the case of an elongated rod-shaped molecule, each molecule is arranged in parallel, with the molecular long axis oriented in a consistent direction. Due to the exclusion volume effect, which is determined by hard-core repulsion, the formation of a liquid crystal phase requires a rod-like molecular shape with a length-to-diameter (*L*/*D*) ratio exceeding 2.75.

Next, the formation of liquid crystals through attractive interactions will be examined. When the molecules are polarized in the long-axis direction, as shown in Fig. 1.11, the molecules are oriented parallel to each other due to the attractive

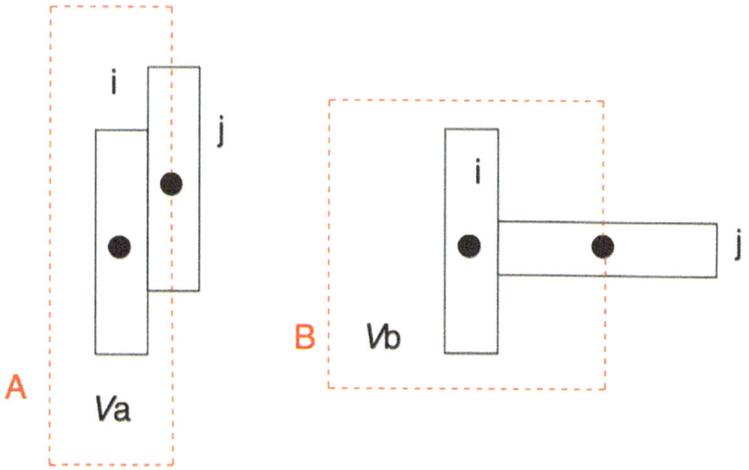

Fig. 1.10 Excluded volume effect of rod-shaped molecules [3]

force originating from dipole–dipole interactions between neighboring molecules. Maier and Saupe expressed the nematic-isotropic liquid phase transition temperature (T_{NI}) from mean-field theory using the following equation:

$$T_{NI} = A/(4.5k_B V^2) \tag{1.10}$$

where A is determined by the molecular nature and is proportional to the square of the anisotropy of the molecular polarization, and V is defined as the molecular volume [11].

Chemists design liquid crystal molecules with these factors in mind [12]. A schematic representation of rod-shaped liquid crystal molecules is depicted in Fig. 1.12. The molecules are composed of a core unit and flexible chains.

As illustrated in Fig. 1.13, lateral substitution has been demonstrated to reduce the nematic-to-isotropic liquid phase transition temperature of rod-like liquid crystals [12]. This phenomenon can be attributed to the reduction in the *L*/*D* ratio.

Next, we will discuss the effect of polar substituents on nematic phase formation. Table 1.1 shows the relationship between the anisotropy of the C–X bond polarization

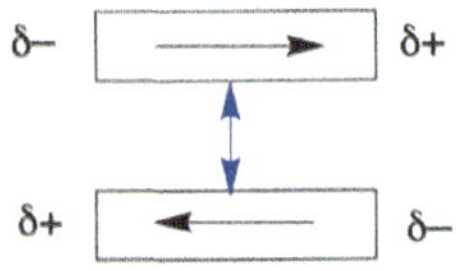

Fig. 1.11 Dipole-dipole interaction between neighboring polarized molecules

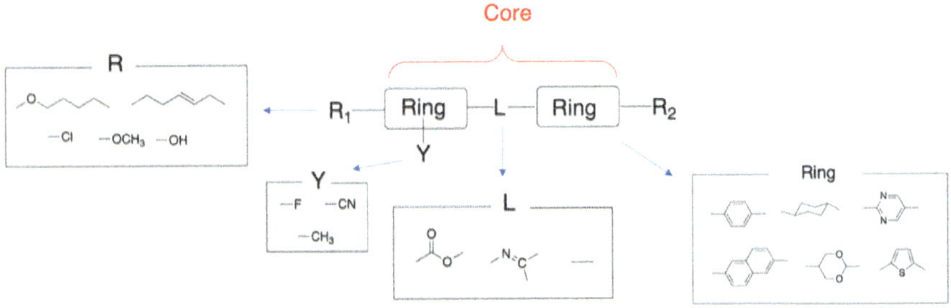

Fig. 1.12 Schematic diagram of rod-shaped liquid crystal molecules

Fig. 1.13 Decrease in the clearing temperatures by the addition of lateral substituents

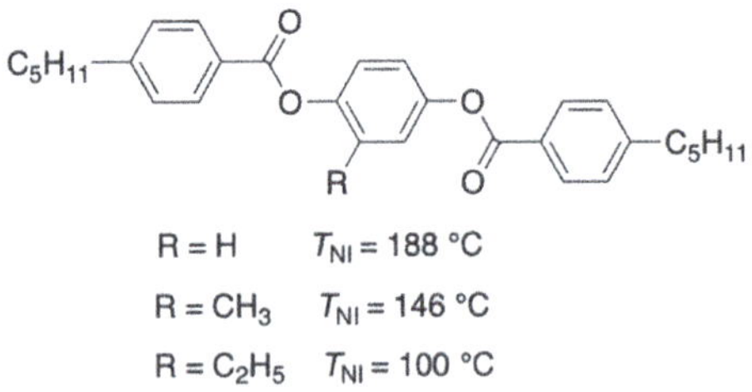

($\Delta\alpha_{C\text{-}X}$) and T_{NI} in rod-like liquid crystals [13]. With two exceptions, T_{NI} increases in response to an increase in $\Delta\alpha_{C\text{-}X}$. This phase transition behavior can be rationalized in terms of Eq. 1.10. The initial exception pertains the instance in which X = Br. In this scenario, the compound that possesses a Br substituent exhibits a lower T_{NI} value in comparison to the compound that possesses a CI substituent. It has been established that the bromine atom is larger than the chlorine atom. Consequently, replacing chlorine with bromine results in an increase in molecular width, leading to a decrease in the *L*/*D* ratio. The second scenario is when X = CN. The compound displays a higher T_{NI} than expected. The following explanation is proposed to elucidate this phenomenon. The molecules that contain a cyano group associate in an antiparallel manner, mimicking a dimer behavior (Fig. 1.14). This process of dimer formation has shown to increase the *L*/*D* ratio.

Two kinds of thermotropic phases are expressed by rod-shaped molecules: nematic phases, in which the positional order is completely lost, and smectic phases, which have a layered structure. The smectic phases are further classified into two categories: those with order in the layer and those without. Figure 1.15 shows a typical nematic phase and smectic A and C phases. In the smectic A phase, the molecules are aligned along the layer normal, while in the smectic C phase, they are tilted. It is noteworthy that both phases exhibit no discernible order within the layer. Disk-shaped molecules also form liquid crystals. The typical phase structures are shown in Fig. 1.16.

The alignment of amphiphilic molecules is attributed to two fundamental forces: the attraction between their hydrophilic groups and the attraction between their hydrophobic groups. These forces give rise to a process of micro-segregation, as illustrated in Fig. 1.17. Amphiphilic thermotropic liquid crystals manifest phase

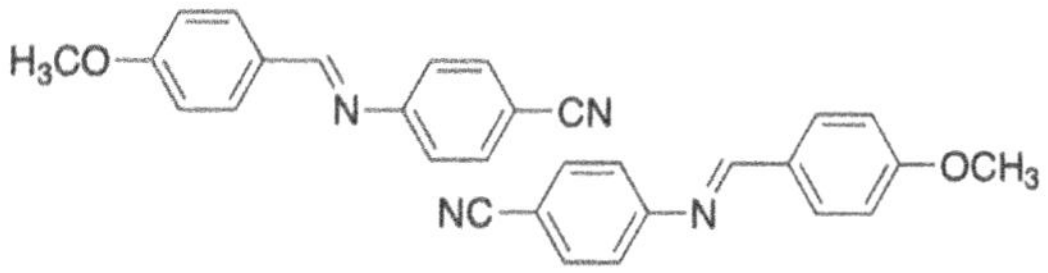

Fig. 1.14 Dimerization of rod-like LCs with a terminal cyano group

Table 1.1 Relationship between the anisotropy of the C–X bond polarization ($\Delta\alpha_{C\text{-}X}$) and T_{NI} in rod-like liquid crystals [13]

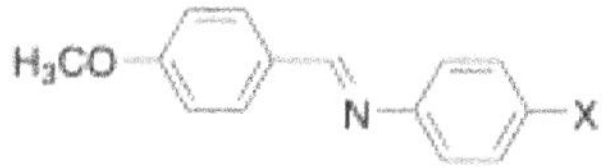

X	$\Delta\alpha_{C\text{-}X}$	T_{NI} /°C
F	0.5	26
CH_3	1.2	38
CI	2.5	47
Br	3.9	34
NO_2	4.0	83
CN	4.4	119

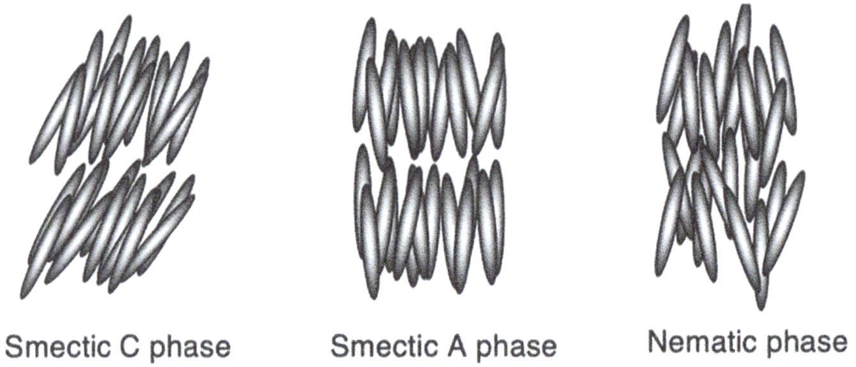

Fig. 1.15 Schematic phase structures exhibited by rod-shaped molecules

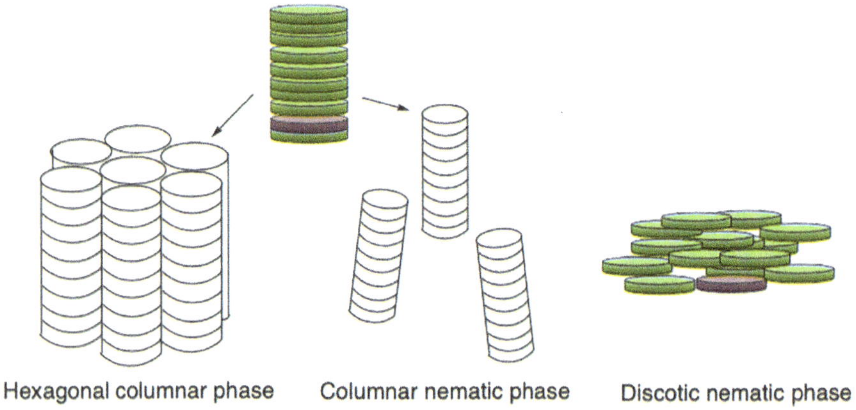

Fig. 1.16 Schematic phase structures exhibited by disk-shaped molecules

structures analogous to those of lyotropic liquid crystals, depending on the ratio of hydrophobic to hydrophilic groups within the molecule [14].

Hydrogen bonding, ionic bonding, and coordinate bonding also give rise to liquid crystal phases. Hydrogen bonding is one of the key interactions for chemical and biological processes. Kato et al. demonstrated that liquid-crystalline supramolecular complexes with well-defined structures are obtained by self-assembly through hetero-intermolecular hydrogen bonding between carboxylic acid and pyridine moieties [15]. The complexation of 4-methoxybenzoic acid and *trans*-1,2-bis(4-pyridyl)ethylene results in the formation of mesogenic complex, which exhibits a nematic phase from 166 to 185 °C (Fig. 1.18) [16, 17]. The complex that has been formed possesses a hydrogen-bonded mesogenic core and has a significantly larger *L/D* ratio in comparison to the nonmesogenic precursors.

Chirality plays an important role in the formation of order in the liquid crystal phase [18–20]. If a molecule possesses an asymmetric carbon (Fig. 1.19a) or an

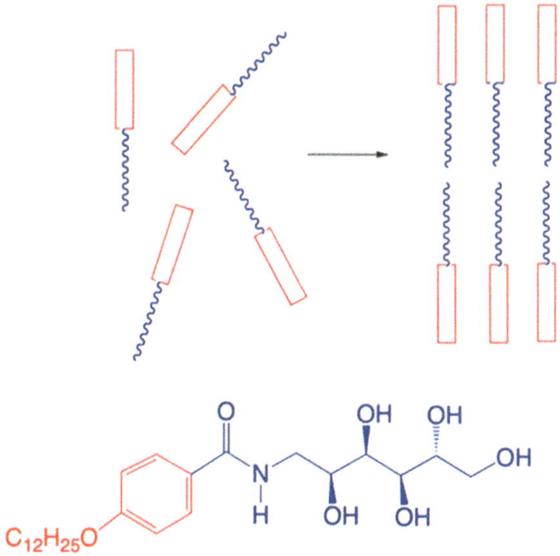

Fig. 1.17 Micro-segregation of amphiphilic molecules

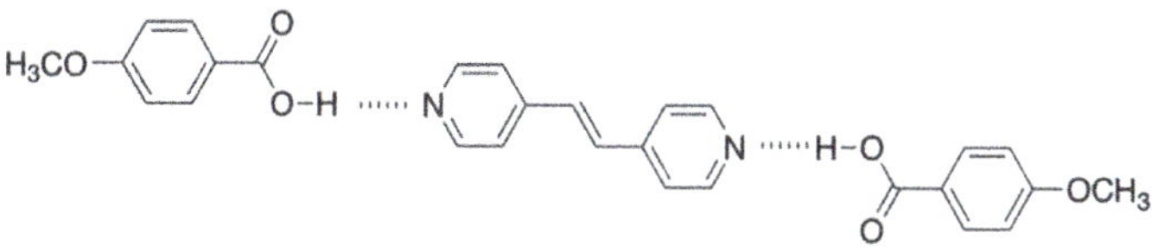

Fig. 1.18 Hydrogen-bonded liquid crystal system

asymmetric axis (Fig. 1.19b), it does not overlap its mirror image. The molecule is chiral. If the system is chiral, helical structures occur in select liquid crystal phases. In the nematic phase doped with a chiral substance, the helix is formed so that the helical axis is perpendicular to the molecular long axis. This is referred to as the chiral nematic (N*) phase or cholesteric (Ch) phase (Fig. 1.20a). The chiral nematic phase appears when the liquid crystal molecules themselves are chiral or when a chiral substance is added into a nematic liquid crystal. In smectic phases, helical structures emerge in the SmC phase, for instance, where the molecular long axis is tilted from the layer normal, and the helical axis is parallel to the layer normal (Fig. 1.20b). The formation of helical structures in the chiral liquid crystal phase is attributable to steric interactions and is obervable in monophilic thermotropic liquid crystals with weak intermolecular interactions. Figure 1.21 shows an example of a chiral liquid-crystalline molecule. The observation of ferroelectricity in liquid crystal was first reported to be observed in the chiral smectic C (SmC*) phase of the chiral liquid crystal **DOBAMBC** by Meyer et al. in 1975 [21].

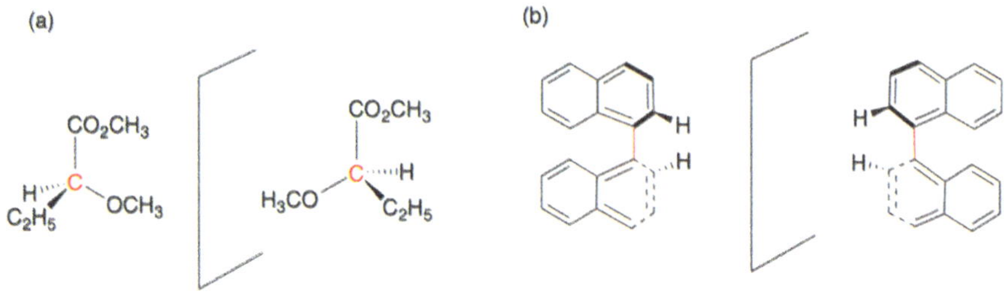

Fig. 1.19 Molecular chirality: **a** Asymmetric carbon (chiral center) and **b** asymmetric axis

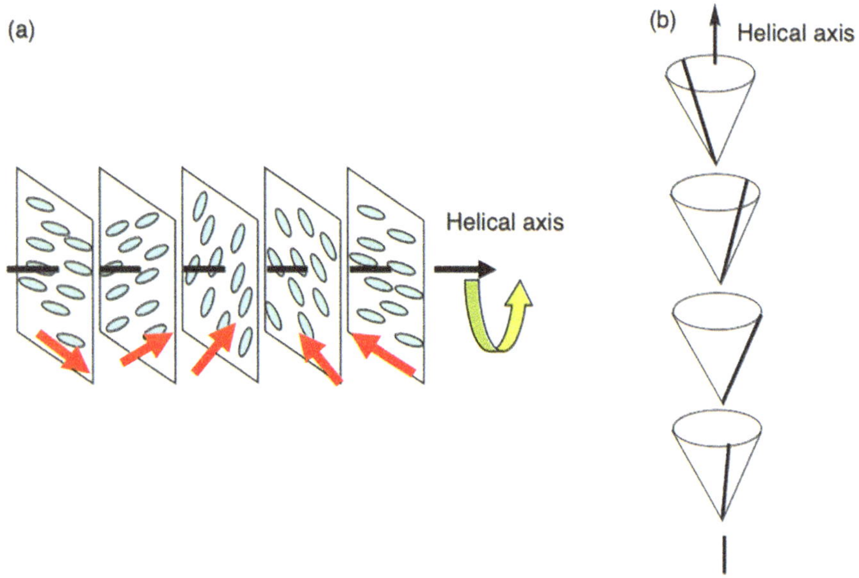

Fig. 1.20 **a** Schematic phase structure of a chiral nematic phase and **b** that of a chiral smectic C (SmC*) phase. The red arrow shows a unit vector (director) that indicates the average orientation direction of the molecular long axis

Fig. 1.21 Molecular structure of chiral liquid crystal **DOBAMBC**

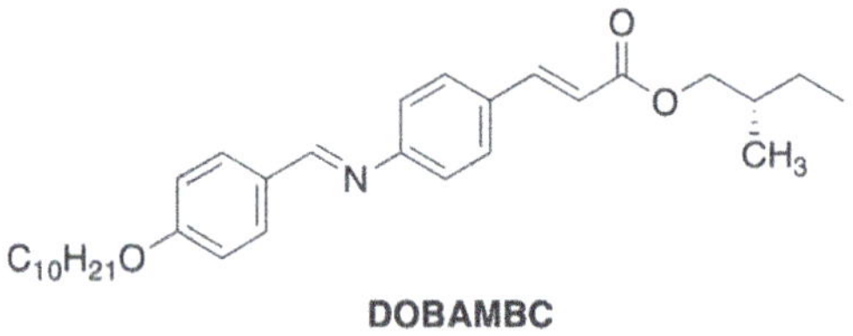

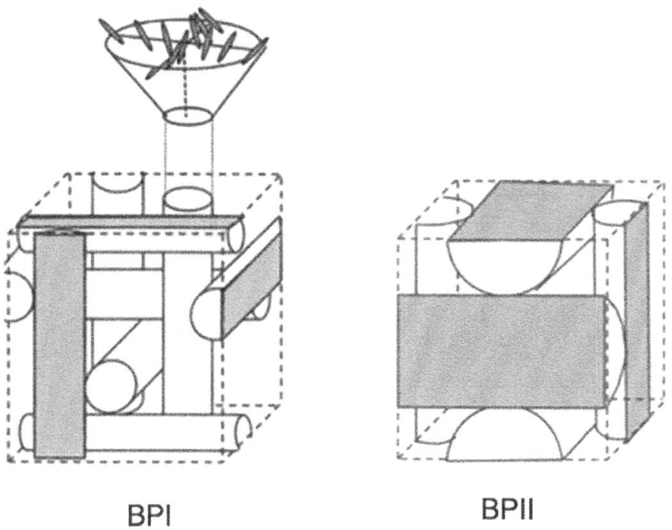

Fig. 1.22 Phase structures of BPI and BPII. Reproduced from [24]

Chirality has been observed to induce certain frustrated phases, such as blue phases [22] and twisted grain boundary (TGB) phases [23]. The manifestation of these frustrated phases results from the competition between the twisting force and the molecular tendency to arrange in a manner that ensures uniform filling of the available space. The blue phases are believed to consist of double twist cylinders (DTCs), and are classifiable into three categories depending on the double-twist-cylinder packing structure: blue phase I (BPI), blue phase II (BPII) and blue phase III (BPIII). BPI and BPII have a body-centered cubic structure and a simple cubic one, respectively. Their models are shown in Fig. 1.22. On the other hand, BPIII is thought to have an amorphous structure. The formation of DTCs inevitably accompanies line defects because the constituent molecules cannot fill the space uniformly.

The TGB phases were predicted by de Gennes [25] and later by Lubensky [26] to be a result of the competition between bend or twist deformations and the desire for molecules to form a layer structure. The first experimental observation of a TGB phase was reported by Goodby et al. [27] Fig. 1.23 shows a schematic model of the TGB phase, resulting from the competition between twist forces and layer formation in the smectic A phase. The TGB phase is referred to as the TGBA phase. It has been demonstrated that the molecules exhibit blocks of SmA molecular order mediated by screw dislocations.

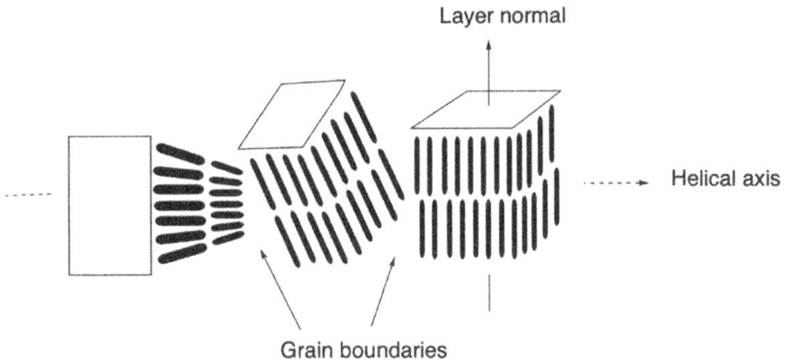

Fig. 1.23 Schematic model of the TGBA phase

1.3 Characterization

The identification of liquid crystal phases and the observation of their structures is primarily facilitated by three method: polarized light microscopy [28, 29], calorimetry, and X-ray diffraction measurements [30]. Polarized light microscopy is a method of analyzing the orientation, order, and defects present in the liquid crystal phase. This analysis is achieved by observing the pattern of the liquid crystal through the use of polarized light. When light in all directions is passed through a polarizing filter, light with a specific oscillation direction (polarization) is transmitted. The phenomenon of passing through two polarizers is illustrated in Fig. 1.24. Devices capable of generating polarized light are designated as polarizers, while those that detect or block polarized light are referred to as analyzers. If the polarizer and analyzer are oriented in a parallel fashion, the polarized light passes through; conversely, if they are oriented in a perpendicular fashion, the polarized light cannot pass through the analyzer.

To illustrate this phenomenon, consider the case of polarized light entering a rod-shaped nematic liquid-crystalline molecule (Fig. 1.25). The refractive index when the photoelectric field is parallel to the molecular short axis is designated as ordinary refractive index (n_o), and the refractive index when it is parallel to the molecular long axis is designated as extraordinary refractive index (n_e). Optical anisotropy, or birefringence, is expressed by the following equation:

$$\Delta n = n_e - n_o. \tag{1.11}$$

In rod-shaped molecules, the electrons cloud typically extends in the direction of the molecular long axis. Consequently, the amplitude of the photoelectric field oriented parallel to the molecular long axis is greater, and the velocity of the extraordinary light is reduced relative to that of the ordinary light. In this instance, Δn is positive.

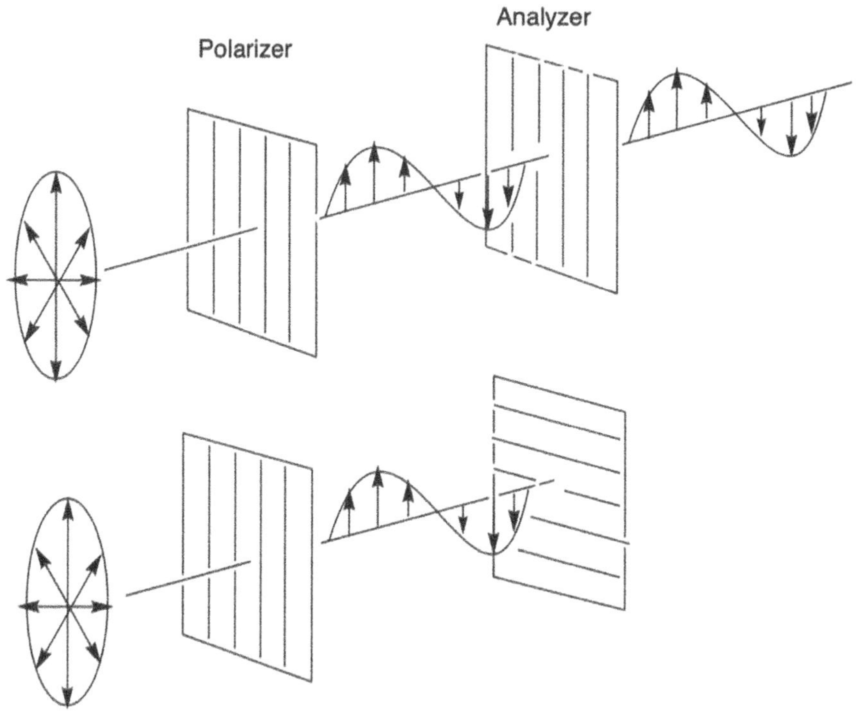

Fig. 1.24 Use of polarization [3]

As illustrated in Fig. 1.26, there is a direct correlation shows between molecular orientation and the progression of polarization. If the molecular long axis is oriented in a parallel (Fig. 1.26a), orthogonal (Fig. 1.26b), or perpendicular (Fig. 1.26c) configuration with respect to the polarizer, the progression of polarized light remains unaltered in direction and is incapable of traversing the analyzer. If the molecular long axis is oriented in parallel with the substrates, yet not aligned parallelly with both the polarizer and the analyzer (Fig. 1.26d), the direction of the polarized light undergoes a progressive alteration along the axis, ultimately passing through the analyzer.

The texture of liquid crystals depends on their orientational state. The standard alignment states of nematic liquid crystals are shown in Fig. 1.27. Figure 1.27a, b show homogeneous alignments. The molecules align in a parallel fashion with respect to the substrates, but not perfectly aligned in one direction (Fig. 1.27a). It is called planar alignment. This phenomenon is typically observed in the absence of specialized surface treatments. If the molecules are oriented so that they are parallel to the substrates and uniformly aligned in one direction (Fig. 1.27b), it is called uniform alignment in this book. If the molecules are oriented perpendicularly to the substates (Fig. 1.27c), the alignment is designated as homeotropic alignment. The arrow in each figure indicates a unit vector, denoting the average orientational

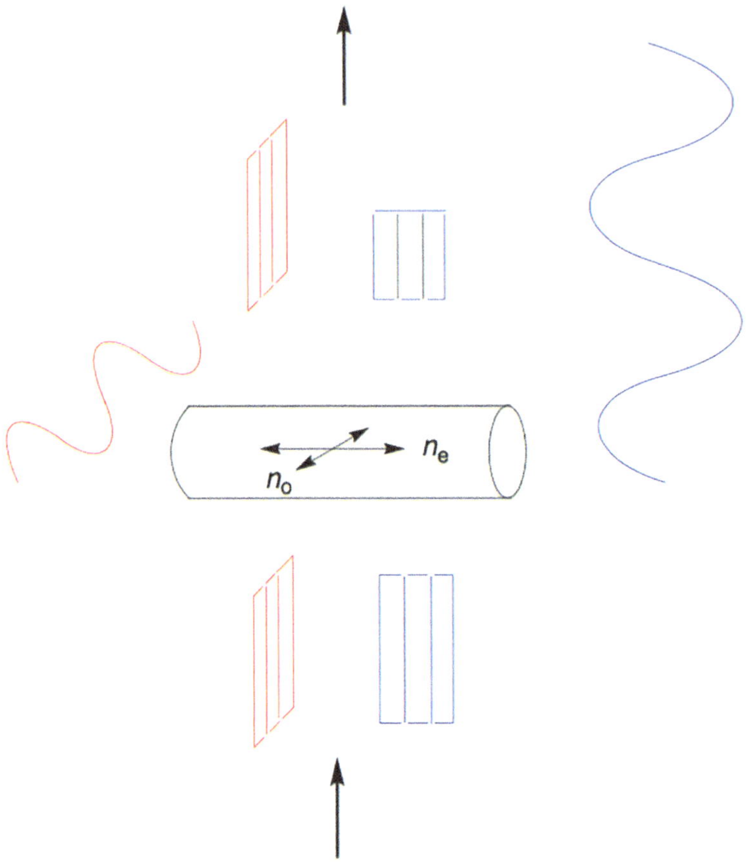

Fig. 1.25 Optical anisotropy of rod-shaped nematic liquid crystals. Blue and red colors show the extraordinary and ordinary lights, respectively. The wavy lines stand for the light waveform, while the rectangles represent the speed of light

direction of the molecular long axis, which is referred to as the orientation vector or director.

A comparison of Fig. 1.27a, b reveals that the directions of their orientational vectors are identical; however, the degree of alignment of the liquid-crystalline molecules differs. The degree to which the liquid-crystalline molecules align along the orientation vector is expressed by the following order parameter S as Eq. 1.12. In this equation θ is the angle between the long axis of individual rod-like liquid-crystalline molecules and the director. The brackets indicate the average value.

$$S = \left(\frac{1}{2}\right) < 3\cos^2\theta - 1 > \tag{1.12}$$

Polarized light microscopy of the planar orientation of liquid crystals provides critical insights into the arrangement of these liquid crystals. To illustrate this point,

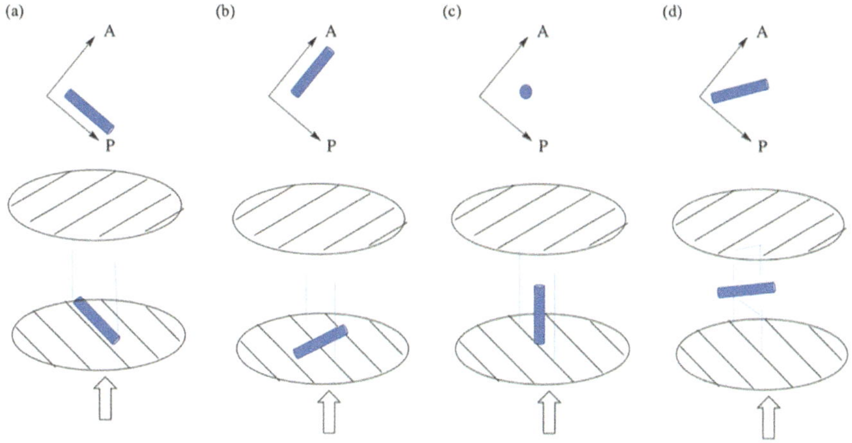

Fig. 1.26 Relationship between molecular orientation and polarization progression

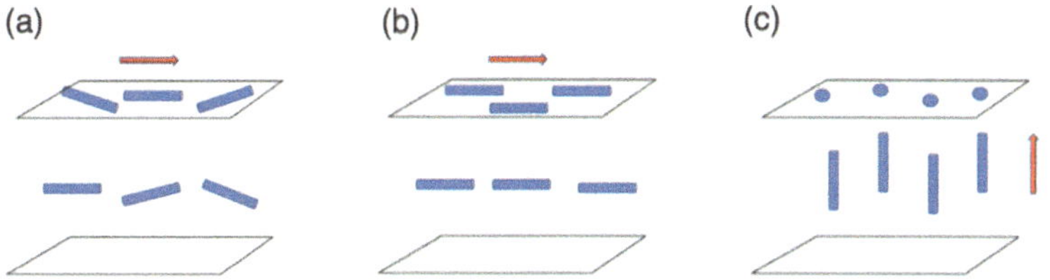

Fig. 1.27 Typical alignment states of nematic liquid crystals. **a** Planar alignment, **b** uniform alignment, and **c** homeotropic alignment. The red arrows show their director

Fig. 1.28 Schlieren texture of a nematic liquid crystal. The photograph was taken in our laboratory

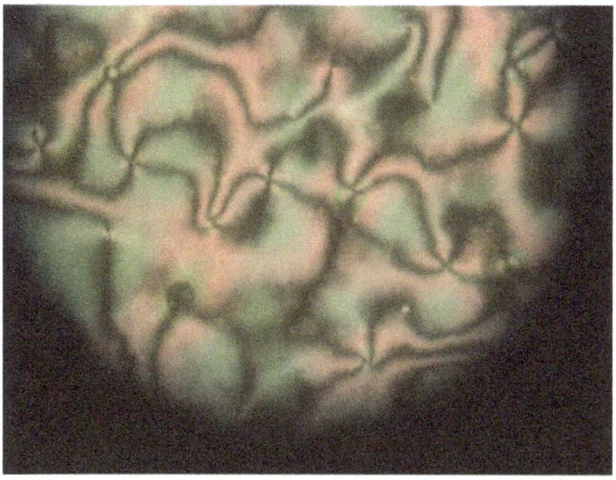

we will examine the schlieren texture (Fig. 1.28) observed in the planar orientation of nematic liquid crystals.

Nematic liquid crystals can be treated as an elastic medium and exhibit three types of deformations: splay, twist, and bend, as shown in Fig. 1.29. The elastic constants corresponding to these types are designated as K_{11}, K_{22}, and K_{33}, respectively.

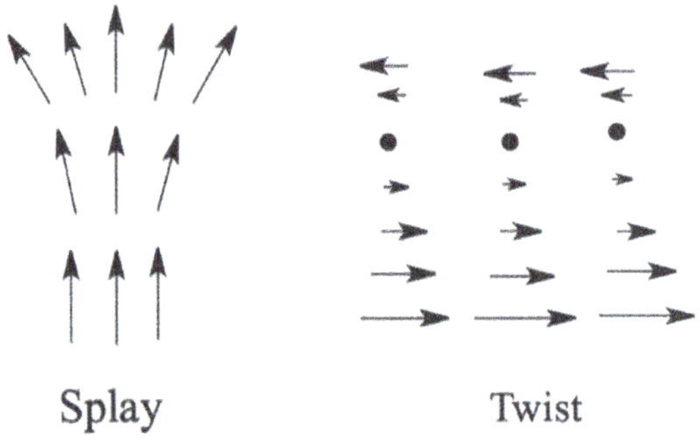

Fig. 1.29 Deformations of nematic liquid crystals

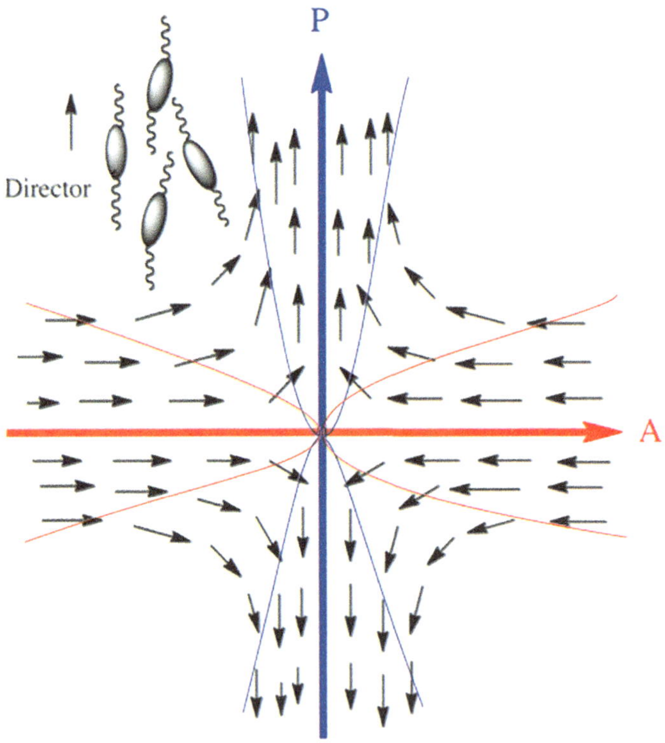

Fig. 1.30 Molecular arrangement near the singularity $s = -1$

In the context of a schlieren texture of a nematic liquid crystal, the presence of four brushes and two brushes meeting is observed. The meeting points of these lines indicate the presence of singularities. The observed type of singularity corresponds to variations in molecular arrangement. The singularities are denoted by specific signs and numerical values. When the polarizer and the analyzer are simultaneously rotated in the same direction, the one in which the brushes rotate in the same direction is considered positive, and the one in which the brushes rotate in the opposite direction is considered negative. The number of brushes is divided by four. Singularities with $s = +1/2, -1/2, +1$, and -1 have been observed in a nematic liquid crystal. Figure 1.30 shows the molecular arrangement in proximity of the $s = -1$ singularity of a nematic liquid crystal. The splay deformation is capable of generating this topology. In the central part, the director is perpendicular to the plane. The black brushes depicted in Fig. 1.28 suggest that their director is oriented parallel to the polarizer or analyzer. In other areas, the director is not parallel to either polarizer or the analyzer, resulting in the visibility of bright textures.

Figure 1.31 presents POM textures and phase models of smectic A and smectic C phases. In the SmA phase (Fig. 1.31a), the planar region exhibits a focal conic texture, while the homeotropic region is completely dark. In the homeotropic region, the optical axis is perpendicular to the glass slides. Consequently, the polarized light progresses without changing its direction and cannot pass through the analyzer (Fig. 1.26c). In regard to the SmC phase (Fig. 1.31b), a focal conic texture is observed in the planar region. Conversely, the homeotropic region manifests a schlieren texture. The molecular tilting produces an orientation vector (the *c*-director) that is oriented in the direction of the tilt. The *c*-director is deformed as the molecule oscillates on the cone, producing the schlieren texture. In the schlieren texture of the SmC phase, singularities consisting of four brushes are observed. These singularities are designate as $s = 1$ or $s = -1$. A smectic C phase possesses two optical axes, thus defining it as biaxial. Conversely, the nematic N and smectic A phases are uniaxial.

Dierking recently presented an over view about the liquid crystal textures in polarizing microscopy [29]. Please see the review.

The formation of a helical macrostructure is a possible outcome of chiral liquid crystals under specific conditions. The helical structure is distinguished by its sense of twist and pitch. According to Davis and Goodby [20], the following description will be provided of the definition of the twist sense. The direction of the helical ordering can be defined by determining the direction in which phase-polarized light is rotated by the helical macrostructure. A rotation of the polarization plane to the left, moving away from an observation point along the helical axis, is denoted as a left-handed helix, and a rotation to the right results in a right-handed helix. In polarimetry experiments, the classification of handedness as dextro- or levorotatory is determined by the direction of the light beam's movement toward the observer. Dextrorotation is defined as a rotation to the right, while levorotation signifies a rotation to the left (Fig. 1.32). Consequently, in this type of experiment, a left-handed helix will yield a dextrorotation, while right-handed helix will produce a levorotation.

The twist sense for a given chiral nematic material can be determined by polarized optical microscopy [20]. A contact preparation of the material of unknown twist

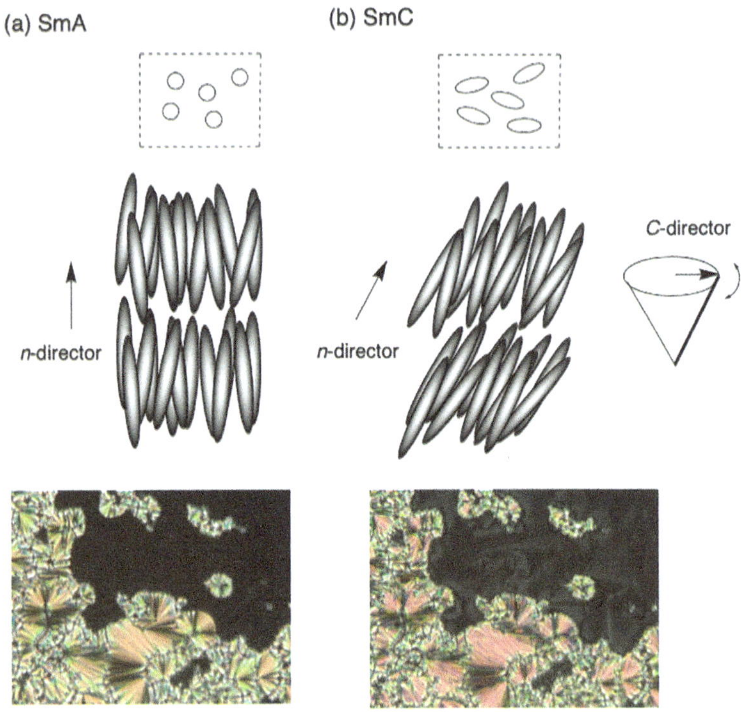

Fig. 1.31 POM textures and phase models of smectic A and smectic C phases. The orientation vector along the molecular long axis is described as *n*-director. These photographs were taken in our laboratory

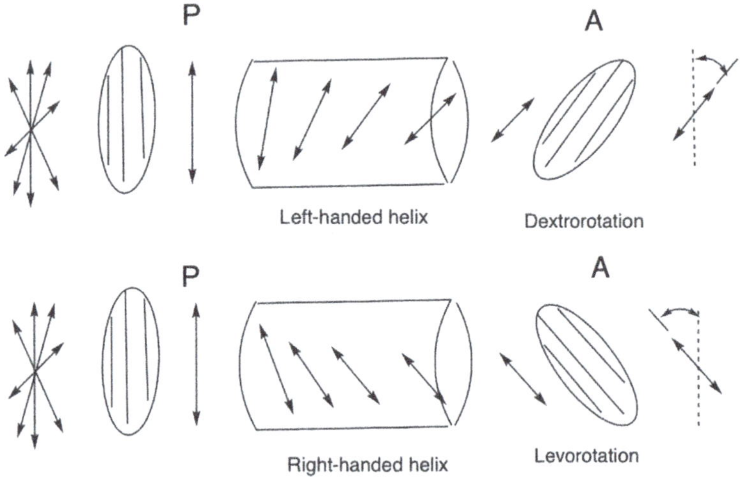

Fig. 1.32 Illustration of the polarity experiments

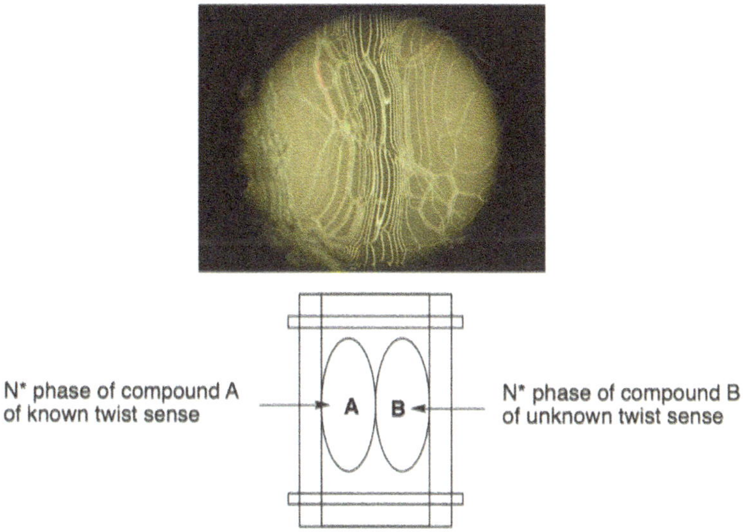

Fig. 1.33 Determination of the twist sense of a chiral nematic phase by polarized optical microscopy

sense and a standard material of known twist sense is made, and the area of contact between the two materials is subsequently observed. Two materials with identical twist senses are expected to demonstrate continuous miscibility across the whole preparation. Conversely, contact preparations involving two materials of opposite twist sense will exhibit two distinct regions of opposite twist sense mediated by a nematic phase with an infinite pitch length (Fig. 1.33).

Figure 1.34 shows the helical structure of a chiral nematic phase and a finger print texture. The twist sense of the N* phase is left-handed. The distance between the adjacent lines in the texture corresponds to a half-pitch length because there is no distinction between the head and tail of the constituent molecule. The measurement

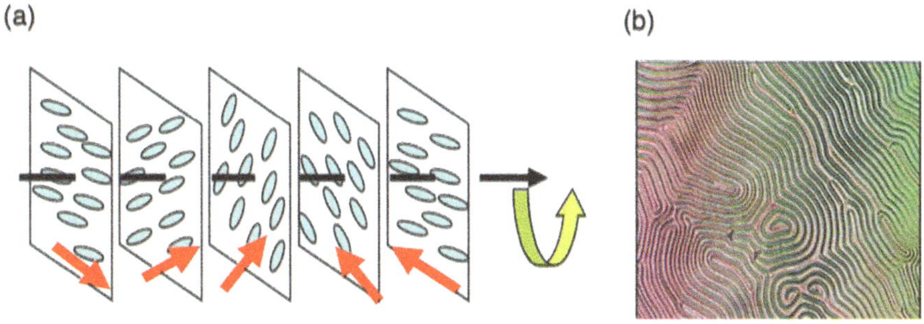

Fig. 1.34 **a** Left-handed helix of a chiral nematic phase and **b** finger print texture of the N* phase

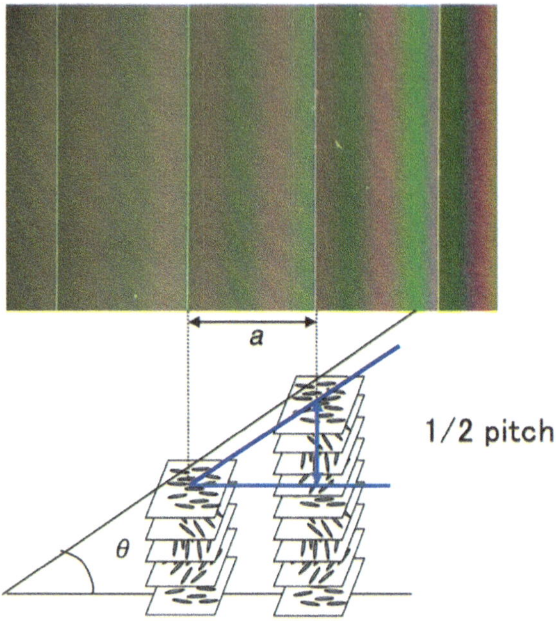

Fig. 1.35 Texture of chiral nematic Grandjean steps in a Grandjean–Cano wedge cell

of pitch length can be facilitated by employing a Grandjean-Cano wedge cell, as illustrated in Fig. 1.35.

As demonstrated in Fig. 1.35, periodic domains are generated depending on the molecular orientation, accompanied by the emergence of a disclination line between these domains. It appears at positions where the length of the helical structure is an integer multiple of half a pitch. The distance a between the two lines is measured under a polarized microscope. The pitch length p is calculated by substituting the distance a and the wedge angle θ into Eq. 1.13.

$$p = 2a\tan\theta \quad (1.13)$$

The phase transition temperature and enthalpy of phase transition are obtained from calorimetry. It is hypothesized that a substance displays a phase sequence of Iso–N–SmA–SmC upon cooling, whereas the crystal melts to the SmA phase then it changes to the N phase upon heating. Figure 1.36 shows an illustration of differential scanning calorimetry (DSC) charts of the substance. The SmC phase does not appear during the heating process. It is a metastable liquid-crystalline phase that is obtained only by cooling below the melting point through the supercooling phenomenon. This phenomenon is referred to as monotropic. Conversely, the N and SmA phases are thermodynamically stable phases that are obtainable by heating. This phenomenon is referred to as enantiotropic. DSC measurements provide a value for the entropy change. Given that the Gibbs energies of the high and low temperature phases are equal at the phase transition temperature, the phase transition entropy is calculated

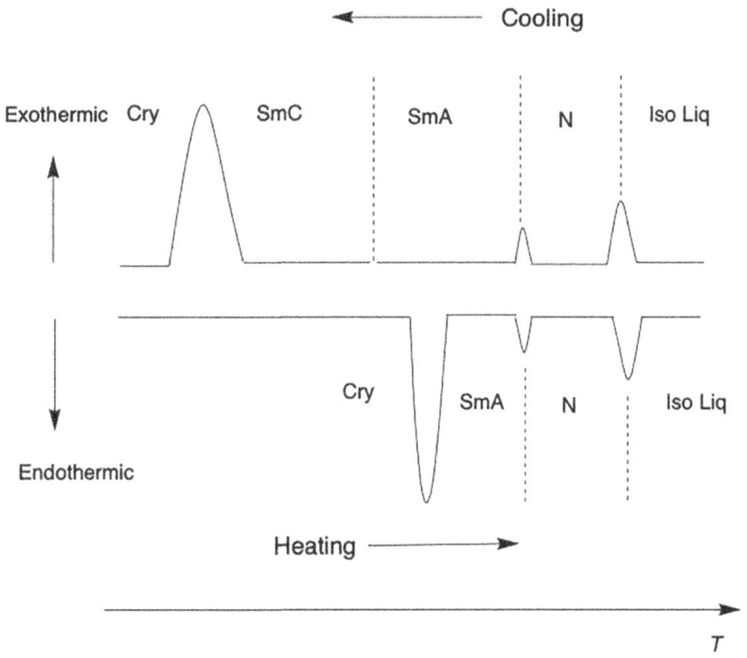

Fig. 1.36 Illustration of differential scanning calorimetry (DSC) charts

using the following equation.

$$\Delta S = \Delta H / T \tag{1.14}$$

Phase transitions can be categorized into two distinct phases. A phase transition that exhibits a discontinuity in the first derivative of the Gibbs free energy is designated as a first-order phase transition. At the first-order phase transition, an enthalpy change occurs, and the high and low temperature phases coexist. The Iso–N and N–SmA phase transitions illustrated in Fig. 1.36 are of the first order type. A phase transition in which a discontinuity appears for the first time in the second derivative of the Gibbs free energy is designated as a second-order phase transition. The SmA–SmC phase transition depicted in Fig. 1.36 is of second order. The system exhibits no transition enthalpy, and the two phases do not coexist.

X-ray diffraction (XRD) measurements are a method of obtaining information regarding the periodic structure of a liquid crystal phase [30]. Smectic phases are characterized by a layered structure, and the periodicity of these phases can be determined through the use of X-ray diffraction measurements. The phenomenon of strong light emission is observed when the path difference of light reflected by neighboring lattice planes is an integer multiple of the light wavelength (see Fig. 1.37). This phenomenon is known as Bragg reflection, and its interference condition is expressed

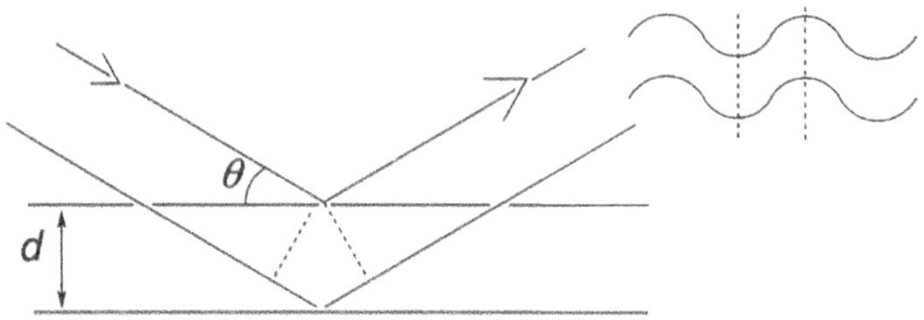

Fig. 1.37 Illustration of Bragg reflection

in Eq. (1.15):

$$n\lambda = 2d\sin\theta \tag{1.15}$$

λ is the wavelength of X-ray, d is the distance between adjacent lattice planes, θ is the angle of incidence, and n is an arbitrary integer.

Figure 1.38a shows a schematic illustration of the XRD pattern of a smectic A phase. A set of spots in the figure corresponds to the period of the layer, and two broad haloes represent the order within the layer is close to that of an isotropic liquid. Figure 1.38b shows that of a smectic C phase.

Figure 1.39 shows a schematic illustration of the XRD profile of the SmA phase. The sharp peak at the small angle corresponds to the layer spacing, and it can be calculated from the angle θ using Eq. 1.14. The broad peak corresponds to the intermolecular distance within the layer. In the SmA and SmC phases, the order in within the layers is liquid-like, resulting in the broad peaks shown in the figure. On the other hand, higher-ordered phases, such as a smectic B phase, manifest sharp peaks.

The correlation length along the layer normal (ξ) was determined using the Ornstein–Zernicke expression as follows. First, the X-ray profile as a function of 2θ was converted to a scattering function of q according to:

$$q = 4\left(\frac{\pi}{\lambda}\right)\sin\theta. \tag{1.16}$$

By fitting the X-ray profiles using the following Lorentzian equation, the correlation length ξ was determined:

$$I(q) = \frac{I_0}{\{1 + (q - q_0)^2\xi^2\}} + \text{background} \tag{1.17}$$

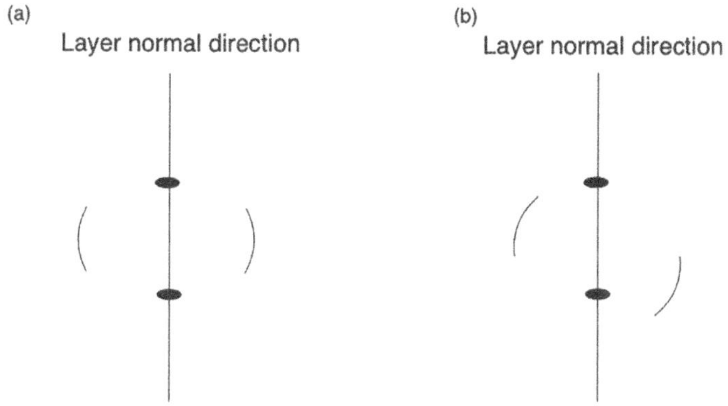

Fig. 1.38 **a** Schematic of the XRD pattern of a smectic A phase and **b** that of a smectic C phase

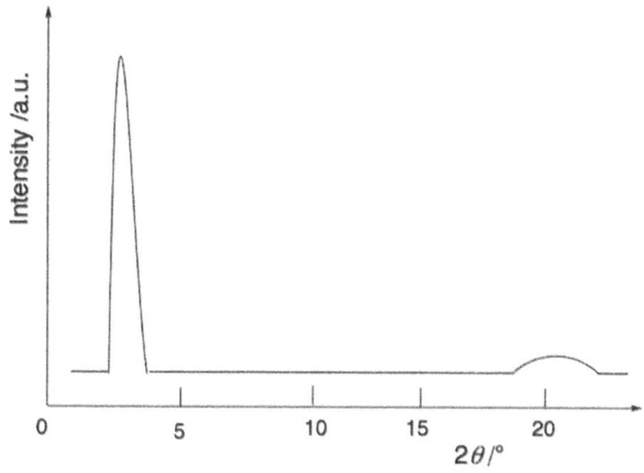

Fig. 1.39 Schematic of the XRD profile of the SmA phase

1.4 Electric Filed Effects

The orientational change of liquid-crystalline molecules due to an electric field based on dielectric anisotropy is described. The application of an electric field induces polarization in the material. The factors that cause polarization can be categorized into three distinct groups: electronic, ionic, and orientational polarization. In the case of liquid-crystalline materials, the orientational polarization occurs. The phenomenon of orientational polarization occurs when the permanent is oriented in the direction of the electric field due to the application of an electric field. The electrical capacitance increases from C to εC when the space between the polar plates of the capacitor is filled with a substance. This rate of change, denoted by ε, is referred to as the dielectric constant of the substrate. In liquid crystal phases such as the nematic phase, the orientation of rod-shaped molecules gives rise to different dielectric constants in the long-axis and short-axis directions of the molecules. In this model, the nematic phase is conceptualized as an elongated ellipsoid, as shown in Fig. 1.40. The dielectric

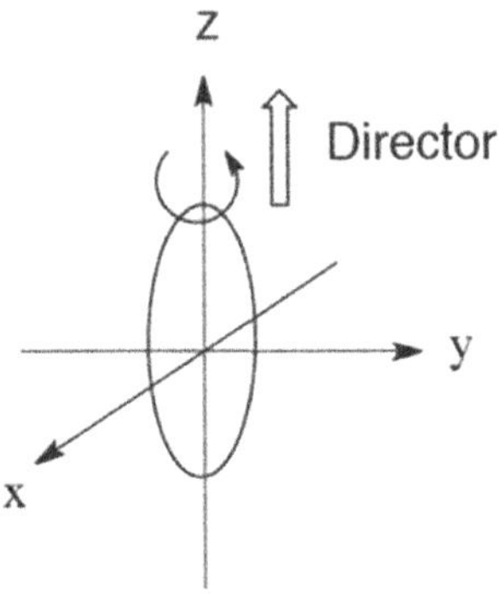

Fig. 1.40 Ellipsoid model of nematic phase

constant along the long axis of a molecule can be expressed as $\varepsilon_{\parallel} = \varepsilon_{zz}$ and the dielectric constant along the short axis as $\varepsilon_{\perp} = 1/2\ (\varepsilon_{xx} + \varepsilon_{yy})$. Dielectric anisotropy is defined as in Eq. 1.18:

$$\Delta\varepsilon = \varepsilon_{\parallel} - \varepsilon_{\perp}. \tag{1.18}$$

Figure 1.41a shows the temperature dependence of the dielectric constant of a nematic liquid crystal with positive dielectric anisotropy. The dielectric anisotropy of this material increases with decreasing temperature [31]. Dielectric anisotropy is proportional to the product of the order parameter and the molecular polarizability. As the temperature decreases, the molecules align in the direction of the orientation vector, thereby increasing the dielectric anisotropy. Figure 1.41b shows the temperature dependence of the dielectric constant of a nematic liquid crystal with negative dielectric anisotropy [32]. The dielectric constant along the long-axis direction of the molecule is smaller than that along the short-axis direction. The dielectric anisotropy is negative. The absolute value of dielectric anisotropy increases with decreasing temperature.

The reorientation process of the director due to an electric or magnetic field is referred to as the Freedericksz transition [33, 34]. The schematic representation of the application of an electric field to the liquid-crystalline molecules situated between the pole plates is depicted in Fig. 1.42. ***E*** is the electric field vector; ***n*** is the director and θ is the angle between ***E*** and ***n***.

The electrical energy density is expressed by the following equation:

$$f_e = \left(-\frac{1}{2}\right)\varepsilon_0\varepsilon_{\perp}E^2 - \left(\frac{1}{2}\right)\varepsilon_0\Delta\varepsilon(\boldsymbol{n} \bullet \boldsymbol{E})^2 = \left(-\frac{1}{2}\right)\varepsilon_0\varepsilon_{\perp}E^2 - (\frac{1}{2})\varepsilon_0\Delta\varepsilon n^2E^2\cos^2\theta. \tag{1.19}$$

When the applied voltage exceeds the threshold voltage, the nematic liquid crystal with positive dielectric anisotropy has a minimum fe when θ is 90°, and is oriented the director is parallel to the electric field direction. Conversely, nematic liquid crystals

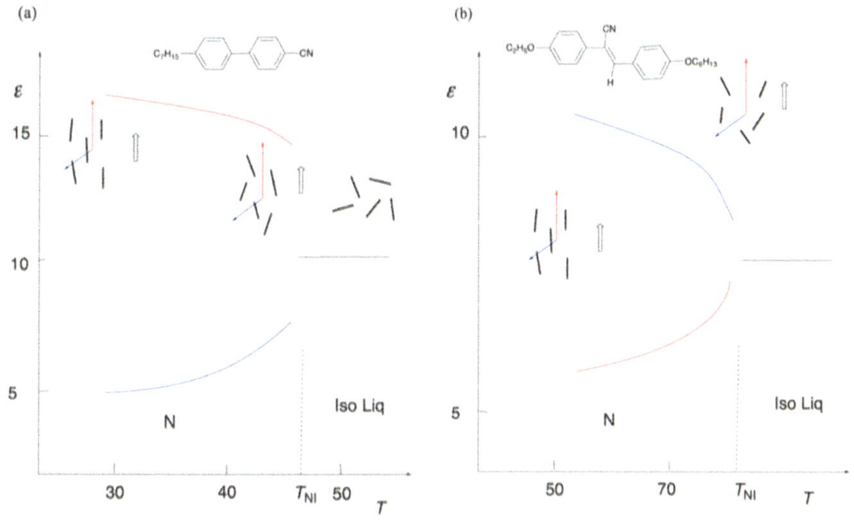

Fig. 1.41 **a** Temperature dependence of the dielectric constant of a nematic liquid crystal with positive dielectric anisotropy [31] and **b** that with negative dielectric anisotropy. The red lines show $\varepsilon_{\parallel}$ and the blue lines show $\varepsilon_{\perp}$ [32]

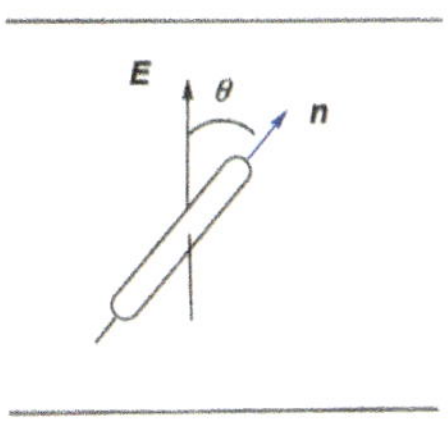

Fig. 1.42 Schematic diagram of the application of an electric field to the liquid-crystalline molecules between the pole plates

with negative dielectric anisotropy exhibit an orientation in which the director is perpendicular to the direction of the electric field.

1.5 Liquid Crystal Displays

Taking a twisted-nematic LCD as an example, the operating mechanism is described below (Fig. 1.43) [35]. Liquid crystalline materials with positive dielectric anisotropy such as 4-cyano-4′-pentylbiphenyl (**5CB**) are used. Two glass substrates are used to seal the liquid crystal. Spacers keep the distance between the substrates at about 5 μm. The polarizer and analyzer are perpendicular to each other. Transparent electrodes are attached to the inside of each substrate. An alignment film is used to orient the liquid crystal. The rubbing direction of the top and bottom alignment films should align with the respective orientations of the polarizer and analyzer. When no voltage is applied, the liquid-crystalline molecules twist 90° up and down. Light passing through the polarizer is twisted 90° by the liquid-crystalline molecules and passes through the analyzer, making the display becomes bright. When voltage is applied, the liquid crystalline molecules change their orientation state so that they become perpendicular to the top and bottom substrates. Consequently, the polarized light continues on its way and cannot pass through the analyzer, resulting in a dark state.

The performance requirements for liquid crystal materials utilized in display applications are subject to variation, depending on the specific application. The physical properties of these materials are typically modified through the incorporation of 10 or more compounds. Organic materials have been demonstrated to be unstable when exposed to light and water, which initially led to their classification as unsuitable for use in electronic materials. However, sufficient reliability has been achieved by encapsulating the liquid crystal material between glass plates and preventing degradation due to light with UV-cut film. The initial step in this research was to identify compounds capable of forming nematic liquid crystal phases at ambient temperature. In order to enhance the response speed, polar groups such as cyano groups,

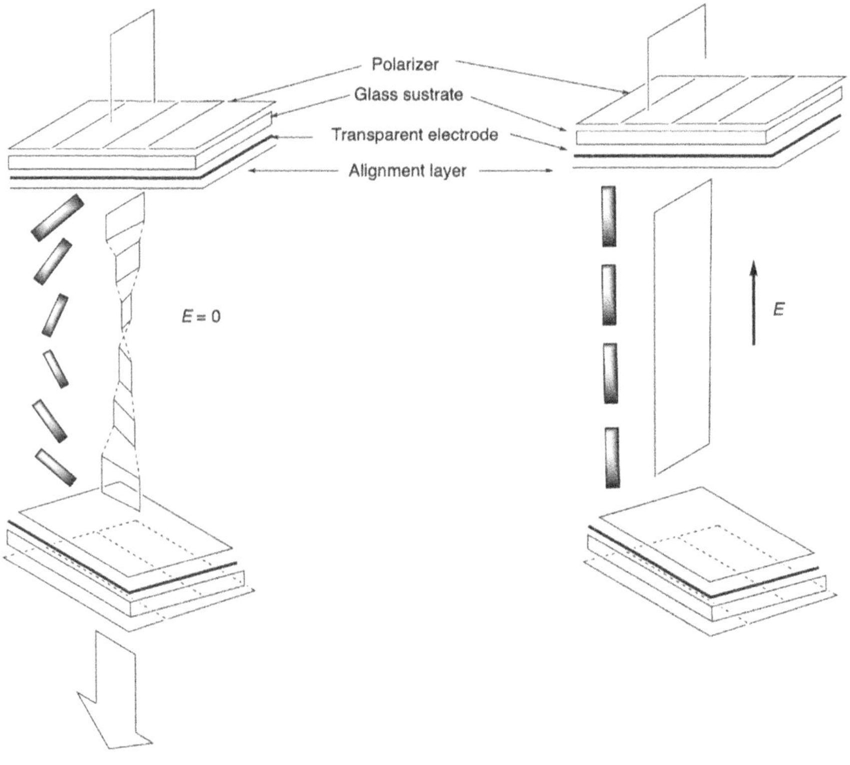

Fig. 1.43 Schematic diagram of twisted-nematic LCD operation

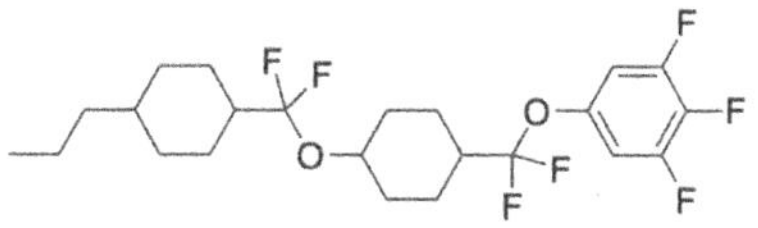

Fig. 1.44 Molecular structure of a nematic liquid crystal for TFT LCDs

which exhibit favorable interactions with electric fields, were incorporated, or cyclohexane rings were substituted for benzene rings to reduce viscosity. Thin-film transistor (TFT) devices are utilized in PC monitors and televisions, which require liquid crystal materials with very low conductivity. The atoms that can be incorporated into molecules are limited to C, H, and F. Fluorine-based liquid crystal materials are often used. Figure 1.44 shows an example of liquid crsystal molecules for TFT-LCDs [36].

The invention of a single molecule can lead to the emergence of a new device, which can engender a substantial ripple effect. Cyanobiphenyl molecules, exemplified by **5CB**, as synthesized by Gray et al., manifest a nematic phase at ambient temperature, exhibit rapid responsiveness to electric fields, and demonstrate notable chemically stability. This development subsequently led to the pervasive utilization of LCDs in clocks and calculations, which, in turn, gave rise to the flat panel displays, such as televisions [37].

1.6 Liquid-Crystalline Gels and Nanoparticles

Nanostructured liquid crystals have been extensively investigated. Liquid-crystalline physical gels and liquid crystals containing nanoparticles are discussed here. Liquid crystals can be gelled by the addition of gelators. The gelator molecules self-aggregate in the liquid crystal through noncovalent interactions. Kato et al. conducted pioneering research on liquid gels [38, 39]. Various liquid-crystalline phases, including nematic, smectic, and discotic phases, can be transformed into their LC physical gels. A schematic representation of the phase-separated structure of a nematic LC gel is depicted in Fig. 1.45 [40].

The utilization of LC gels has been demonstrated to enhance the functionality of the original LC. Nematic LC gels were injected into TN cells to investigate the impact of gel incorporation on electro-optical response. The LC gels containing 0.4 wt% of a gelator exhibited a rise response time and threshold voltage that were less than half of those observed in the original LC. The authors discuss that the dispersed aggregates may weaken the interactions between the LC molecules and the surface of the alignment layer [41]. LC gels are a class of functional soft materials that find application in a variety of fields, including displays, electro-optical memory, semiconductors, sensors, and drug delivery.

Nanoparticles (NPs) are generally in the range of 1–100 nm in size and have the capacity to form the supramolecular assemblies depending on the morphology of the host liquid crystal [42–44]. In order to enhance the solubility of metal NPs in liquid crystals and to functionalize them, a variety of chemical reactions can

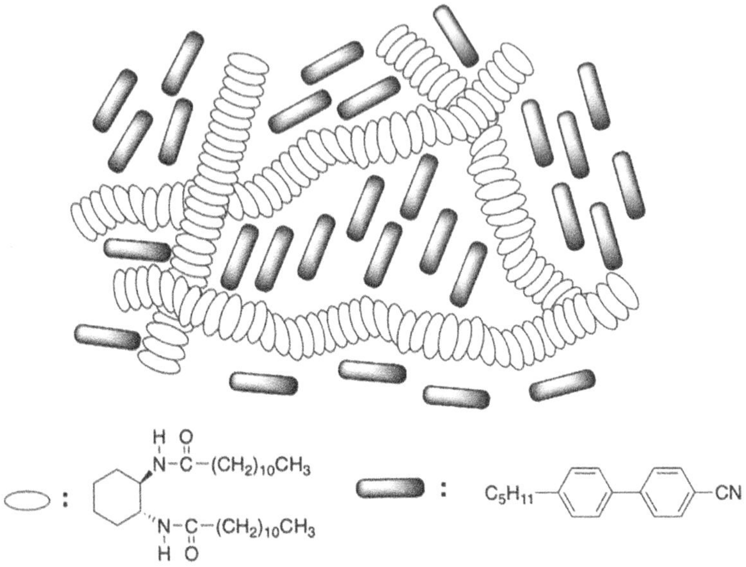

Fig. 1.45 Schematic illustration of a nematic LC gel [40]

be performed. Figure 1.46 shows a schematic representation of mixed-monolayer protected gold NPs [45]. A two-step ligand-exchange process was employed to provide them. Such protected metal particles may exhibit liquid-crystalline properties in isolation; however, they significantly impact the physical properties of the mixed system through interactions with liquid-crystalline molecules. The use of doping inorganic nanoparticles has been documented to enhance the performance of liquid crystal displays [43]. The presence of MgO and SiO_2 in the nanoparticles has been demonstrated to reduce the threshold volage of TN LCD [46]. Nematic liquid crystals doped with low concentrations of ferroelectric nanoparticles such as $Sn_2P_2S_6$ or $BaTiO_3$, have been observed to exhibit a decrease in the switching voltage and an increase in the isotropic–nematic phase transition temperature [47–49]. These particles can generate local electric fields within the LC system, thereby enhancing the intermolecular interactions [50].

Carbon nanotubes (CNTs) are a class of extraordinary anisotropic nanomaterials with diameters measured in nanometers and lengths ranging from micrometers to centimeters. CNTs are seamless cylinders of carbon atoms and can exhibit remarkable properties such as tensile strength and high electric conductivity applying to the direction of the tube axis. The organization of CNTs for to demonstrate their anisotropic properties is a matter of significant importance [51]. The alignment methods of CNTs employing liquid crystals have been the subject of extensive investigation. Dierking et al. reported that CNTs dispersed in a nematic liquid crystal are aligned along the director of the liquid crystal [52]. The LC director field acts as a template for nanotube ordering.

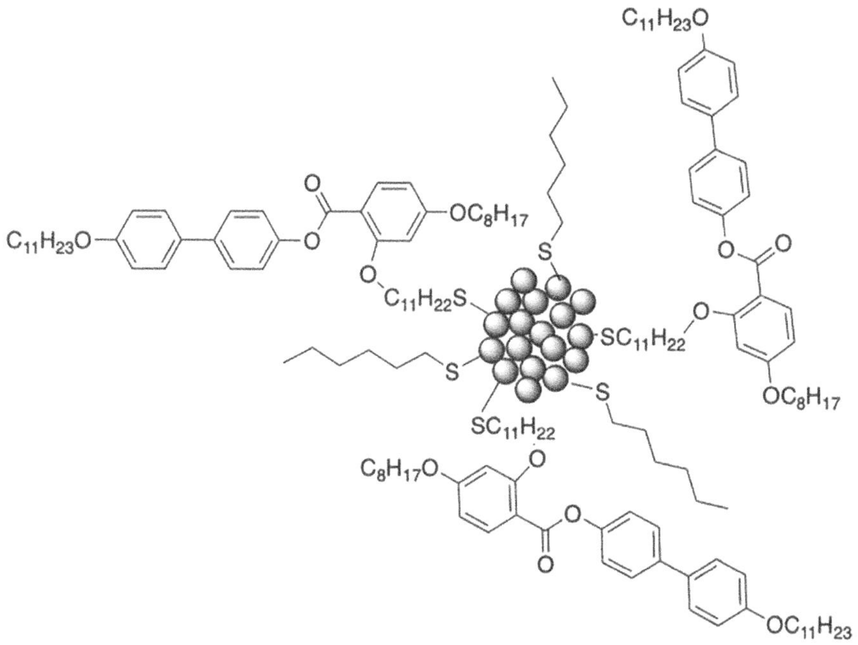

Fig. 1.46 Schematic representation of the protected gold nanoparticles [45]

1.7 Liquid-Crystalline Supermolecules

Molecules in liquid crystal phases possess dual faces, manifesting as static and dynamic images (Fig. 1.47). The static molecular structure can be readily delineated, while the dynamic properties, such as rotation around the long axis, reorientation of the long axis, internal motions, and cooperative motions, are more challenging to conceptualize. Chemists design thermotropic LC molecules with the anisotropic intermolecular interactions, namely, hard-core repulsion and electrostatic interactions. It is generally accepted that visualizing the static state of the molecule is the norm. From a dynamic perspective, the correlation of molecular motion neighboring molecules considerably affects long-range order in the liquid-crystalline phases.

Liquid-crystalline supermolecules, comprising multiple mesogenic units linked by spacers, have been developed. Liquid crystal dimers serve as the simplest exemplification of these phenomena. These molecules are attractive due to their ability to exhibit different properties from the corresponding low-molecular-weight liquid crystal [53]. The transition properties depend on the parity of methylene number in the spacer. The overall molecular shape is estimated based on a static description, namely a thermodynamically stable state. The even-numbered dimer possesses a linear configuration, while whereas the odd-numbered dimer exhibits a bent shape as shown in Fig. 1.48.

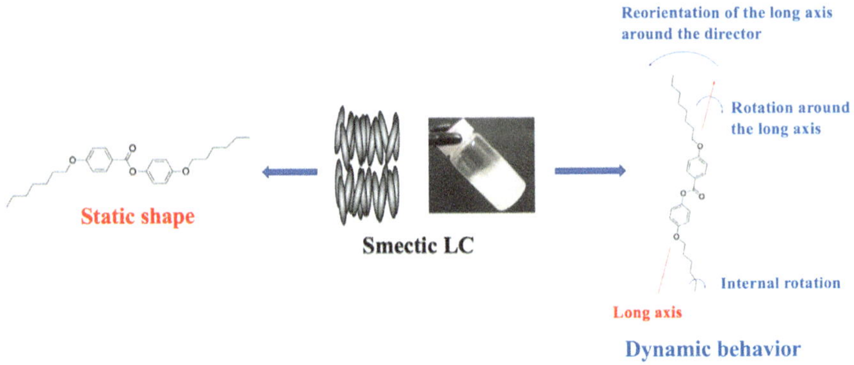

Fig. 1.47 Molecules in LC phases have two faces, i.e., static and dynamic images

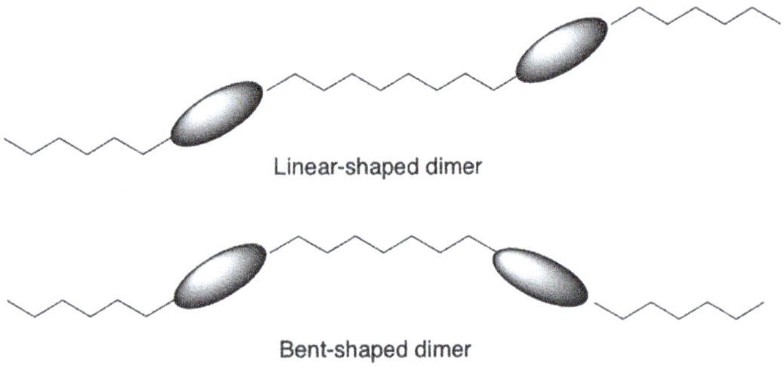

Fig. 1.48 Liquid crystal dimers: the overall molecular shape depends the parity of methylene number in the spacer

We have designed flexible liquid-crystalline supermolecules. These molecules produce characteristic anisotropy of molecular motion in the liquid crystalline phase. In order to stabilize the system, the flexible molecules have the capacity to form thermodynamically unstable conformations. The following H-shaped compound (Fig. 1.49) [54] will be introduced. The liquid-crystalline supermolecule displays an electro-optical switching in the isotropic phase above the nematic phase. Figure 1.50a shows the optical texture of the compound in the absence of an electric field. A completely dark state was obtainable in the isotropic liquid at T_{NI} + 0.5 °C. Figure 1.50b shows that with an electric field of 18 V μm^{-1}. A bright state was obtained. The compound induced a significantly higher level of transmittance and exhibited a broader temperature range of switching than **6OCB**. They organize the microscopic nematic order with a long-range coherence length of the molecules in the isotropic liquid (Fig. 1.51).

Liquid-crystalline supermolecules, designed according to this methodology, exhibit a short-range order within a molecule and possess the capacity to generate

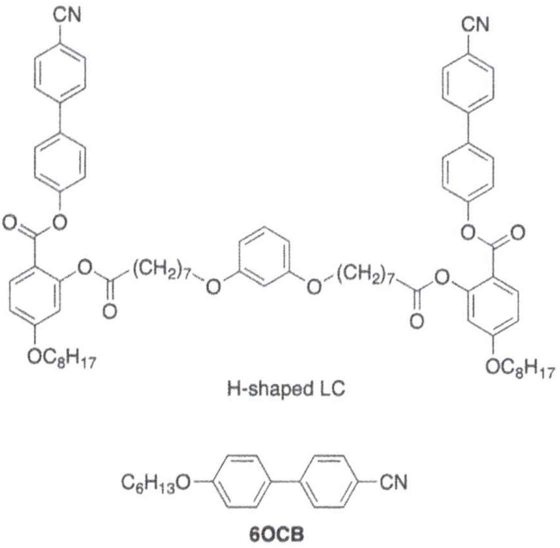

Fig. 1.49 Molecular structures of the H-shaped LC and **6OCB**

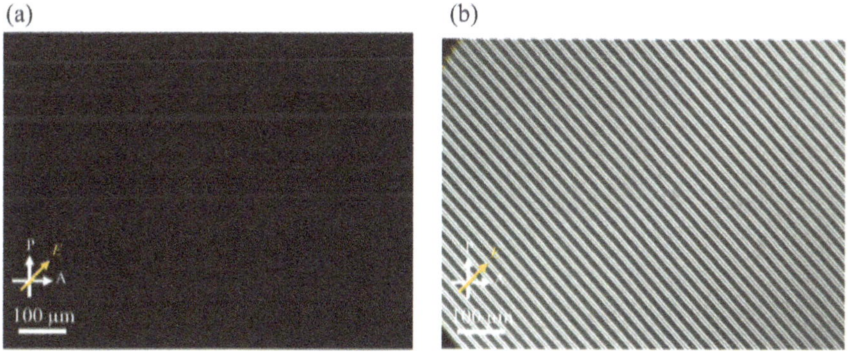

Fig. 1.50 **a** Optical texture of the H-shaped compound in the isotropic liquid at T_{NI} + 0.5 °C; **b** that with an electric field of 18 V μm^{-1} at a frequency of 10 Hz, which was set at 45° to the polarising axes of both polarizers. The stripes in **b** are electrodes. Reprinted with permission from [54]. Copyright (2019) Taylor & Francis Ltd.

liquid-crystalline phases with a hierarchical structure [55]. Even familiar liquid-crystalline materials when observed from a new angle can show us new worlds.

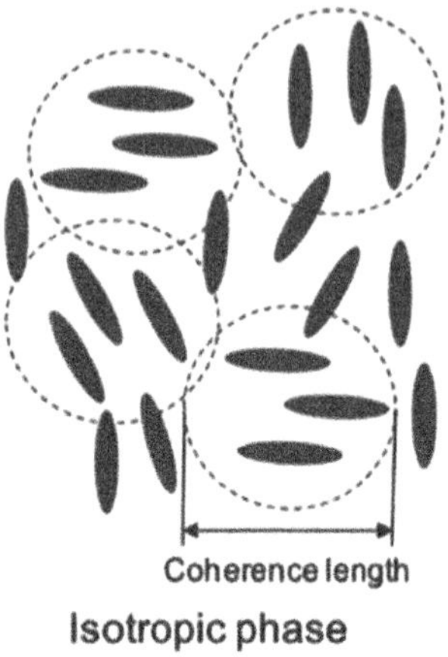

Fig. 1.51 Molecules with coherence length in the isotropic phase. Reprinted with permission from [54]. Copyright (2019) Taylor & Francis Ltd.

References

1. Dunmur, D., Sluckin, T.: Soap, Science, and Flat-Screen Tvs: A History of Liquid Crystals. Oxford University Press, Oxford, UK (2011)
2. Gray, G.W., Harrison, K.J., Nash, J.A.: New family of nematic liquid crystals for displays. Electronic. Lett. **9**, 130–131 (1973)
3. Yoshizawa, A.: Liquid crystalline materials. In: Nomura, M., Suzuka, T. (eds.) Modern Industrial Chemistry, pp. 194–201. Kodansha Scientific, Tokyo (2004)
4. Yoshizawa, A.: Research and development of liquid-crystalline supramolecular assemblies as anticancer drugs. In: Goel, N., Kumar, N. (eds.) Pharmaceutical Applications of Supramolecules, pp. 89–112. Springer, Cham, Chap. 4 (2022)
5. Fairhurst, C.E., Fuller, S., Gray, J., Holmes, M.C., Tiddy, G.J.T.: Lyotropic surfactant liquid crystals. In: Demus, D., Goodby, J.W., Gray, G.W., Spiess, H.W., Vill, V. (eds.) Handbook of Liquid Crystals, vol. 3, 1st edn, pp. 341–392. Wiley-VCH Verlag GmbH, Weinheim, Germany (1998)
6. Hyde, S.T.: Identification of lyotropic liquid crystalline mesophases. In: Holmberg, K. (ed.) Handbook of Applied Surface and Colloid Chemistry, pp. 299–332. John Wiley & Sons Ltd., USA (2001)
7. Israelachvili, J.: The science and applications of emulsions–an overview. Colloids Surf. A **91**, 1–8 (1994)
8. Lee, W.B., Mezzenga, R., Fredrickson, G.H.: Anomalous phase structure in lyotropic liquid crystals. Phys. Rev. Lett. **99**, 187801 (2007)
9. Huang, Y., Gui, S.: Factors affecting the structure of lyotropic liquid crystals and the correlation between structure and drug fusion. RSC Adv. **8**, 6978–6987 (2018)
10. Onsager, L.: The effects of shape on the interactions of colloidal particles. Ann. N. Y. Acad. Sci. **51**, 627–659 (1949)
11. Maier, W., Saupe, A.: Eine einfache molekulare Theories des nematischen kristallinflüssigen Zustandes. Zeitschrift Naturforschung, **13a**, 564–566 (1958)
12. Demus, D.: Chemical structure and mesogenic properties. In: Demus, D., Goodby, J.W., Gray, G.W., Spiess, H.W., Vill, V. (eds.) Handbook of Liquid Crystals, 1st edn, vol. 1, pp. 133–187. Wiley-VCH Verlag GmbH, Weinheim, Germany (1998)
13. Funakoshi, K., Hoshino, N., Matsunaga, Y.: Effects of terminal substituents on mesomorphic properties: phenyl 444-X-substituted phenyliminomethyl)benzoate and its isomeric compounds. Mol. Cryst. Liq. Cryst. 197–206 (1994)

14. Tschierske, C.: Micro-segregation, molecular shape and molecular topology-partners for the design of liquid crystalline materials with complex mesophase morphologies. J. Mater. Chem. **11**, 2647–2671 (2001)
15. Kato, T.: Hydrogen-bonded systems. In: Demus, D., Goodby, J.W., Gray, G.W., Spiess, H.W., Vill, V. (eds.) Handbook of Liquid Crystals, 1st edn, vol. 2B, pp. 969–979. Wiley-VCH Verlag GmbH, Weinheim, Germany (1998)
16. Kato, T., Wilson, P.G., Fujishima, A., Frechet, J.M.: Hydrogen-bonding liquid crystals. A novel mesogen incorporating nonmesogenic 4,4′-bipyridine through selective recognition between hydrogen bonding donor and acceptor. Chem. Lett. **19**, 2003–2006 (1990)
17. Kato, T., Frechet, J.M., Wilson, P.G., Saito, T., Uryu, T., Fujishima, A., Jin, C., Kaneuchi, F.: Hydrogen-bonded liquid crystals. Novel mesogens incorporating nonmesogenic biphyridyl compounds through complexation between hydrogen-bond donor and acceptor moieties. Chem. Mater. **5**, 1094–1100 (1993)
18. Goodby, J.W.: Chirality in liquid crystals. J. Mater. Chem. **1**, 307–318 (1991)
19. Kitzerow, H.-S., Bahr, C. (eds.) Chirality in Liquid Crystals. Springer-Verlag, New York, NY, USA (2001)
20. Davis, E.J., Goodby, J.W.: Symmetry and chirality in liquid crystal. In: Goodby, J.W., Collings, P.J., Kato, T., Tschierske, C., Gleeson, H.F., Raynes, P. (eds.) Handbook of Liquid Crystals, vol. 1, 2nd edn, pp. 197–230. Wiley-VCH Verlag GmbH, Weinheim, Germany (2014)
21. Meyer, R.B., Liebert, L., Strzelecki, L., Keller, P.: Ferroelectric liquid crystals. J. Phys. Lett. **36**, L69–L71 (1975)
22. Crooker, P.P.: Blue phases. In: Kitzerow, H.-S., Bahr, C. (eds.) Chirality in Liquid Crystals, pp. 186–222. Springer-Verlag, New York, NY, USA (2001)
23. Goodby, J.W.: Twist grain boundary and frustrated liquid crystal phases. Curr. Opin. Colloid Interface Sci. **7**, 326–332 (2002)
24. Yoshizawa, A.: Material design for blue phase liquid crystals and their electro-optical effects. RSC Adv. **3**, 25475–25497 (2013)
25. De Gennes, P.G.: An analogy between superconductors and smectics A. Solid State Commun. **10**, 753–756 (1972)
26. Renn, S.R., Lubensky, T.C.: Abriksov dislocation lattice in a model of the cholestericto–smectic-A transition. Phys. Rev. A **38**, 2132–2147 (1988)
27. Goodby, J.W., Waugh, M.A., Stein, S.M., et al.: Characterization of a new helical smectic liquid crystal. Nature **337**, 449–452 (1989)
28. Cowling, S.J.: Optical microscopy studied of liquid crystals. In: Goodby, J.W., Collings, P.J., Kato, T., Tschierske, C., Gleeson, H.F., Raynes, P. (eds.) Handbook of Liquid Crystals, vol. 1, 2nd edn, pp. 263–300. Wiley-VCH Verlag GmbH, Weinheim, Germany (2014)
29. Dierking, I.: Liquid crystal textures: an overview. Liq. Cryst. **52**, 1474–1506 (2025)
30. Agra-Kooijman, D.M., Kumar S.: X-ray scattering investigations of liquid crystals. In: Goodby, J.W., Collings, P.J., Kato, T., Tschierske, C., Gleeson, H.F., Raynes, P. (eds.) Handbook of Liquid Crystals, vol. 1 , 2nd edn, pp. 301–338. Wiley-VCH Verlag GmbH, Weinheim, Germany (2014)
31. Ratna, B.R., Shashidhar, R.: Dielectric properties of 4’-*n*-alkyl-4-cyanobiphenyls in their nematic phases. Pramana **6**, 278–283 (1976)
32. De Jeu, W.H., Van Der Veen, J.: Instabilities in electric fields of a nematic liquid crystal with large negative dielectric anisotropy. Phys. Lett. A **44**, 277–278 (1973)
33. Freedericksz, V., Repiewa, A.: Theoretisches und Experimentelles zur Frage nach der Natur der anisotropies Flüssigkeiten. Zeit. für Phys. **42**, 532–546 (1927)
34. Blinov, L.M.: Behavior of liquid crystals in electric and magnetic field. In: Demus, D., Goodby, J.W., Gray, G.W., Spiess, H.W., Vill, V. (eds.) Handbook of Liquid Crystals, vol. 1, 1st edn, pp. 477–534. Wiley-VCH Verlag GmbH, Weinheim, Germany (1998)
35. Schadt, M., Helfrich, W.: Voltage-dependent optically activity of a twisted nematic liquid crystal. Appl. Phys. Lett. **18**, 115–118 (1971)
36. Kubo, T., Goto, Y.: JP 5962164.

37. Kobayashi, S., Miyama, T., Akiyama, H., Ikemura, A., Kitamura, M.: Development of liquid crystal displays and related improvements to their performance. Proc. Jpn. Acad. Ser. B **98**, 493–516 (2022)
38. Kato, T., Hirai, Y., Nakaso, S., Moriyama, M.: Liquid-crystalline physical gels. Chem. Soc. Rev. **36**, 1857–1867 (2007)
39. Yabuuchi, K., Kato, T.: Liquid-crystalline gels. In: Goodby, J.W., Collings, P.J., Kato, T., Tschierske, C., Gleeson, H.F., Raynes, P. (eds.) Handbook of Liquid Crystals, vol. 6, 2nd edn, pp. 3–25. Wiley-VCH Verlag GmbH, Weinheim, Germany (2014)
40. Kato, T., Kutsuna, T., Hanabusa, K., Ukon, M.: Gelation of room-temperature liquid crystals by the association of a *trans*-1,2-bis(amino)cyclohexane derivative. Adv. Mater. **10**, 606–608 (1998)
41. Mizoshita, N., Hanabusa, K., Kato, T.: Fast and high-contrast electro-optical switching of liquid-crystalline physical gels: formation of oriented microphase-separated structures. Adv. Func. Mater. **13**, 313–317 (2003)
42. Bisoyi, H.K., Kumar, S.: Liquid-crystal nanoscience: an emerging avenue of soft self-assembly. Chem. Soc. Rev. **40**, 306–319 (2011)
43. Umadevi, S., Ganesh, V., Hegmann, T.: Nanoparticles: additives and building blocks for liquid crystal phases. In: Goodby, J.W., Collings, P.J., Kato, T., Tschierske, C., Gleeson, H.F., Raynes, P. (eds.) Handbook of Liquid Crystals, vol. 6, 2nd edn, pp. 27–76. Wiley-VCH Verlag GmbH, Weinheim, Germany (2014)
44. Chen, C.-H., Dierking, I.: Nanoparticles in thermotropic and lyotropic liquid crystals. Front. Soft Matter (2025). https://doi.org/10.3389/frsfm.2024.1518796
45. Cseh, L., Mehl, G.H.: The design and investigation of room temperature thermotropic nematic gold nanoparticles. J. Am. Chem. Soc. **128**, 13376–13377 (2006)
46. Haraguchi, F., Inoue, K., Toshima, N., Kobayashi, S., Takato, K.: Reduction of the threshold voltages of nematic liquid crystal electrooptical devices by doping inorganic nanoparticles. Jpn. J. Appl. Phys. **46**, L796–L797 (2007)
47. Reznikov, Yu., Buchnev, O., Tereshchenko, O., Reshetnyak, V., Gluschchenko, A, West, J.: Ferroelectric nematic suspension. Appl. Phys. Lett. **82**, 1917–1919 (2003)
48. Ouskova, E., Bunchnev, O., Reshetnyak, Y., Reznikov, Y., Kresse, H.: Dielectric relaxation spectroscopy of a nematic liquid crystal doped with ferroelectric $Sn_2P_2S_6$ nanoparticles. Liq. Cryst. **30**, 1235–1139 (2003)
49. Kaczmrek, M., Buchnev, O., Nanhadkumar, I.: Ferroelectric nanoparticles in low refrective index liquid crystals for strong electro-optic response. Appl. Phys. Lett. **92**, 103307 (2008)
50. Lopatina, L.M., Selinger, J.V.: Theory of ferroelectric nanoparticles in nematic liquid crystals. Phys. Rev. Lett. **102**, 197802 (2009)
51. Lagerwall, J., Scalia, G.: Carbon nanotubes in liquid crystals. In: Goodby, J.W., Collings, P.J., Kato, T., Tschierske, C., Gleeson, H. F., Raynes, P. (eds.) Handbook of Liquid Crystals, vol. 6, 2nd edn, pp. 99–137. Wiley-VCH Verlag GmbH, Weinheim, Germany (2014)
52. Dierking, I., Scalia, G., Morales, P., LeClere, D.: Aligning and reorienting carbon nanotubes with nematic liquid crystals. Adv. Mater. **16**, 865–869 (2004)
53. Imrie, C.T., Luckhurst, G.R.: Liquid crystal dimers and oligomers. In: Goodby, J.W., Collings, P.J., Kato, T., Tschierske, C., Gleeson, H.F., Raynes, P. (eds.) Handbook of Liquid Crystals, vol. 7, 2nd edn, pp. 137–210. Wiley-VCH Verlag GmbH, Weinheim, Germany (2014)
54. Sayama, S., Muraki, R., Yoshizawa, A.: H-shaped liquid crystals inducing nematic order in the isotropic liquid. Liq. Cryst. **46**, 1756–1762 (2019)
55. Yoshizawa, A.: Unconventional liquid crystal oligomers with a hierarchical structure. J. Mater. Chem. **18**, 2877–2889 (2008)

Chapter 2
Hierarchical Structures and Dissipative Structures Driven by Smectic Layer Fluctuations

Abstract Liquid crystals, which possess properties of both liquids and solids, inevitably exhibit fluctuations. Frustrated liquid crystal phases with hierarchical structures, including cybotactic nematic phases, modulated and incommensurate smectic phases, and bicontinuous cubic phases, are of particular interest as they arise from fluctuations in the layered structure. Furthermore, pattern formation as a non-equilibrium order driven by fluctuations is one of the most compelling research areas in the field of soft mater. The present chapter is concerned with the design of frustrated layer order derived by smectic layer fluctuations. Liquid-crystalline supermolecules are characterized by a structural arrangement in which multiple mesogens are interconnected by flexible spacers. It is evident that the subjects under scrutiny are distinguished not only by their distinctive shape but also by an inherent dynamic order within their molecular composition. The designed supermolecules induce the fluctuations in the layer structure, thereby producing not only the hierarchical liquid crystal phases but also the pattern propagation in a smectic phase. In addition to the thermotropic liquid crystal systems, it is demonstrated that an anionic hybrid surfactant possessing a **6OCB** group exhibits the periodic formation and breakdown of lamellar aggregates in water.

Keywords Liquid crystals · Supermolecules · Fluctuation · Cybotactic nematic phases · Cubic phases · Modulated smectic phases · Dissipative structures · Pattern formation · Rayleigh–Bénard convection · Hybrid surfactants

2.1 Introduction

Liquid crystals combine the anisotropy of solids with the fluidity of liquids. Therefore, fluctuations are a phenomenon characteristic of liquid crystals and their phase transitions. The nematic phase exhibits orientational fluctuations. Some smectic

Supplementary Information The online version contains supplementary material available at https://doi.org/10.1007/978-981-92-2524-8_2.

A. Yoshizawa, *Liquid-Crystalline Supermolecules*, The Materials Research Society Series, https://doi.org/10.1007/978-981-92-2524-8_2

phases show fluctuated layer structures, such as modulated and incommensurate smectic phases. Orientational fluctuations increase near the phase transition [1, 2]. For an example of the pre-transitional behavior, smectic-like layered clusters are repeatedly generated and annihilated in a nematic phase in the vicinity of the nematic–smectic phase transition. Fluctuations in the layer structure give rise to a liquid crystal phase with a hierarchical structure. A cybotactic nematic phase (Ncyb) is a frustrated LC phase where short-range ordered smectic clusters form within the long-range nematic structure [3–6]. Smectic fluctuations in which the smectic-like layer is either normal or tilted, related to the nematic director $\boldsymbol{n}$, exist (Fig. 2.1) [7, 8]. The layer fluctuations in smectic phases produce various frustrated layer structures such as modulated smectic phases [9–12]. Figure 2.2 shows the illustration of the modulated SmA phase [13]. Interactions between molecules in adjacent layers can form a bilayer structure. If the monolayer and bilayer structures coexist, the layer deformation can appear in the crossover region [14]. Bicontinuous cubic phases as shown in Fig. 2.3 are also thought to form through fluctuations in the layer structure. They have three-dimensional interpenetrating networks [15, 16]. They have attracted much attention due to the intriguing nanoscale structure [16–19], but for use in applications such as 3D electronic or ionic conductors [20, 21]. Apart from these equilibrium structures, dissipative structures named as order through fluctuation [22, 23] are an important research topic in the field of soft mater. Figure 2.4 shows the polarized optical texture of a striped pattern propagation observed for a smectic phase. This chapter shows some researches on the pattern formation in both thermotropic and lyotropic systems. We discuss the dynamic behavior of liquid-crystalline supermolecules in the formation of these patterns.

2.2 Molecular Design of Liquid-Crystalline Supermolecules

Typical nematic and smectic liquid-crystalline molecules have a rod-like shape. This molecular design is based on the excluded volume effect determined by hard-core repulsion. On the other hand, a lot of unconventional liquid crystals being different from a rod-like shape have been reported. Some of them are classified as supermolecules in which several or many mesogenic units are connected via a flexible spacer have been investigated (Fig. 2.5) [12, 25–27].

Liquid crystal dimers and trimers are the simplest structures of the supermolecules [12, 26]. An overall molecular shape of the LC dimer depends on the parity of the methylene number of the flexible spacer. The even-numbered LC dimer forms a linear shape in which two mesogenic cores are almost parallel whereas the odd-numbered LC dimer forms a bent shape in which two mesogenic cores are inclined each other (Fig. 2.6). The transition properties of linear dimeric liquid crystals depend on the length and parity of the flexible spacer. The linear liquid-crystalline dimers show pronounced odd–even effects of the spacer on the phase transition behavior are shown; the even-numbered dimer shows the higher Iso–N transition temperature than

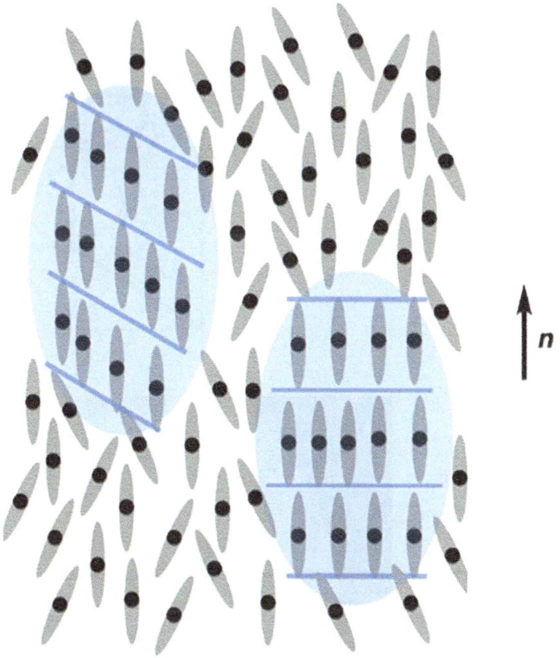

Fig. 2.1 Sketch of a nematic phase exhibiting skewed (left) and normal (right) smectic fluctuations. Reproduced from [8] with permission from the Royal Society of Chemistry

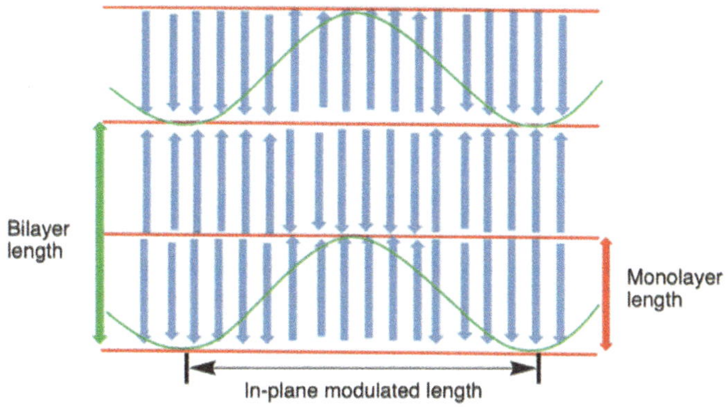

Fig. 2.2 Sketch of a modulated smectic A phase. Reproduced from [13] with permission from the Royal Society of Chemistry

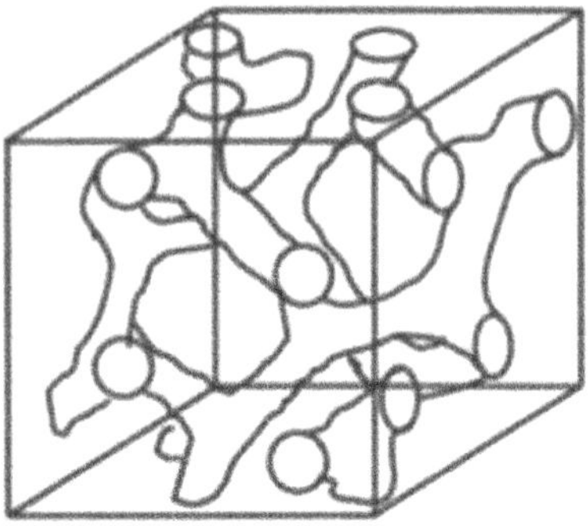

Fig. 2.3 Sketch of a bicontinuos cubic phase

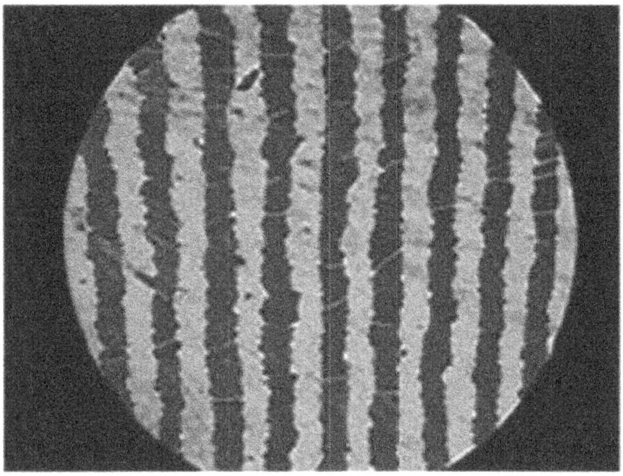

Fig. 2.4 Photomicrograph of the striped pattern propagation in a smectic phase. Reproduced from [24] with permission from the Royal Society of Chemistry

the odd-numbered dimer. This phenomenon is explainable in terms of a length-to-diameter ratio. On the other hand, some odd-numbered LC dimers and trimers exhibit spontaneous mirror symmetry breaking in achiral liquid crystal phases. Furthermore, the twist-bend nematic phase (N_{TB}) showing chiral conglomerates was found for 1,7-bis(4-cyanobiphenyl-4′-yl)heptane [28, 29, also see Chap. 5].

Non-symmetric dimers possessing different cores, for example, cyanobiphenyl and phenylpyrimidine unnits, can exist two kinds of smectic layer structures, i.e., intercalated and interdigitated SmA phases. Their driving forces are specific core–core interactions depending on the lengths of the spacer and the terminal alkyl chain. When the spacer is longer than sum of the terminal alkyl chain and the cyano group lengths, the core–core interactions between unlike cores produce an intercalated SmA phase (Fig. 2.7a). On the other hand, when the spacer is shorter than the terminal alkyl chain, the core–core interactions between same cores form an interdigitated SmA phase (Fig. 2.7b).

According to the definition by Saez and Goodby [25], the difference between super- and supra-molecular liquid crystals is described below. "A supermolecule is a giant molecular entity that is made up of covalently bonded identifiable molecular

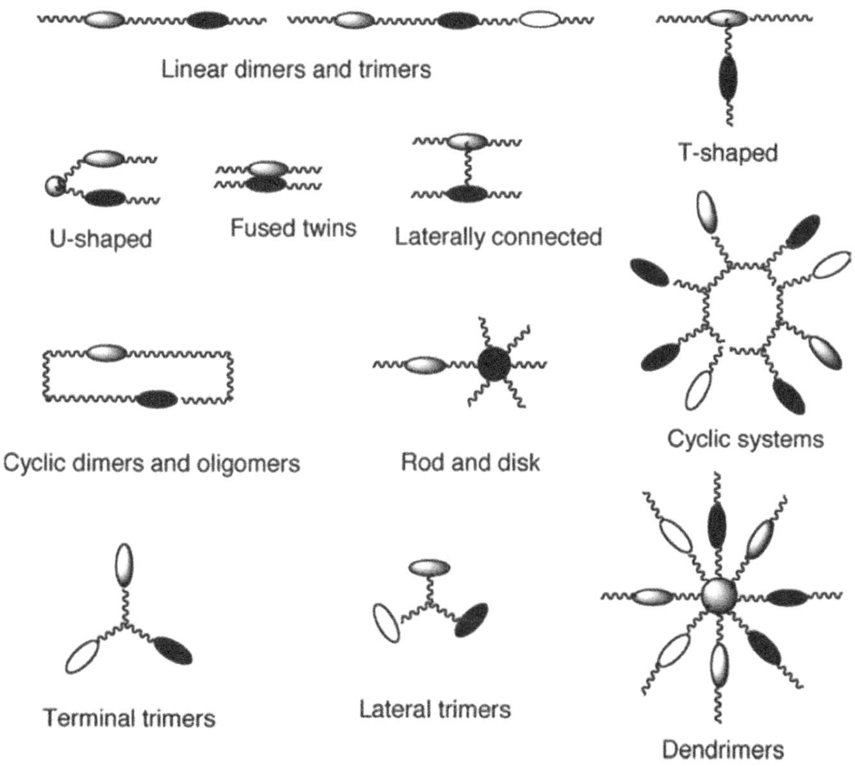

Fig. 2.5 Types of liquid-crystalline super-molecules. Reproduced from [24] with permission from the Royal Society of Chemistry

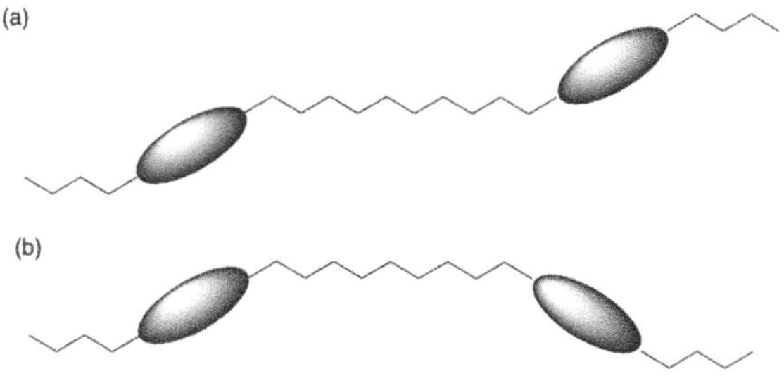

Fig. 2.6 Schematic images of the even-numbered LC dimer (**a**) and the odd-numbered LC dimer (**b**)

units; thus it is similar in constitution to that of a tertiary structure of a protein. A supramolecular system, on the other hand, is a self-assembled, non-covalently bonded entity where complete molecular units are brought together through non-covalent forces to create a complex structure similar in constitution to that of a quaternary structure of a protein." Liquid-crystalline supermolecules possess internal order, and this order changes depending on how the constituent mesogens connect to each other. As the length of the linking unit decreases, the degrees of freedom of

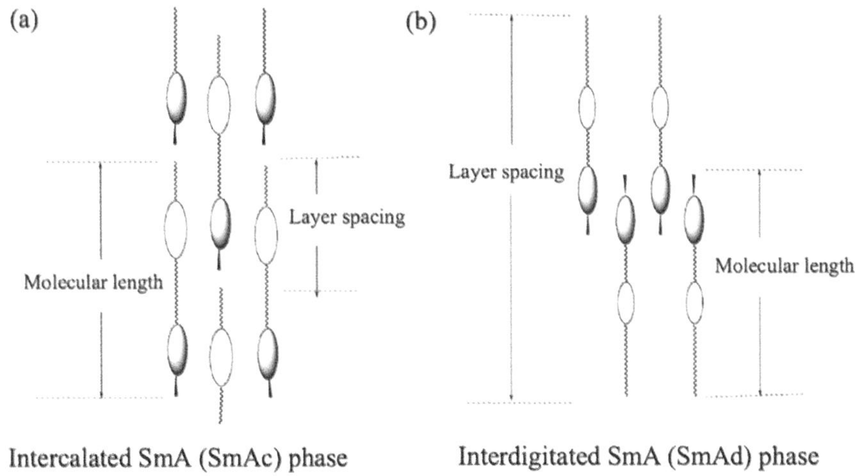

Fig. 2.7 Schematic representations for **a** intercalated SmA and **b** interdigitated SmA structures. Reproduced from [24] with permission from the Royal Society of Chemistry

each mesogen diminish, and the gross molecular shape is reflected in the formation of the mesophase. On the other hand, as the length of the linking unit increases, the properties of each mesogen influence the overall properties of the supermolecular liquid crystals. Various structure–property relationships in supramolecular liquid crystals are shown in the above review. Consider a supermolecule where a scaffold is located at the molecular center and rod-like shaped mesogens are attached to its periphery via spacers. The orientation of the attachment can markedly influence the type of mesophases formed. When mesogens are connected laterally to the central scaffold, the supermolecule tends to form a nematic phase; conversely, when they are connected terminally, it tends to form a smectic phase (Fig. 2.8) [25]. Saez and Goodby designed "Janus" liquid-crystalline supermolecules, which possess two distinct faces within a single molecule [30, 31]. The Janus compound shown in Fig. 2.9 is based on a central scaffold linked possessing end-on three cyanobiphenyl units (smectic preferring) and side-on three chiral phenyl benzoate units (chiral nematic preferring). The compound actually exhibited both chiral nematic and chiral smectic C phases.

We surmised that inducing layer fluctuation produces the above mentioned hierarchical and/or dissipative structures. In order to realize this hypothesis, we have designed various types of liquid crystalline supermolecules [24, 32]. The background is described below. Our ^{13}C NMR studies on smectic liquid crystals reveal the dynamic behavior of each molecule in the layer structure; intralayer cooperative motion and interlayer correlation exist [33]. By connecting two mesogens within or between the layers, we reasoned that these liquid-crystalline supermolecules could amplify either the cooperative motion within the layer or the correlation between the adjacent layers (Fig. 2.10). If so, the U-shaped molecules stabilize the layered structure and the linear-shaped molecules affect the layer correlation. Therefore,

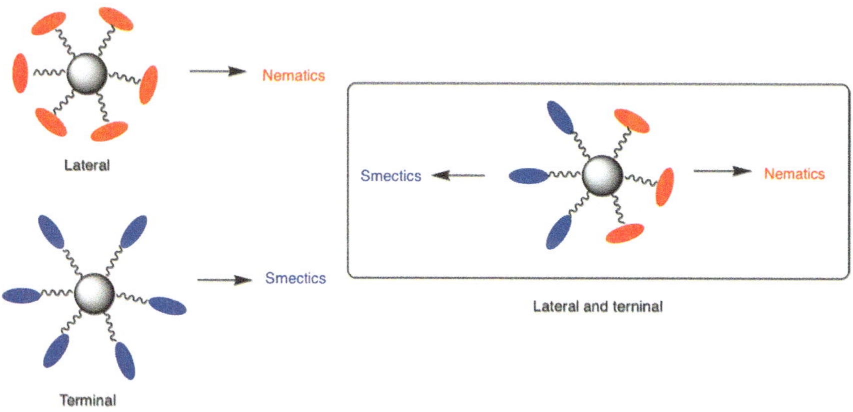

Fig. 2.8 The effect of difference in attaching orientation of rod-shaped mesogens to a central scaffold on mesophase formation

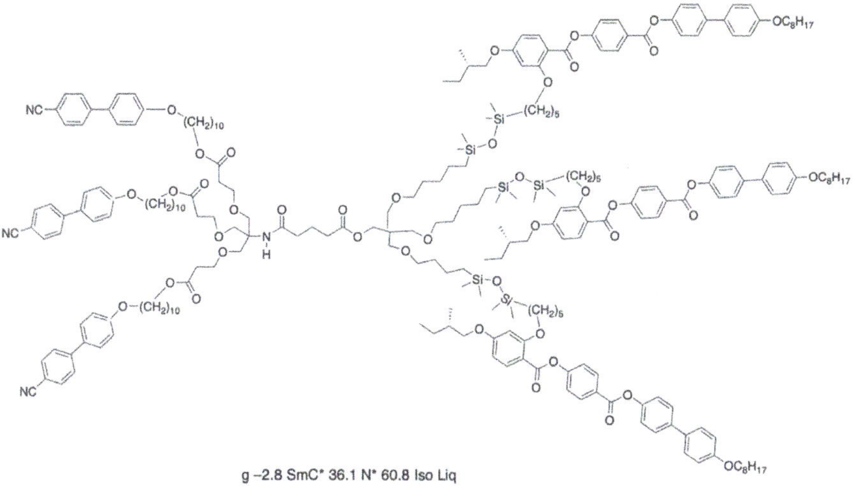

Fig. 2.9 The molecular structure and phase transition temperatures (°C) of "Janus" supermolecule

these liquid crystal oligomers could induce the layer fluctuation in smectic liquid crystals. This chapter shows how the designed liquid-crystalline supermolecules induce the layer fluctuations to produce the frustrated hierarchical and/or dissipative structures. Furthermore, we have demonstrated that some liquid crystal oligomers produce a dynamic chiral origin. A binaphthyl derivative stabilized blue phases [34] (see Chap. 4). The two mesogens might form a twisted conformation. An achiral odd-numbered linear trimer exhibited spontaneous mirror symmetry breaking in the soft-crystalline layered phase [35] (see Chap. 5). The trimer might adopt a metastable twisted conformer to induce the helical order.

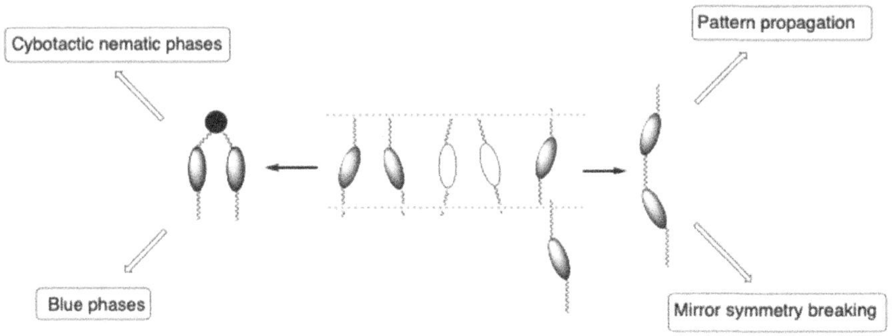

Fig. 2.10 Design concept of the liquid-crystalline supermolecules

2.3 Hierarchical Structures

2.3.1 Cybotactic Nematic Phases

The molecules in nematic phases spontaneously align parallel to each other along the director. The nematic phases are fluid like the isotropic liquid, so they exbibit no positional order but have long-range orientational order [36, 37]. In addition to the conventional nematic phase, unusual nematic phases, such as biaxial nematic (Nb) [38–41] and cybotactic nematic (Ncyb), have been reported. Furthermore, twist bend nematic (Ntb) phases [28, 42] and ferroelectric nematic phases [43–47] have recently been discovered. Nematic liquid crystals are the most widely known and common type of liquid crystal, yet new discoveries continue to be made. With respect to the existence of cybotactic clusters of the Ncyb phases, De Vries observed the short-range order in nematic phases by using X-ray scattering measurements [3, 4]. The Ncyb phases are classified depending on how they appear [6, 48]. Type I: They appear as a pre-transitional order in the low temperature range of the nematic phase. They can be said as fluctuation-induced Ncyb phases and usually observed for rod-shaped molecules. Type II: They appear as structural features in the entirety of the nematic range and do not necessarily have a low-temperature smectic phase. They are observed for bent-core LCs [49, 50]. The fluctuation of cybotactic clusters was detected by dielectric spectroscopy in the Ncyb phase of bent-core LCs [51, 52]. Interestingly, some liquid crystal dimers exhibited Ncyb phases [53] (Fig. 2.11). Both the dimers form the Ncyb phase independent of the parity of the methylene number of the spacer, indicating that the appearance of the Ncyb phase is attributed not to the rigid bent-shape but to the flexible supermolecular dimeric system. The cybotactic order was also observed for a main-chain LC polymer with bent-core repeat unit in the nematic phase [54]. Nasir and Rahman summarized structure–property relationships in Ncyb phases in their review [48]. Recently, it has been discovered that these cluster structures are involved in the formation of various frustrated liquid crystal phases. The cluster formation can play an important role in the appearance of the

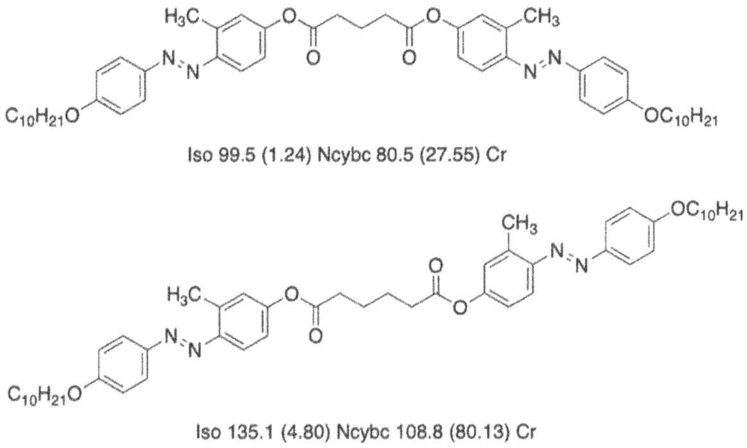

Fig. 2.11 The molecular structures, transition temperatures (°C) and enthalpies (kJ mol^{-1}, in parentheses) of transition on the first cooling for the liquid crystal dimers [53]. Ncybc: Cybotactic nematic phase composed of smectic C-like nanoclusters

biaxial nematic phase [6, 40, 55–57]. Bent-core LCs exhibiting the Ncyb phase are known to stabilize blue phases [58] (see Chap. 4). Furthermore, ferroelectric-like polar switching has been observed in some Ncyb phases [52, 59, 60]. Interestingly, Nishikawa et al. found that the nano-cluster formation is related to the appearance of the nematic ferroelectric order [61] (see Chap. 3).

The fluctuation-induced Ncyb (Type I) phases appear in a narrow temperature range above the N–Sm phase transition. There are only few reports on broadening the temperature range [24]. We demonstrated that U-shaped and hydrogen-bonding systems stabilize the Ncyb phases [8, 62]. Their molecular structures and phase transition temperatures are shown in Fig. 2.12.

We designed the U-shaped compound **BHPDFB** possessing two 2,3-difluoro-1,4-diphenylbenznen units (Fig. 2.12a) As shown in Fig. 2.10, the liquid-crystalline supermolecule has an intra-molecular lateral order, which is expected to stabilize a layer structure. **BHPDFB** exhibited N and SmC phases. Figure 2.13 shows the X-ray diffraction patterns of **BHPDFB** in the N and SmC phases. Clear but diffuse scattering appeared beside the meridian in the N phase (Fig. 2.13b), indicating that the cybotactic clusters exist. Furthermore, the layered cluster form an SmC-like skewed structure. The temperature-dependent XRD measurements reveal that the Ncyb phase began to appear at $T–T_{NC}$ = +20 K (Fig. 2.14). The layer spacing continuously decreases from 31.7 to 29.8 Å as the temperature decreases (Fig. 2.14a). The layer spacings are shorter than the estimated molecular length of 33 Å. Therefore, the molecular long axis is tilted with respect to the layer normal as the same as a typical monolayered SmC phase and the tilt angle continuously increases as the temperature decreases. On the other hand, the discontinuous increase in the correlation length

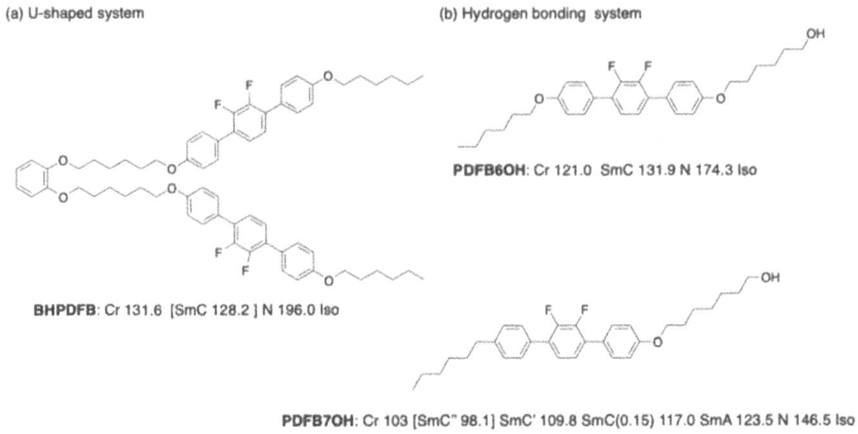

Fig. 2.12 The molecular structures and transition temperatures (°C) for **BHPDFB**, **PDFB6OH**, and **PDFB7OH**. Square brackets denote a monotropic transition

corresponds to the Ncyb–SmC transition (Fig. 2.14b). **BHPDFB** possessing an intramolecular lateral order is thought to induce short-lived nano layered structures in the N phase as shown in Fig. 2.15.

We designed the on/off switching of interlayer interactions by using the hydrogen bonding system (Fig. 2.12b) [8, 62]. As shown in Fig. 2.16, the Ncyb phase of **PDFB6OF** appeared over 25 K above the N–SmC transition. Interestingly, the layer spacing discontinuously decreases at the Ncyb–SmC transition as decreasing temperature. A typical SmA–SmC transition shows the same behavior. In the Ncyb phase, the layer spacing increases with decreasing temperature, after which the temperature dependence disappears. This is also observed for an SmA phase. The layer spacing of 27 Å in the Ncyb phase is shorter than the molecular length of 29 Å. The difference

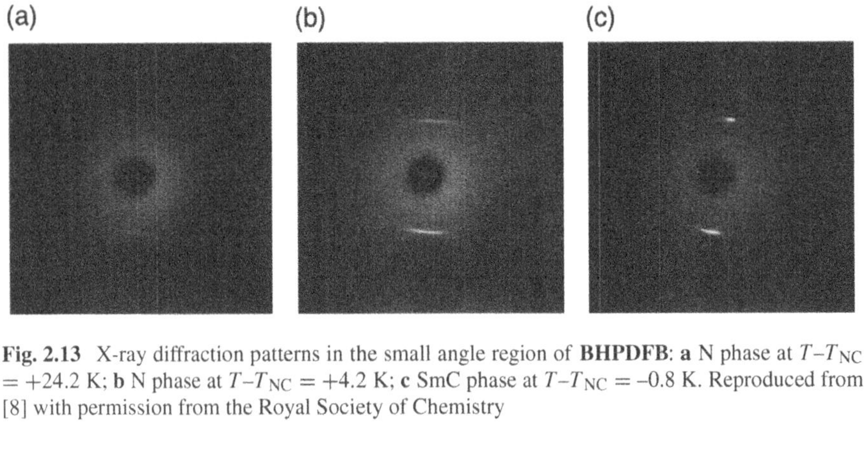

Fig. 2.13 X-ray diffraction patterns in the small angle region of **BHPDFB**: **a** N phase at T–T_{NC} = +24.2 K; **b** N phase at T–T_{NC} = +4.2 K; **c** SmC phase at T–T_{NC} = –0.8 K. Reproduced from [8] with permission from the Royal Society of Chemistry

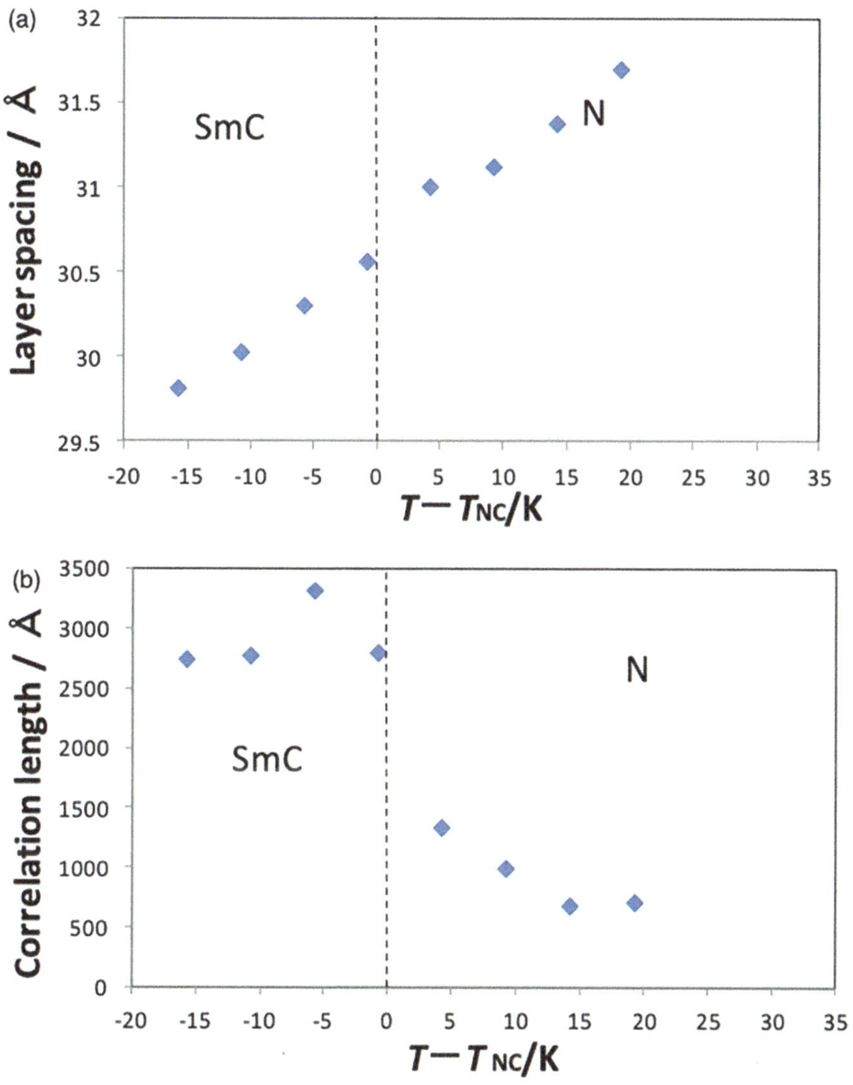

Fig. 2.14 Temperature dependence of **a** the layer spacing and **b** the correlation length of **BHPDFB** in the N and SmC phases. Reproduced from [8] with permission from the Royal Society of Chemistry

between them can be explained by the penetration of the alkyl chains between adjacent layers. Therefore, the Ncyb phase of **PDFB6OF** has a monolayered SmA-like cluster.

Compound **PDFB7OH** exhibited the Ncyb phase and three smectic C phases [60]. The phase structures of the three SmC phases as follows [63]. The SmC and the SmC″ phases have monolayer and bilayer structures, respectively. Interestingly, an in-plane modulated layer structure was observed for the SmC′ phase. The layer modulation

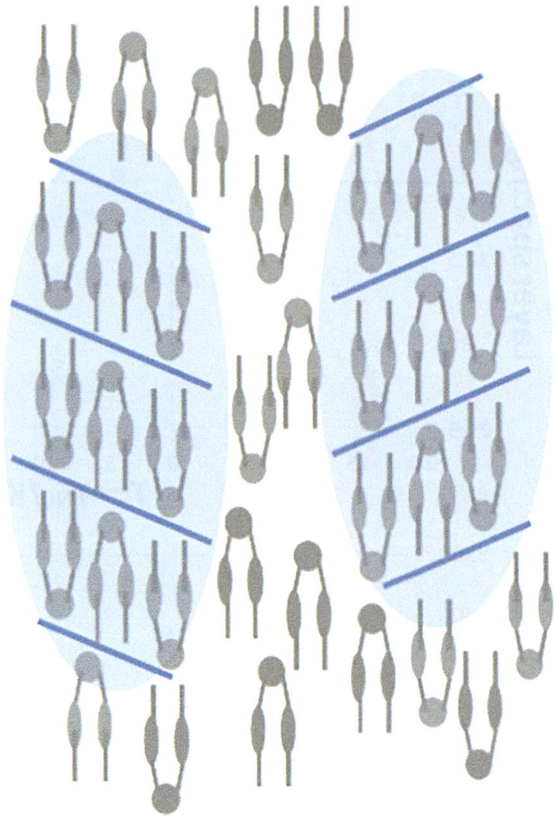

Fig. 2.15 Model for the Ncyb phase of **BHPDFB**. Reproduced from [8] with permission from the Royal Society of Chemistry

results from the competition between the monolayer ordering in the SmC phase and the bilayer ordering in the SmC″ phase. On the other hand, the difluoroterphenyl derivative possessing two terminal hydroxyl groups shown in Fig. 2.17 exhibited the conventional N phase instead of the Ncyb phase [8]. The inter-layer interactions are uniformly strong due to the hydrogen bonding so that the layer fluctuation cannot be induced in the N phase.

We discuss how the hydrogen bonding organizes the layer fluctuation. The Ncyb of **PDFB6OF** has a monolayered SmA-like cluster as described above. Micro-segregation prefers a bilayer-like ordering in which the molecules are aligned in parallel within each layer as shown in the right side of Fig. 2.18. There are two types of interlayer interactions. The one is strong interaction through the hydrogen bonding. The other is weak dispersion force due to penetration of the alkyl chains between adjacent layers. This difference in inter-layer interaction might destabilize a long-range smectic order. On the other hand, packing entropy prefers a monolayer structure in which the molecules are aligned in an anti-parallel manner as shown in the left side of Fig. 2.18. There is no difference in the inter-layer interactions, which might stabilize the monolayer layer structure. Competition between the monolayer

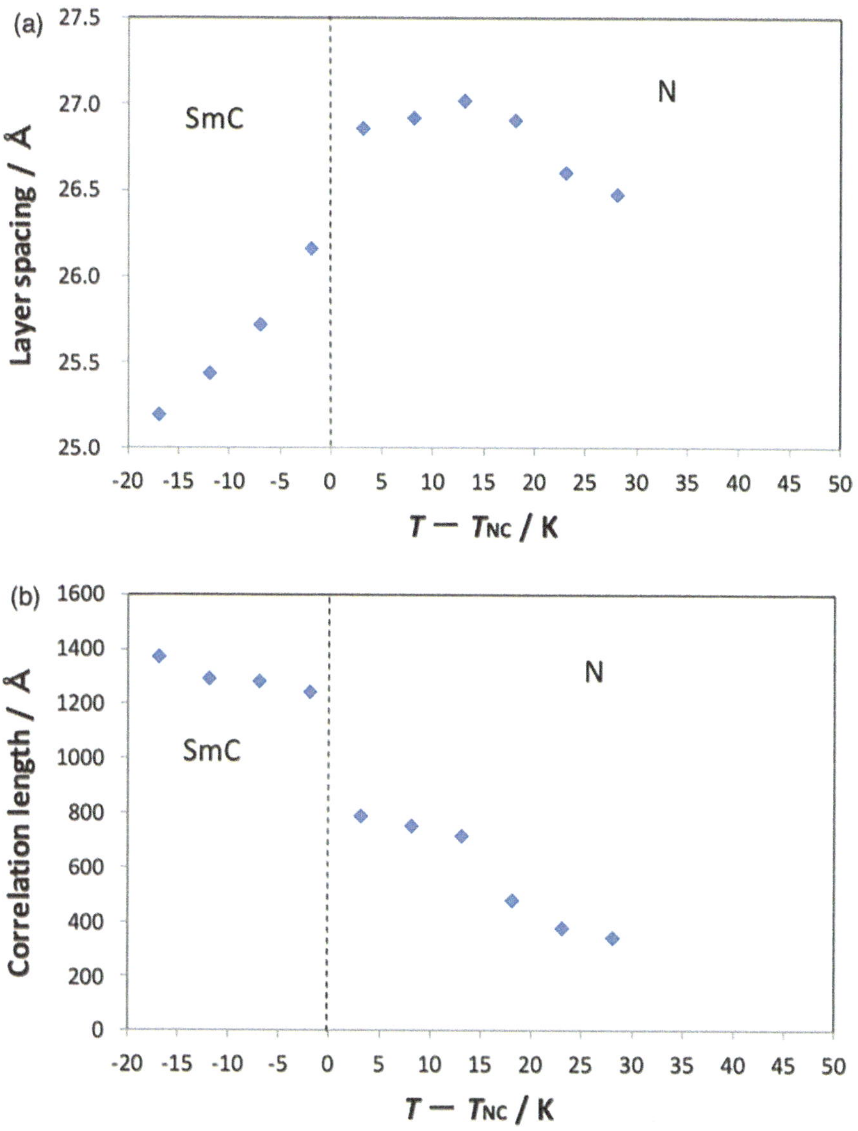

Fig. 2.16 Temperature dependence of **a** the layer spacing and **b** the correlation length of **PDFB6OF** in the N and SmC phases. Reproduced from [8] with permission from the Royal Society of Chemistry

and bilayer characteristics in the monolayer structure can induce the layer fluctuation to organize the Ncyb phase.

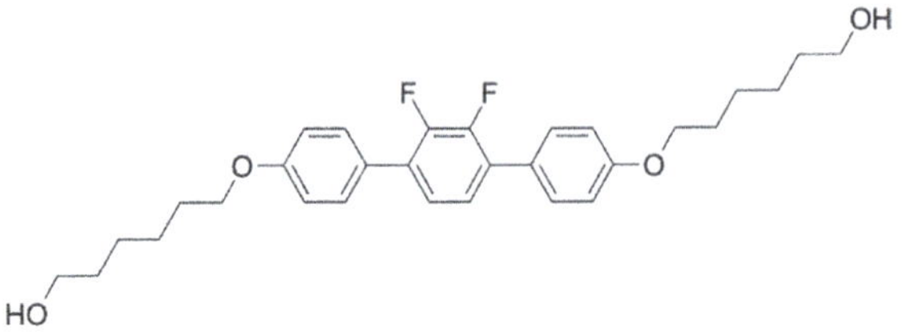

Fig. 2.17 The molecular structure and transition temperatures (°C) for the difluoroterphenyl derivative [8]

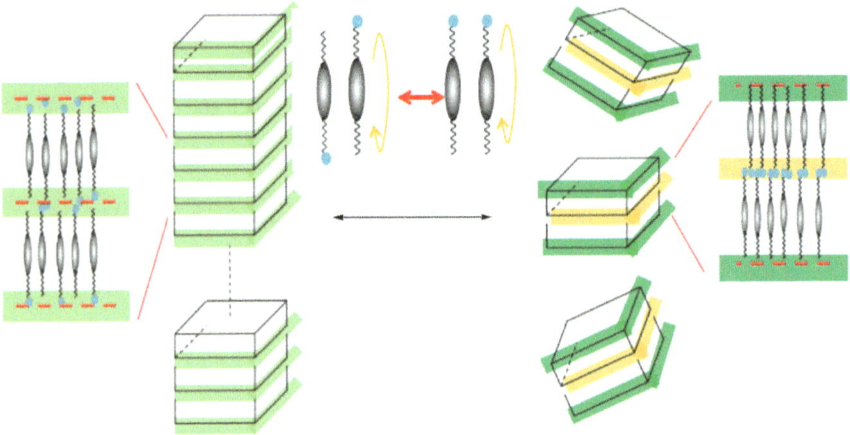

Fig. 2.18 Possible model for the formation of the Ncyb phase of **PDFB6OH**. Reproduced from [8] with permission from the Royal Society of Chemistry

2.3.2 Modulated Smectic Phases

Competition between layered structures with different periodicities gives rise to frustrated smectic phases such as modulated and incommensurate phases [11, 64]. Additional small-angle Brag peaks are observed away from the meridian in the modulated phases [11]. Those peaks serve as evidence for the existence of transverse layer modulation (see Fig. 2.2). The modulated phases have been usually observed for smectic phases composed of polar molecules. Dipole–dipole interactions between the antiparallel molecular pairs in adjacent layers drive the head-to-head dimerization, producing the bilayer structure. The bilayer structure competes against a monolayer structure to induce the layer modulation. The small angle Brag peaks correspond to the in-plane modulation periodicity. On the other hand, the incommensurate phases exhibit several small-angle Brag peaks along the meridian, revealing that layer structures with competing layer spacings coexist. Modulated structures are thought to occur as escape from incommensurability [65]. The modulated smectic phases have

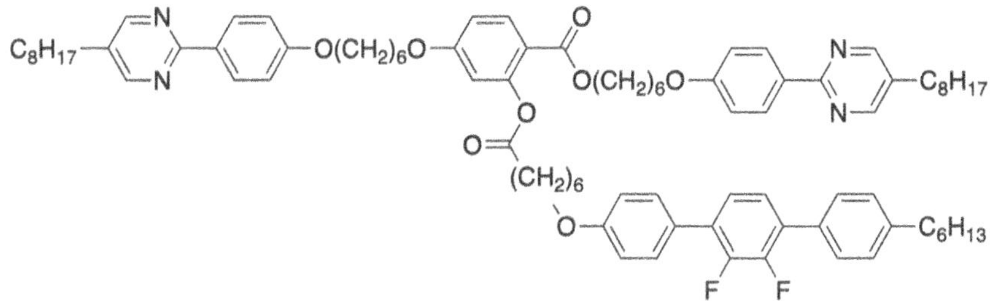

BPYDFB: Cry 113.3 [$\mathrm{Sm}\tilde{\mathrm{C}}$ 83.8 (1.0) SmCanti 87.1 (0.8) SmA 95.7 (1.9) N 104.2 (3.9)] Iso

Fig. 2.19 The molecular structure, transition temperatures (°C), and enthalpies (kJ mol^{-1}, in parentheses) of transition for **BPYDFB** [75]. Square brackets denote a monotropic transition

been observed in biforked mesogens [66], dimers [67–69], liquid crystals with a terminal hydroxyl group [63], bent-shaped molecules [70], mesogens with terminal fluorinated chains [71], and polymeric mesogens [72–74].

Kashima et al. reported that a taper-shaped liquid-crystalline trimer **BPYDFB** shows a modulated SmC phase. Two phenylpyrimidine cores and a difluoroterphenyl core are connected to 2,4-hydroxybenzoic acid via flexible spacers within the liquid-crystalline supermolecule. It exhibited a phase sequence of isotropic liquid–nematic–smectic A–anticlinic smectic C (SmCanti)–modulated smectic C ($\mathrm{Sm}\ \tilde{\mathrm{C}}$) (Fig. 2.19).

The layer spacing of 28.4 Å in the SmA phase was less than half its molecular length of 61 Å, indicating that the SmA phase has an intercalated structure. The polarized optical texture of the SmA phase in the homeotropically aligned region was completely dark, revealing that the director is parallel to the layer normal. Cooling from SmA phase, birefringence in the homeotropically region increased. This texture change is usually observed for an SmA–SmC phase transition. But two brushes in addition to four brushes were detected. The two brushes are characteristic to an anticlinic structure where molecules are tilted in opposite directions between adjacent layers [76]. This phase was identified as SmCanti phase. The SmCanti phase exhibited the scattering corresponding to a layer spacing of 27.8 Å in the two-dimensional X-ray diffraction pattern in the small-angle region (Fig. 2.20a). Therefore, the SmCanti phase also has an intercalated structure. On further cooling, another phase transition was detected by both POM and DSC measurements. The two brushes disappeared. The molecules are thought to reorganize from anticlinic to synclinic orientation. Additional small-angle diffraction spots appeared symmetrically away from the meridian (Fig. 2.20b), which are characteristic of a modulated smectic phase. Therefore, the low-temperature synclinic phase was identified as modulated smectic C ($\mathrm{Sm}\ \tilde{\mathrm{C}}$) phase. The $\mathrm{Sm}\ \tilde{\mathrm{C}}$ phase has three periodicities: the intercalated layer spacing of 27.2 Å, the monolayer spacing of 54.4 Å, and the in-plane modulated length of 115 Å.

Let us discuss how the liquid-crystalline supermolecule **BPYDFB** exhibits the SmCanti and the $\mathrm{Sm}\ \tilde{\mathrm{C}}$ phases. Figure 2.21 shows MOPAC models for (a) T-shaped

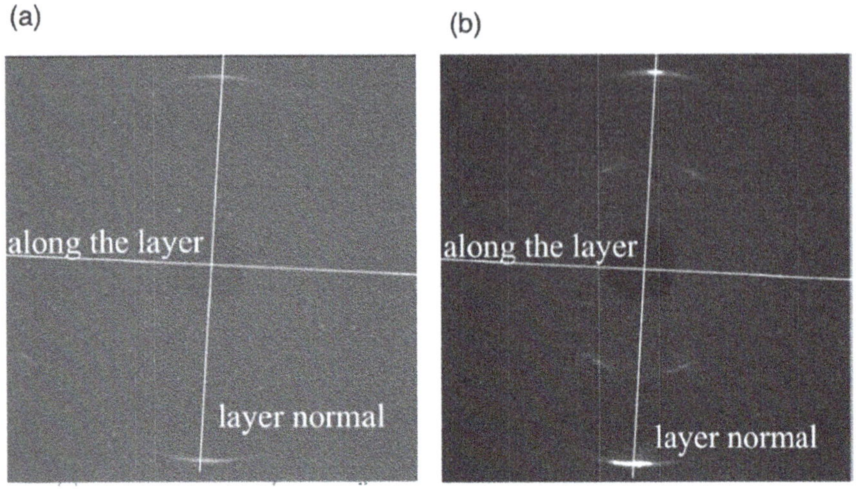

Fig. 2.20 X-ray diffraction patterns in the small-angle region of **BPYDFB** in **a** the SmCanti phase at T–T_{NA} = –9.5 K and **b** the Sm $\tilde{C}$ phase at T–T_{NA} = –13.5 K. Reprinted with permission from [75]. Copyright (2014) Taylor & Francis Ltd.

and (b) taper-shaped conformers of **BPYDFB**. According to the MOPAC calculations, the taper-shaped conformer is stabler than the T-shaped one by 17 kJ mol^{-1} [77]. We assume that **BPYDFB** forms a taper shape in the liquid-crystalline phases. The dipole–dipole interactions between phenylpyrimidine cores in adjacent molecules may induce their conformational change from straight to bent, producing the SmA–SmCanti phase transition as shown in Fig. 2.22.

Then we discuss the SmCanti–Sm $\tilde{C}$ phase transition. The conformational change from bent to straight and the molecular packing in an antiparallel manner are thought to organize the synclinic monolayer structure as shown in the right side of Fig. 2.23. The dynamic intramolecular order of the flexible supermolecule and the intermolecular core–core interactions between unlike cores may drive the reorganization. The intercalated layer structure with a spacing of 27.2 Å competes against the monolayer

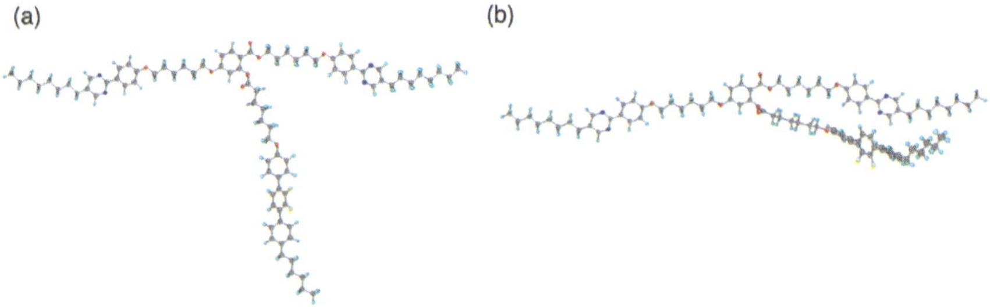

Fig. 2.21 MOPAC models for **a** T-shaped and **b** taper-shaped conformers of **BPYDFB** [77]

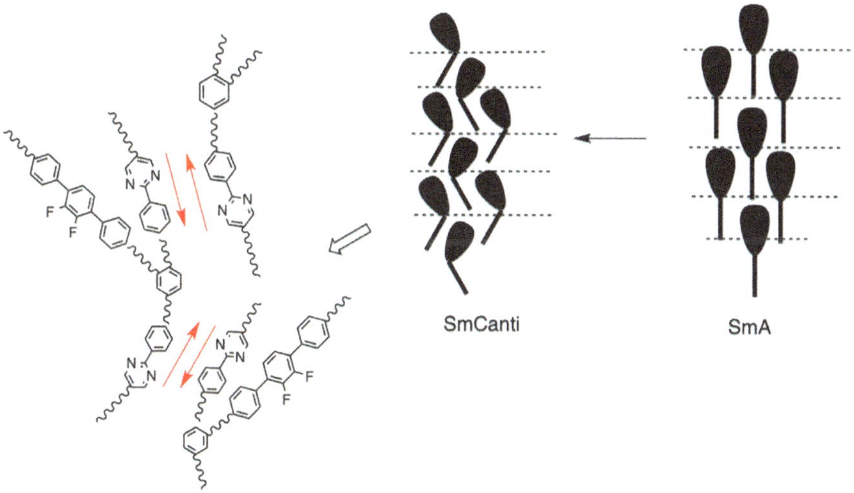

Fig. 2.22 Molecular organization models for the intercalated SmA and SmCanti phases of **BPYDFB**. Red arrows indicate the dipole moment of each phenylpyrimidine unit. Reprinted with permission from [75]. Copyright (2014) Taylor & Francis Ltd.

structure with a spacing of 54.4 Å, inducing the in-plane modulation in the Sm $\tilde{C}$ phase as shown in the left side of Fig. 2.23.

Next, we show a system that incorporates two types of competition (Fig. 2.24). The designed system is as follows. It incorporates competition between a monolayer structure (Type I) and an interdigitated bilayer structure (Type II) for an amphiphilic

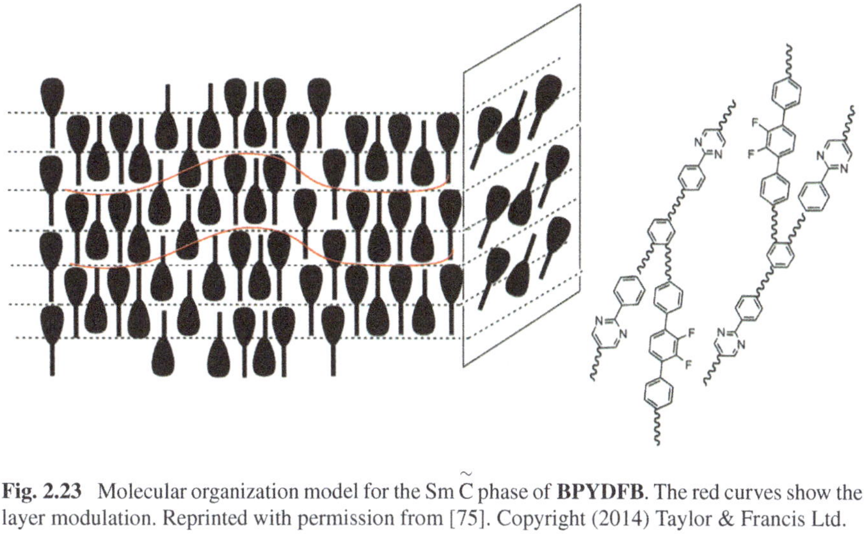

Fig. 2.23 Molecular organization model for the Sm $\tilde{C}$ phase of **BPYDFB**. The red curves show the layer modulation. Reprinted with permission from [75]. Copyright (2014) Taylor & Francis Ltd.

smectic liquid crystal. The competition is coupled with competition between nematic and smectic structures for a mixture of conventional rod-like nematic liquid crystals (Type III) [13]. Figure 2.25 shows the molecular structures of those compounds with their phase transition temperatures.

The amphiphilic compound **HDFCBH** possessing a cyanobiphenyl core and a semiperfluorinated alkyl chain exhibited interdigitated SmA and SmC phases. Figure 2.26 presents the model for the SmA phase. The layer spacing is 51.4 Å whereas the molecular length is 34.5 Å. F-aggregation [78] and dipole–dipole interactions between cyanobiphenyl units stabilize the interdigitated structure (Type II). An equimolecular mixture of nematic liquid crystals **4OCB** and **8OPPY4O** were prepared (**Mixture A**). It showed only smectic phases instead of the N phase. The phase transition temperatures were Iso 98.0 °C SmA 24.2 °C Sm E. The SmA phase has a monolayer structure stabilized by core–core interactions between unlike cores as shown in Fig. 2.27 (Type III).

HDFCBH was mixed with **Mixture A** in various ratios. The phase diagram between them is shown in Fig. 2.28. Interestingly, a mixture containing 20 mol%

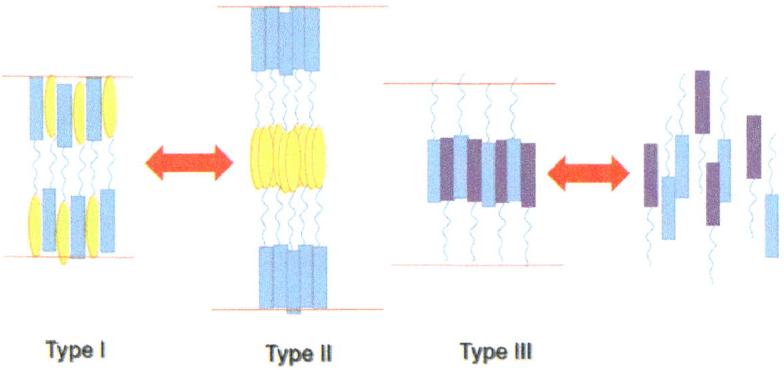

Fig. 2.24 Design concept of the ternary system. Amphiphilic smectic liquid crystals form a monolayer structure (Type I) or an interdigitated bilayer structure (Type II). Nematic liquid crystals with different cores can induce a smectic phase via interactions between unlike cores (Type III). Reproduced from [13] with permission from the Royal Society of Chemistry

Fig. 2.25 The molecular structures and transition temperatures on cooling (°C) for compounds **HDFCBH**, **4OCB**, and **8OPPY4O** [13]

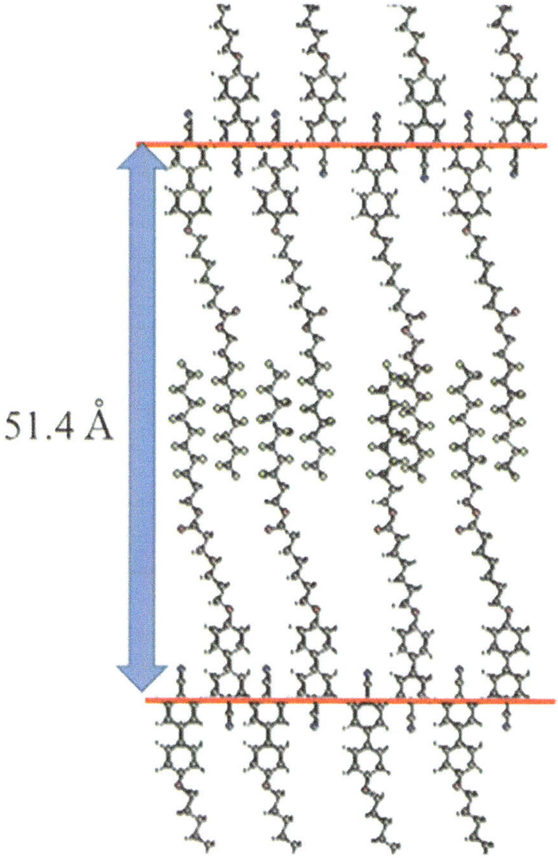

Fig. 2.26 Model for the SmA phase of **HDFCBH**. The layer spacing obtained by XRD studies is indicated. The molecular length is estimated as 34.5 Å. Reproduced from [13] with permission from the Royal Society of Chemistry

of **HDFCBH** exhibited complex phase transitions as follows: Iso 102.3 °C Iso + SmA 97.5 °C N 95.6 °C ~ 92.2 °C X 72.8 °C Y 60.7 °C Z. Figures 2.29 and 2.30 show their optical textures and the cooling DSC thermogram, respectively. On cooling from the isotropic liquid, the SmA phase appeared and kept coexisting with the isotropic liquid until the appearance of the N phase. According to the XRD measurements, the SmA phase has a monolayer structure. The layer spacing was 34.2 Å, which is almost equal to the molecular length of **HDFCBH** (34.5 Å). The amphiphilic compound **HDFCBH** forms the monolayer structure (Type I) instead of the interdigitated structure (Type II). **4OCB** and **8OPPY4O** molecules might enter the layer structure of **HDFCBH**, as shown in Fig. 2.31. Unexpectedly, the N phase appeared below the SmA phase. The N phase is not a re-entrant phase. The N phase changed to an unidentified X phase. The N–X did not accompany enthalpy change. A fan-shaped texture (Fig. 2.29c) was observed in the planar alignment region, whereas a homeotropically aligned sample in the X phase showed a completely dark texture. The POM textures suggest that the X phase is a uniaxial SmA-like phase. However, the X phase showed no periodicity corresponding to the layer spacing. Furthermore,

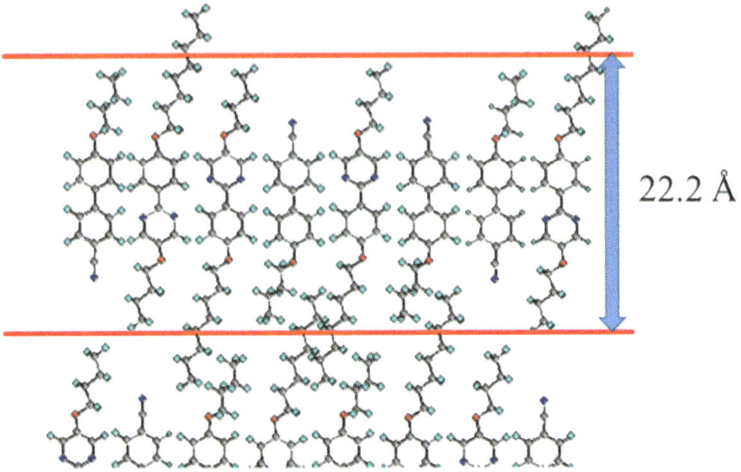

Fig. 2.27 Model for the SmA phase of **Mixture A**. The layer spacing obtained by XRD studies is indicated. The molecular lengths of **4OCB** and **8OPPY4O** are estimated as 16.3 Å and 25.8 Å, respectively. Reproduced from [13] with permission from the Royal Society of Chemistry

the fluidity the X phase exhibited the same fluidity as the N phase. The X–Y transition was detected by POM but did not accompany enthalpy change. The shear stress discontinuously increased at the phase transition. The Y phase is a modulated phase with three periodicities: the intercalated monolayer spacing of 29 Å, the interdigitated bilayer spacing of 57 Å, and the in-plane modulated length of 150 Å (Fig. 2.32a).

The X phase shows a smectic A-like texture but does not have a layer structure. Let us discuss how the X phase occurs. On cooling from the coexistent state of the isotropic liquid and the SmA phase, **4OCB** and **8OPPY4O** in the isotropic

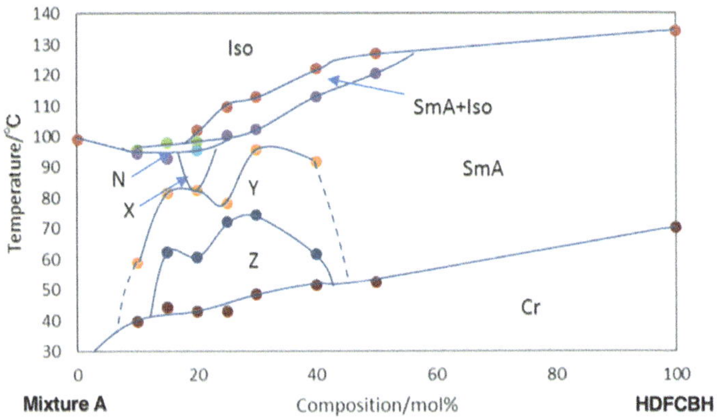

Fig. 2.28 Phase diagram between **Mixture A** and **HDFCBH**. Reproduced from [13] with permission from the Royal Society of Chemistry

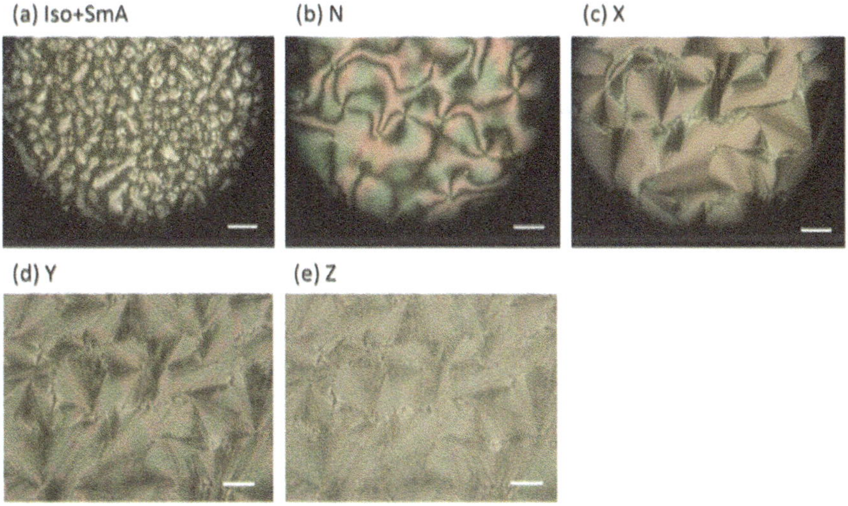

Fig. 2.29 Polarized optical textures of a mixture containing 20 mol% of **HDFCBH**. White bars represent 100 μm. Reproduced from [13] with permission from the Royal Society of Chemistry

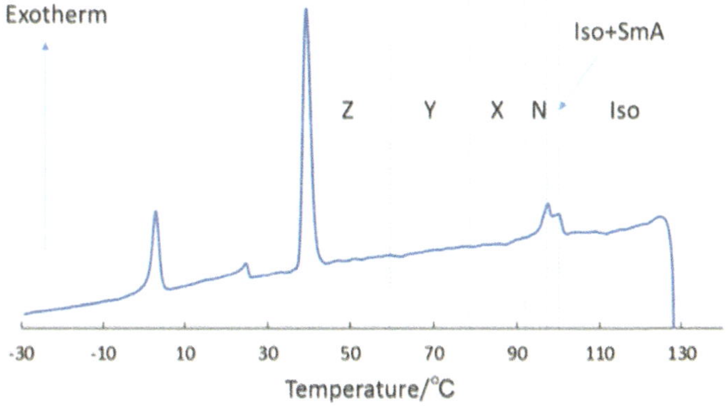

Fig. 2.30 DSC thermogram of the mixture. Reproduced from [13] with permission from the Royal Society of Chemistry

liquid tend to form the monolayer structure. Due to the competition among the interdigitated layer structure of **HDFCBH** (Fig. 2.26), the monolayer structure of **Mixture A** (Fig. 2.27), and the monolayer structure of **HDFCBH** mixed with **4OCB** and **8OPPY4O** (Fig. 2.31), a stable layer structure where the constituted molecules easily accommodate cannot form. Thus, the nematic phase appeared. After further cooling, the molecules in the N phase are thought to form a modulated smectic phase to escape from incommensurability. However, the molecules are difficult to find a solution. Therefore, the three different layer structures (Type I, Tye II, and

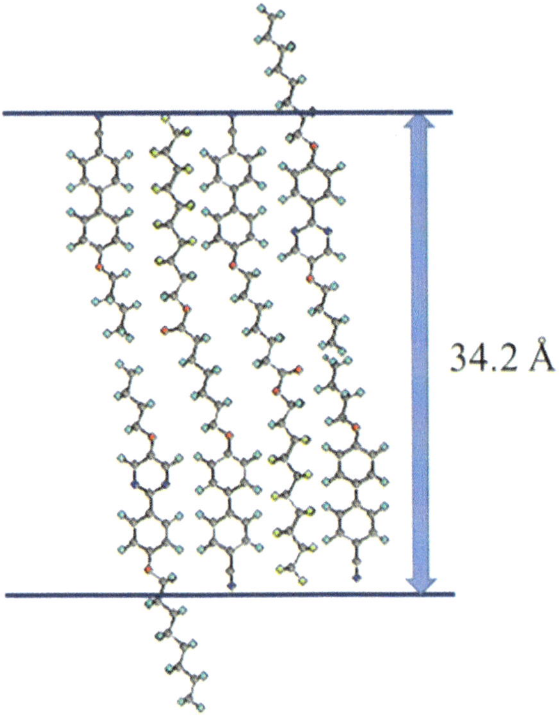

Fig. 2.31 Molecular model for the SmA phase in the Iso + SmA coexistence region of the mixture containing **HDFCBH** (20 mol%), **4OCB** (40 mol%), and **8OPPY4O** (40 mol%). The layer spacing is indicated. Reproduced from [13] with permission from the Royal Society of Chemistry

Type III) with short-range orders might repeatedly be generated and annihilated to produce the X phase. The X–Y phase transition can be explained as follows. **4OCB** and **8OPPY4O** are originally nematic liquid crystals. They are difficult to induce the SmA phase (Type III) in the ternary system. Both Type I and Type II structures can coexist in the Y phase due to the layer modulation that is stabilized by **4OCB** and **8OPPY4O** (Fig. 2.32b). The frustrated X phase exhibits the spatiotemporal fluctuations of the constituted molecules among three different layer periodicities. The competition between the monolayer and the interdigitated bilayer structures for the amphiphilic molecules **HDFCBH** prefers to form a layer modulation. However, the formation is disturbed by the microscopic monolayer structure of **Mixture A** to produce the X phase. Then the modulated layer structure is stabilized by the nematic liquid crystals **4OCB** and **8OPPY4O** in the Y phase.

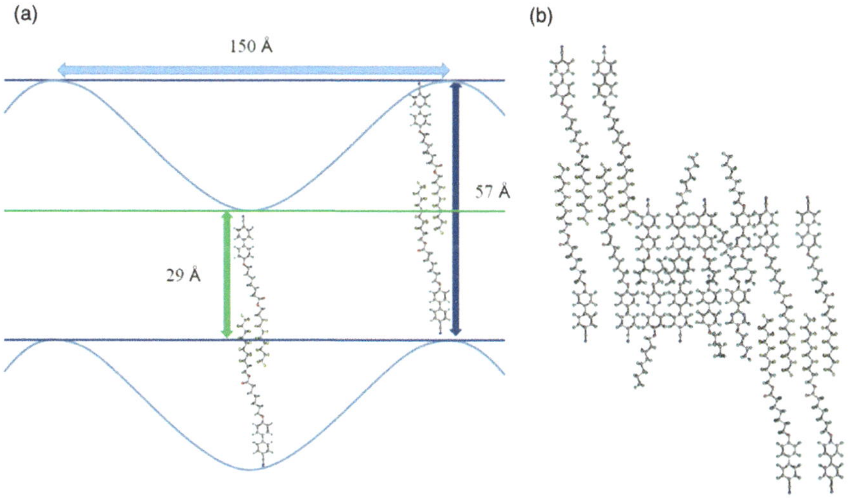

Fig. 2.32 **a** Molecular arrangements for the periodicities in the Y phase. Monolayer length, interdigitated bilayer length, and in-plane modulated length are indicated. **b** Modulated bilayer molecular organization consisting of the interdigitated packing of the semifluorinated molecules with the nematic liquid-crystalline molecules. Reproduced from [13] with permission from the Royal Society of Chemistry

2.3.3 Incommensurate Smectic Phases

Prost and Barois proposed a theoretical description of incommensurate SmA phases [10]. Leadbetter et al. reported the experimental evidence of the incommensurate smectic phase in the smectic E phase of n-alkyl cyanoterphenyls [79, 80]. The frustrated smectic phases were observed in polar liquid crystals. The coexistence of two independent layer periodicities is discussed in terms of the competition between the two incommensurate periodicities, i.e., the molecular length and the length of a pair formed by dipolar association. Figure 2.33 shows the illustration of the incommensurate SmA phase structure formed by polar liquid crystals. in which the monolayer SmA and the bilayer smectic SmA periodicities coexist along the layer normal.

Hardouin et al. reported the existence of incommensurate SmA (SmAinc) phases in a dimesogenic liquid crystal without strong polarized head [81] and then investigated the influence of the spacer length on the SmAinc phase formation [82]. Figure 2.34 shows the molecular structure of the dimesogenic compound (**KI*n***). The phase sequence of **KI5** was as follows: Cry 85 °C SmAinc 144.5 °C SmC* 149 °C SmAinc 151.5 °C SmC* 165 °C TGB 168 °C N* 191 °C BP 192 °C Iso liq. This compound exhibited incommensurate smectic A phases in addition to commensurate chiral smectic C phases. Furthermore, **KI5** showed chiral frustrated phases, i.e., the twist grain boundary (TGB) phase and the blue phase (BP). Although **KI4**, **KI7**, and **KI10** exhibited the TGB and BP, they did not show any incommensurate

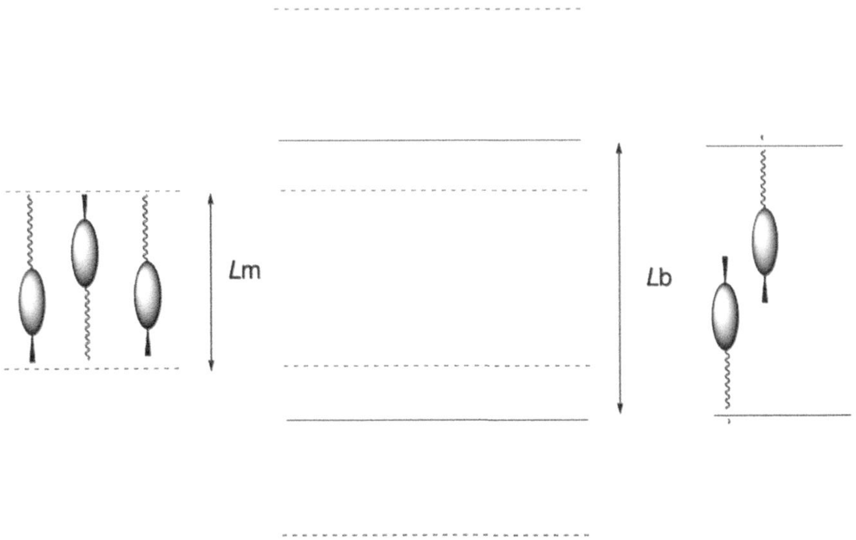

Fig. 2.33 Illustration of the incommensurate SmA phase of a polar liquid crystal. The monolayer and bilayer periodicities coexist along the layer normal. *Lm* shows the layer spacing of the monolayer SmA structure and *Lb* shows that of the bilayer SmA structure

smectic phase. The spacer length plays an important role in the molecular packing for the SmAinc phase formation.

We designed a λ-shaped liquid crystal trimer that has two different non-symmetric dimeric molecular configurations, a straight configuration (Fig. 2.35a) and a bent one (Fig. 2.35c), in a single molecule [83]. It also has a linear dimer and a U-shaped dimer within the structure. We prepared a homologous series of λ-shaped liquid crystal trimers consisting of one cyanobiphenyl unit and two phenylpyrmidine groups **CBOPB*n*** (n = 4–9, 12). The molecular structure is shown in Fig. 2.36. **CBOPB5** only exhibited the SmAinc phase. The phase transition temperatures and Δ*S*/R (in parenthesis) on cooling were as follow: Iso 183 °C (2.1) N 177 °C (0.3) SmA 157 °C (0.2) SmAinc. The melting point was 150 °C. **CBOPB7** and **CBOPB9** showed an anticlinic SmC phase instead of the SmAinc phase. The other trimers

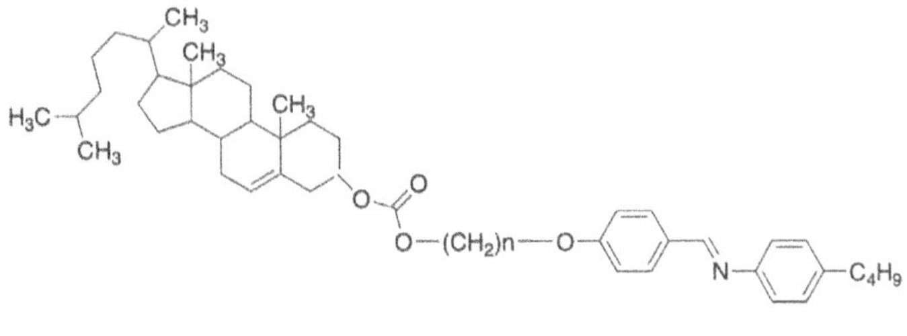

Fig. 2.34 The molecular structure of **KI*n***

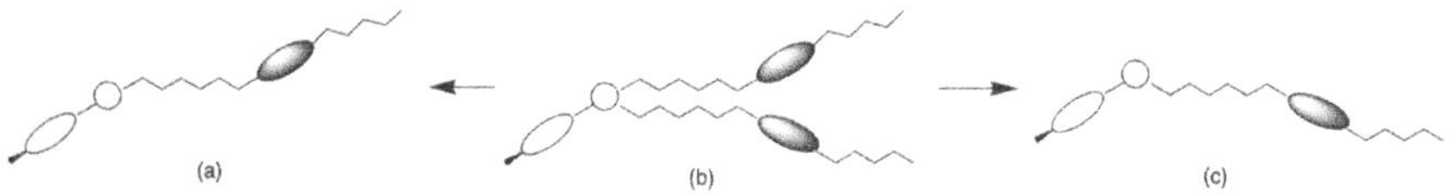

Fig. 2.35 Design concept of the λ-shaped liquid crystal trimer

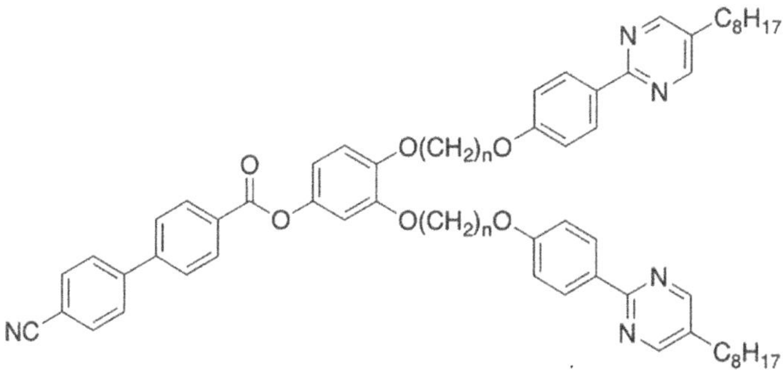

CBOPB*n*

Fig. 2.36 The molecular structure of **CBOPB*n***

showed the SmA phase and/or the N phase. As the same as observed for **KI*n***, the incommensurate phase only appears for an intermediate spacer length ($n = 6$), but not for shorter or longer spacers.

The XRD profile of **CBOPB6** exhibited three sharp reflections, $2\theta = 1.62°, 2.20°$, and 3.82° in the uniaxial smectic phase below the SmA phase (Fig. 2.37a), corresponding respectively to spacings of about 55, 40, and 23 Å. The 2D pattern of the aligned sample showed that three reflection peaks in the small angle region are positioned in one direction (Fig. 2.37b). Therefore, the smectic phase was characterized to be an incommensurate SmA phase that has three periodicity waves along the layer normal. The coexistence of three different molecular packing, corresponding to the monolayer, intercalated and interdigitated structures as shown in Fig. 2.38, can be proposed as a possible model of the SmAinc phase. Both the intercalated and interdigitated models have unfavorable free volumes, which may help the coexistence of three independent structures. The monolayer structure (Fig. 2.38a) is entropy-driven. Dipole–dipole interactions stabilize the entropy unfavorable structures. The interactions between the unlike cores, i.e., cyanobiphenyl and phenylpyrmidine units, may form the intercalated structure (Fig. 2.38b), whereas those between the same cores may produce the interdigitated structure (Fig. 2.38c).

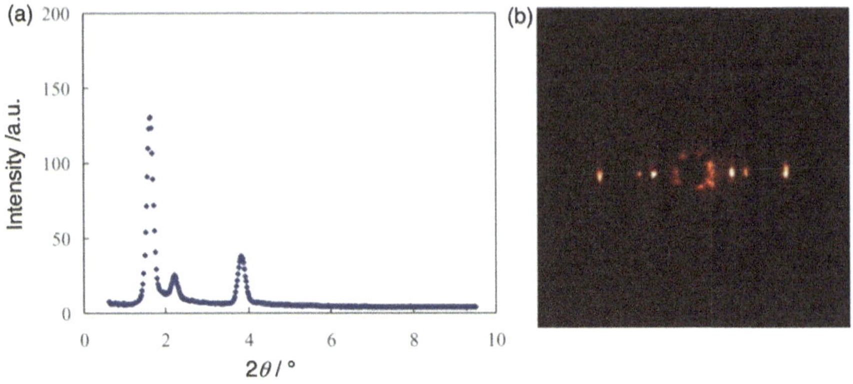

Fig. 2.37 **a** X-ray diffraction profile and **b** pattern of **CBOPB6** in the SmAinc phase at 148 °C. Reproduced from [83] with permission from the Royal Society of Chemistry

2.3.4 Bicontinuous Cubic Phases

We surmise that biscontinuous cubic phases appear to result from layer fluctuation. They are commonly observed for lyotropic liquid crystals in surfactant–solvent systems [84, 85]. In the system of an amphiphilic liquid crystal and water, the typical phase is lamellar–bicontinuous cubic (Cub_V)–hexagonal columnar (Col_h)–micellar cubic (Cub_I) (Fig. 2.39) as the volume fraction of water is increased. This phase transition originates from the increased curvature of the layer, which maximizes the contact surface area between the hydrophobic group and water. With respect to the molecular shape, the mechanism of the structural transitions can be explained by the critical packing parameter (CPP) (see Chap. 1). As the molecular shape changes from rod-like to fan-shaped, the curvature of the layer increases. There are three types of bicontinuous cubic phases in lyotropic liquid crystals: *Ia*3*d* double gyroid phases, *Im*3*m* triple-network phases, and *Pn*3*m* double diamond phases. They are also observed in thermotropic cubic phases [15]. Gray et al. observed the first thermotropic cubic phase in nitro-substituted carboxylic acids [86]. Since the discovery, lots of thermotropic cubic liquid crystals have been reported [15, 87]. Interestingly, some amphiphilic compounds exhibit the lamellar, Cub_V and Col_h phases depending on the molecular structure. Tetracatenar mesogens shows the structural phase changes from SmC to Cub_V and then to Col, as the terminal chains increase (Fig. 2.40a) [88]. Polyhydroxy derivatives exhibit a phase sequence of SmA–Cub_V–Col_h–Cub_I as the lipophilic regions are increased (Fig. 2.40b) [89]. If we consider the hydrophobic alkyl chains of those compounds to act as the solvent, both cases can be explained by the phase transition mechanism of lyotropic liquid crystals. Furthermore, with respect to the polyhydroxy derivatives, their phase transitions is also explainable in terms of CPP. The structure–property relationships in thermotropic bicontinuous cubic phases are described in some reviews [15, 16]. Recently, the spontaneous

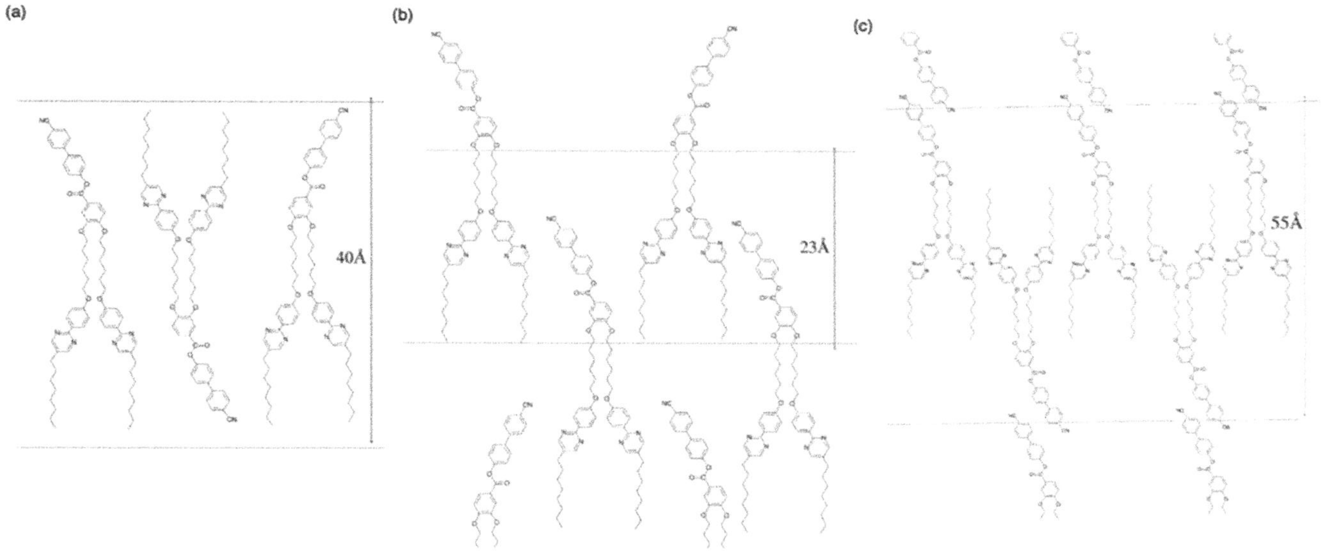

Fig. 2.38 Possible models of **a** monolayer, **b** intercalated and **c** interdigitated structures for the SmAinc phase of **CBOPB6**. Reproduced from [83] with permission from the Royal Society of Chemistry

mirror symmetry breaking has been observed in thermotropic cubic phases [90–93]. Dressel and Tschierske et al. reported chiral conglomerates in the bicontinuous cubic phase with *Im*3*m* symmetry formed by achiral multichain-terminated diphenyl-2,2′-bithiophes [18].

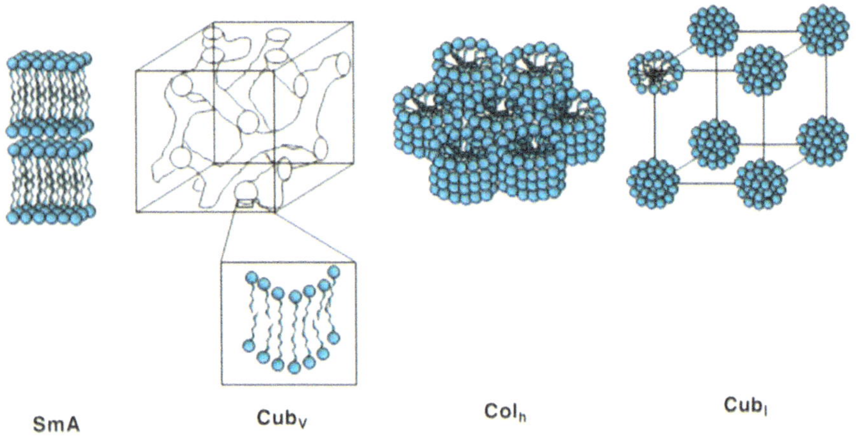

Fig. 2.39 Schematic sketches for morphologies of lamellar (SmA), bicontinuous cubic (Cub$_V$), hexagonal columnar (Col$_h$) and micellar cubic (Cub$_V$) phases. Reprinted with permission from Ref. [4, Chap. 1]. Copyright [2022] Springer Nature

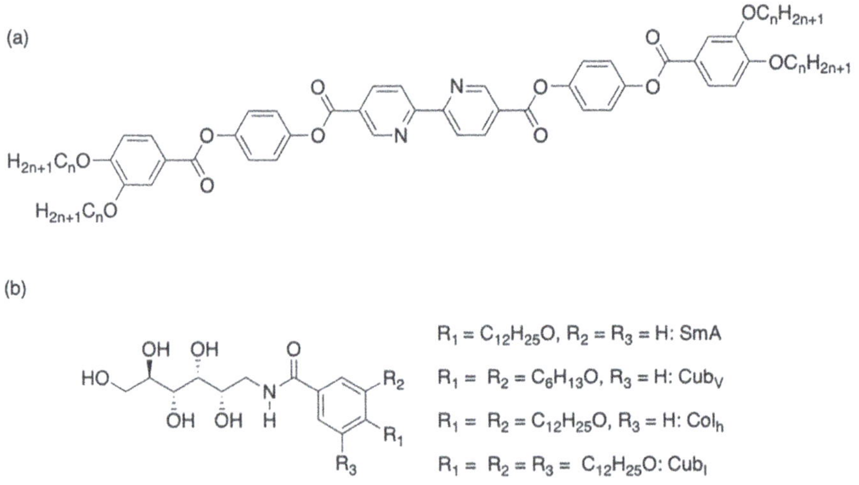

Fig. 2.40 Chemical structures of thermortropic liquid-crystalline compounds: **a** tetracatenar molecule [88] and **b** polyhydroxy derivatives [89]

As described above, a lamellar-to-bicontinuous cubic phase transition might be induced by microscopic fluctuation within the layer. The structural fluctuations were observed in the vicinity of the lamellar-to-cubic phase transition in the lyotropic system of hexaethylene glycol mono-*n*-dodecyl ether and water [94]. X-ray scattering and NMR studies suggest that the fluctuations correspond to local fragmentations of the bilayers.

Yamaguchi et al. demonstrated the appearance of a fluctuation-induced Cub_v phase in liquid-crystalline supermolecules composed of a semiperfluorinated alkyl chain and cyanobiphenyl units [95]. Their molecular structures and phase transition properties are shown in Fig. 2.41. Compound **HDFCBP-1** showed only an SmC phase. Compound **HDFCBP-2** possessing two cyanobiphenyl moieties showed SmA and Col_h phases. Unexpectedly, compound **HDFCBP-3** possessing three cyanobiphenyl moieties exhibited a Cub_v phase instead of the Col_h phase. These amphiphilic oligomers showed a phase sequence of SmC (lamellar)–Col_h–Cub_v when the numbers of cyanobiphenyl moieties were increased. The sequence of Col_h–Cub_v cannot be explained by above mentioned mechanisms, i.e., the interface curvature between two incompatible regions and the CPP theory.

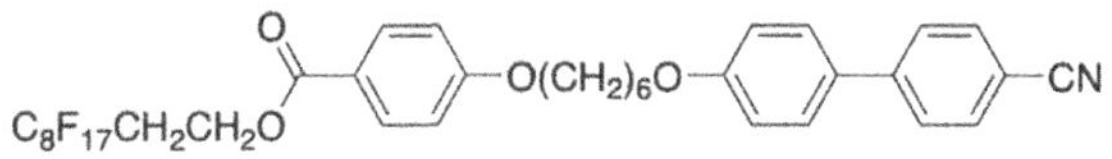

HDFCBP-1: Cr 110.3 (39.4) SmC 139.1 (11.5) Iso

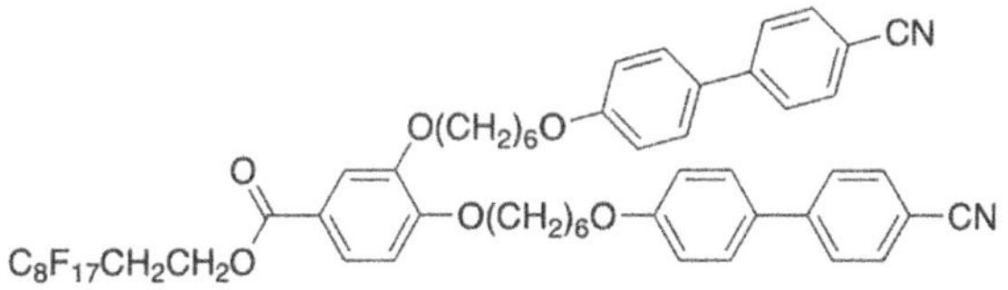

HDFCBP-2: Cr 62.2 (18.7) [glass 10.1] Col_h 115.7 (3.4) SmA 169.6 (10.3) Iso

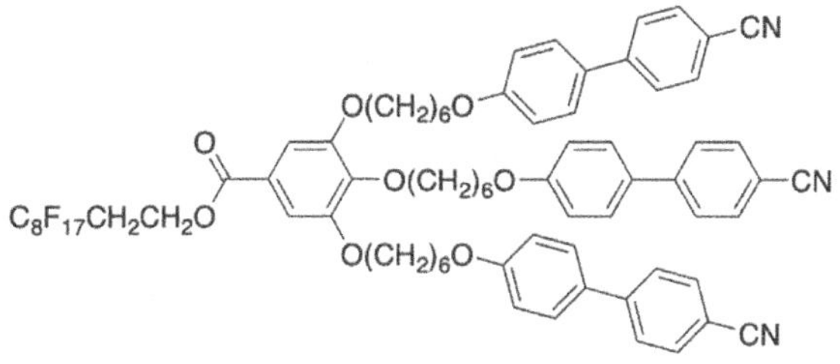

HDFCBP-3: Cr 83.2 (34.5) [glass 9.0] Cub_v 135.9 (3.5) SmA 154.2 (5.4) Iso

Fig. 2.41 The molecular structures, transition temperatures (°C), and enthalpies (kJ mol^{-1}, in parentheses) of transition for **HDFCBP-1**, **HDFCBP-2**, and **HDFCBP-3**. Square brackets denote a monotropic transition

We focus on the Cub_V formation of compound **HDFCBP-3**. The SmA-to-Cub_V transition of **HDFCBP-3** was observed by using polarized optical microscopy. Optically isotropic domains appeared and grew rapidly in the fan texture of the SmA phase (Fig. 2.42). Those domains were highly viscous. These are characteristic to a thermotropic cubic phase. The X-ray diffraction profiles provide the structural information of the SmA and cubic phases. The SmA phase exhibited slightly diffuse two peaks in the small-angle region, corresponding to the periodicities of 55.8 and 45.0 Å (Fig. 2.43a). The periodicity lengths are longer than the molecular length of 36.6 Å, indicating that the SmA phase has an interdigitated structure. A few small reflections can be seen between the two reflections, which might result from perforated lamellae. A diffuse asymmetric tail is spread around the Brag peaks, which is attributed to the undulation of lamellae [96]. These results indicate that fluctuations exist in the interdigitated lamellar structure. With respect to the Cub_V phase (Fig. 2.43b), the observed reflections reveal that the Cub_V phase belongs to a space group $Ia3d$, which represents a bicontinuous cubic phase constructed from two interpenetrating networks of branched cylinders.

An undulatory lamellar deformation mode for a nonequilibrium lamellar structure was reported to exist as the unstable transition state between the lamellar and the Gyroid in some block copolymers [97–99]. Imai et al. observed fluctuations of lamellar structure prior to a lamellar-to-gyroid phase transition in a nonionic surfactant system [100]. They proposed an explanation for the fluctuation as follows: amplitude-modulation fluctuations occur in lamellae and they develop to a perforation layer. We discuss how the SmA-to-Cub_V phase transition occurs in **HDFCBP-3**. Amphiphilic liquid crystal oligomers possessing phenylpyrimidine moieties instead of the cyanobiphenyl moieties exhibited a phase sequence of SmC–SmA–Col as increasing phenylpyrimidine moieties [101]. The peripheral three cyanobiphenyl moieties play an important role in the appearance of the Cub_V phase. The incompatibility between the semiperfluorinated chain and the cyanobiphenyl moieties produces micro-segregation in each layer of **HDFCBP-3**. The competition of the steric effect

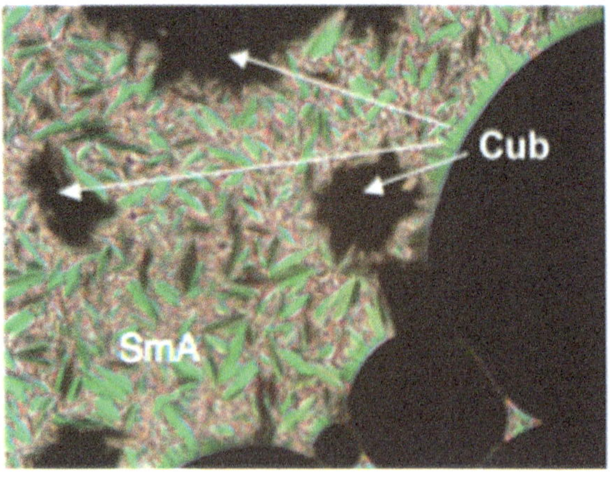

Fig. 2.42 Polarized optical texture of the SmA–Cub_V transition for **HDFCBP-3**. Reprinted with permission from [95]. Copyright (2006) American Chemical Society

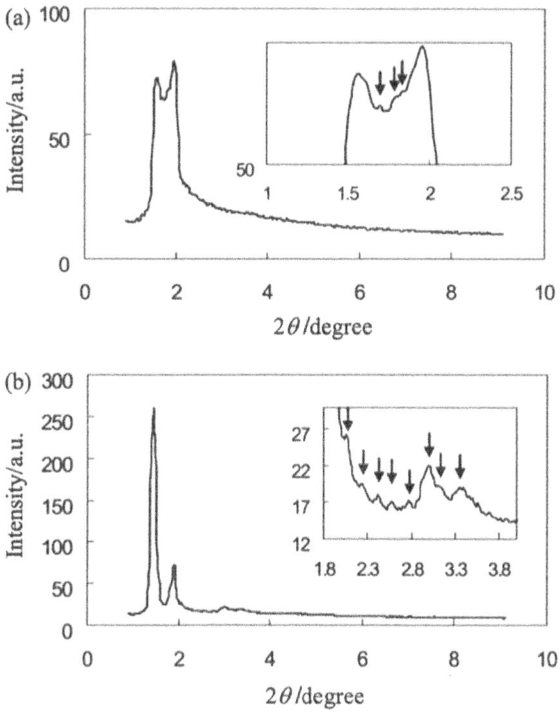

Fig. 2.43 X-ray diffraction profiles in the small-angle region of **HDFCBP-3** in **a** the SmA phase at 150 °C and **b** the Cub_V phase at 100 °C; expansions are included for clarity. Reprinted with permission from [95]. Copyright (2006) American Chemical Society

attributable to the pronounced taper shape and the segregation effect might give rise to the fluctuation of the lamellar structure (Fig. 2.44). The fluctuation of the interdigitated lamellar structure is likely amplified by antiparallel dipole–dipole interactions between cyanobiphenyl moieties in adjacent layers, which perforates the lamellar to form the bicontinous cubic phase.

In order to support the proposed mechanism, Yamaguchi et al. designed an amphiphilic liquid crystal hexamer possessing six cyanobiphenyl moieties and a semiperfluorinated alkyl chain [102]. The hexamer corresponds to a dimer of **HDFCBP-3** and forms a dumbbell shape. The steric effect attributable to the taper shape of **HDFCBP-3** is considered to be inhibited. The molecular structure and transition temperatures of the liquid crystal hexamer are shown in Fig. 2.45. The hexamer showed monotropic liquid crystal phases and exhibited a SmA-to-SmA′ transition instead of the SmA-to-Cub_V phase transition observed for **HDFCBP-3**.

Figure 2.46 shows XRD profiles in the small angle region in the N, SmA, and SmA’ phases for the LC hexamer on cooling from the isotropic liquid. A very weak and broad peak appeared around $2\theta = 2°$ at 110 °C in the N phase. The peak slightly became sharp at 100 °C, suggesting the formation of cybotactic clusters. Therefore,

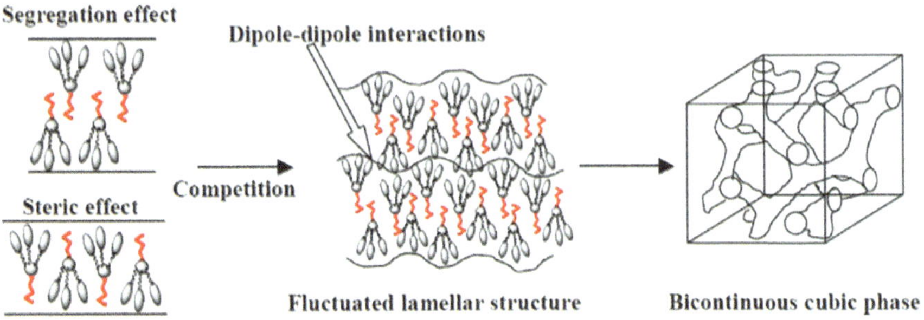

Fig. 2.44 Schematic illustration of the SmA-to-Cub_V transformation of **HDFCBP-3**. Reprinted with permission from [101]. Copyright (2010) Taylor & Francis Ltd.

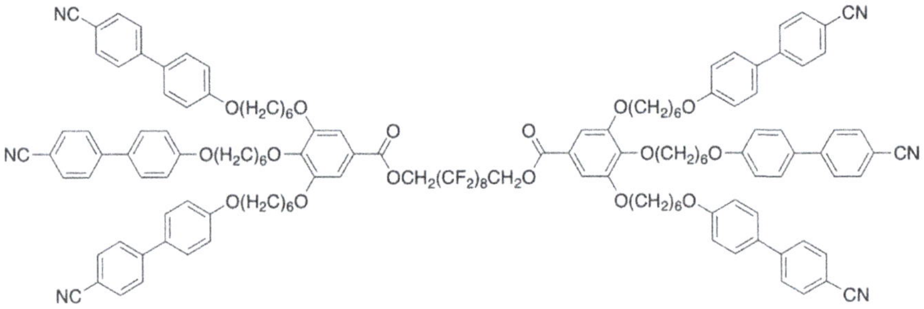

Cr 122.0 (41.2) [SmA' 46.8 (3.7) SmA 101.8 (–[a]) N 121.4 [1.4] Iso

Fig. 2.45 Molecular structure, phase transition temperatures (°C), and associated transition enthalpies (kJ mol^{-1} in brackets) of the LC hexamer. Square brackets indicate monotropic phase transitions. [a]Too small to be detected

the N-to-SmA phase transition does not accompany an enthalpy change. On cooling to the SmA phase at 85 °C, the LC hexamer exhibited a sharp peak around $2\theta = 2°$, corresponding to the layer spacing of 44.3 Å. In the SmA' phase at 39 °C, the peak disappeared and a new peak appeared around $2\theta = 4.9°$, corresponding to the layer spacing of 17.9 Å. The SmA-to-SmA′ transition accompanied a discontinuous and marked decrease in layer spacings.

The layer spacings in the SmA and SmA′ phases are shorter than the elongated molecular length of 65 Å. We proposed molecular organization models for the SmA and SmA′ phases as shown in Fig. 2.47. The high-temperature SmA phase has a bilayer structure resulting from the segregation effect. The molecules are deformed to be organized into a layer structure in which the cyanobiphenyl moieties and the semiperfluorinated chains are segregated (Fig. 2.47a). The low-temperature SmA′ phase has an intercalated structure derived from the steric effect. The molecules adopt an elongated conformation to fill the space (Fig. 2.47b). The phase transition is attributable to the marked conformation change of the LC hexamer.

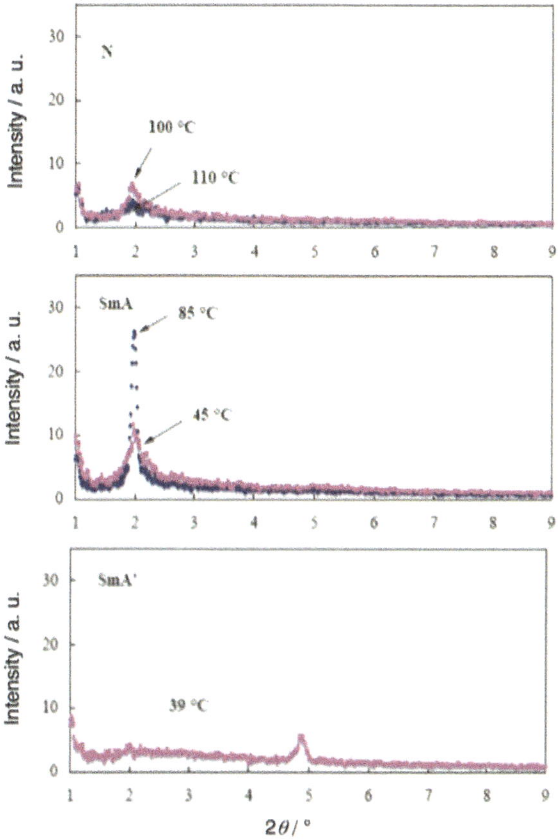

Fig. 2.46 X-ray diffraction profiles in the N, SmA, and SmA′ phases for the LC hexamer. Reprinted with permission from [102]. Copyright (2007) American Chemical Society

2.4 Dissipative Structures: Order Through Fluctuation

2.4.1 Pattern Formation and Propagation

We show that some liquid-crystalline supermolecules induce a dissipative pattern propagation in this section. Prigogine describes the dissipative structures as follows: "An open system *out of equilibrium* may be associated with dissipative structures when specific constrains are satisfied. Dissipative structures are associated with an entirely different ordering principle, which may be called order through fluctuation. They arise from the amplification of fluctuations resulting from an instability of the *thermodynamic branch*" [23]. The dissipative structures appear beyond a certain critical distance from equilibrium. Their typical examples are shown in Rayleigh–Bénard convection [103–105] and Belousov–Zhabotinsky reaction [106–108].

The pattern formation in liquid crystals is one of the attractive dissipative structures and has been actively investigated [109, 110]. Because liquid crystals are ideal

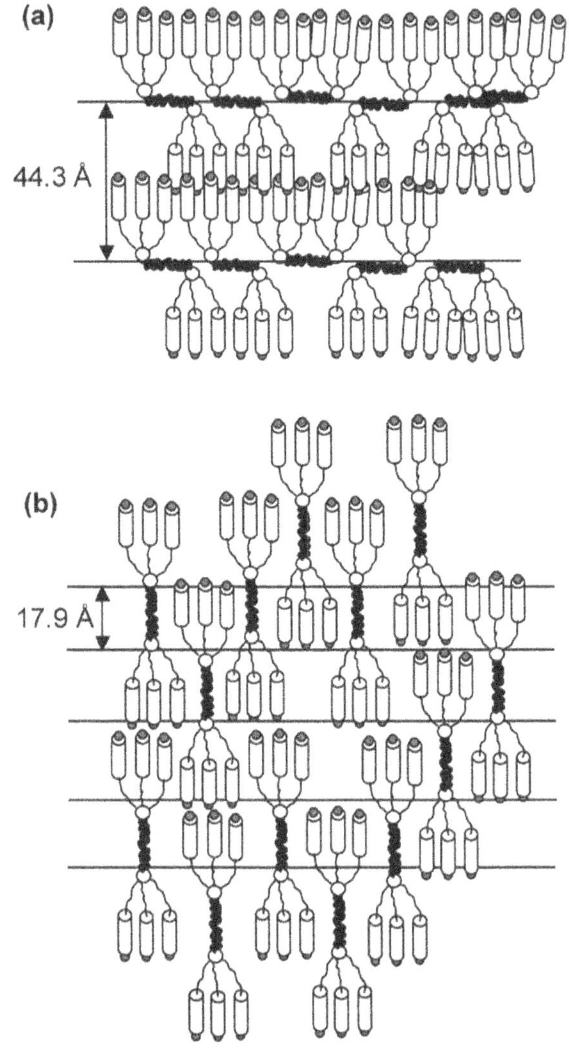

Fig. 2.47 Possible molecular arrangements in the **a** SmA and **b** SmA′ phase of the LC hexamer. Reprinted with permission from [102]. Copyright (2007) American Chemical Society

model systems for the study of dissipative structures [109]. External stimuli such as an electric field, flow or heat cause a liquid-crystalline system to deviate from equilibrium. The pattern formation has been observed in some nematic liquid-crystalline systems under electric or magnetic fields [111–115]. There are several types of electric field-induced patterns. Among them, the electroconvection (EC) is the well-known phenomenon. It is also known as the electro-hydrodynamic instability. A typical example of EC is as follows. A nematic liquid crystal with negative dielectric anisotropy ($\Delta\varepsilon < 0$) and positive conductivity anisotropy ($\Delta\alpha > 0$) is planarly aligned in a thin cell. When an applied electric filed (E) exceeds a threshold value, the homogenous state of the liquid-crystalline molecules becomes unstable towards distortion. The periodic tilt distortion of the director produces the striped pattern (Fig. 2.48a). This pattern formation was explained by the Carr-Helfrich feedback mechanism [116–118]. According to Eber et al. [115], we apply the mechanism to the phenomenon (see Fig. 2.48b). Spatial director tilt modulations may always be present in a planar nematic due to thermal fluctuations. Because the director is tilted and the conductive anisotropy is not zero, the electric current has a nonzero component perpendicular to E, which leads to a spatial separation of charges. A material flow is induced due to the Coulomb force on charges. The flow forms vortices, which exert a viscous torque on the director. The viscous torque is generated to balance the nematic elastic torque, stabilizing the distorted director distribution. The striped pattern emerges. The pattern formation under an electric field is also described in the other types: the transients during the Freedericksz transition and the flexoelectric domains [115]. Front propagation in smectic liquid crystals has attracted much attention. Finger fronts emerged in the SmA phase near the SmA to N* phase transition under an electric field [119]. This pattern formation occurs during the Freedericksz transition. Tang et al. found that flexoelectricity-induced solitary waves in nematic liquid crystals exhibit "butterfly"-like or "bullet"-like structures [120].

We reported the pattern propagation in a smectic liquid-crystalline phase in 2006 [121]. The smectic phase was the interdigitate anticlinic smectic C (SmCanti) phase of a binary system of the unsymmetric dimers. Their molecular structures and phase transition temperatures are shown in Fig. 2.49. They have odd-numbered spacers with different lengths. Figure 2.50 shows the striped pattern of an equivalent-weight mixture of **12PY7OCB** and **12PY9OCB** contained in a homeotropically aligned cell. The discussion on the fluctuation-induced hierarchical structures has just given a new

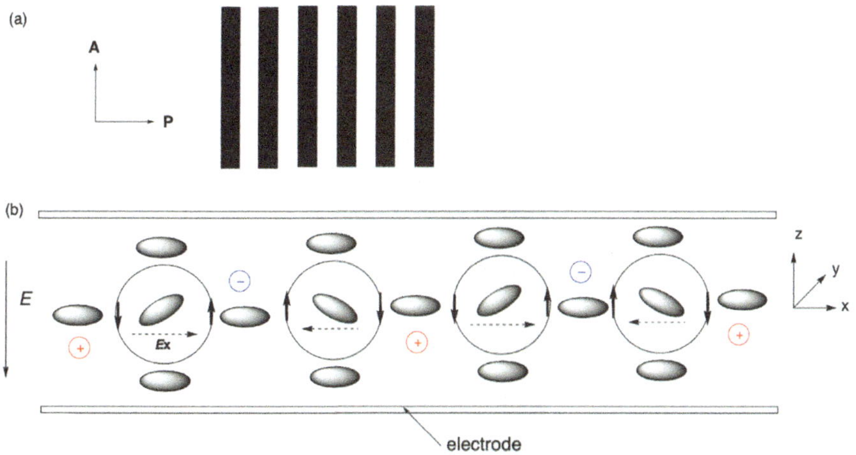

Fig. 2.48 **a** Illustration of the striped pattern. **b** Schematic diagram for the Carr-Helfrich feedback mechanism. Rods represent nematic director. The red and blue circles show + and – space charge clouds, respectively

interpretation of the pattern propagation. The mixture exhibited phase transitions as follows: Iso 120 °C N 109 °C interdigitated SmA 86 °C interdigitated SmCanti [121].

The pattern propagation was observed in thin samples with a wedge-like thickness gradient when the temperature of the sample was varied. It can be seen in Video S1 (see Electronic Supplementary Material). In order to induce a gradient in the thickness of the sample, we slightly pressed a certain position of the sample. Therefore, the sample thickens from that pressed position towards the periphery. When the sample was cooled from the SmA phase to the SmCanti phase, the beginning of the pattern emerged repeatedly around that thinner position of the sample. They spread out in concentric circles to form the periodic pattern (see Video S2a). When the sample was heated in the SmCanti phase, the waves converged from the periphery of the pattern towards the center, where they disappeared (see Video S2b). Figure 2.51 shows the pattern propagation in the periphery region on cooling. The pattern of dark

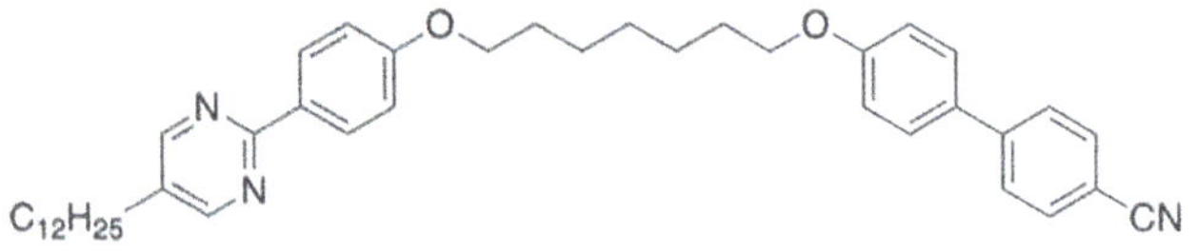

12PY7OCB: Cr 120 [SmA 116] N 121 Iso

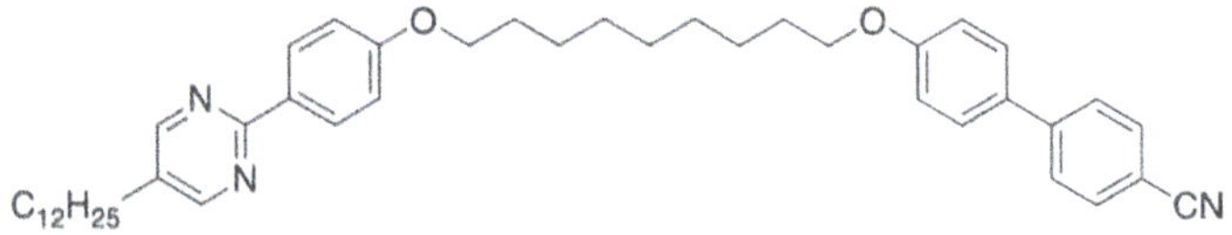

12PY9OCB: Cr 113 [SmA 102] N 122 Iso

Fig. 2.49 The molecular structures and transition temperatures (°C) of **12PY7OCB** and **12PY9OCB** [121]

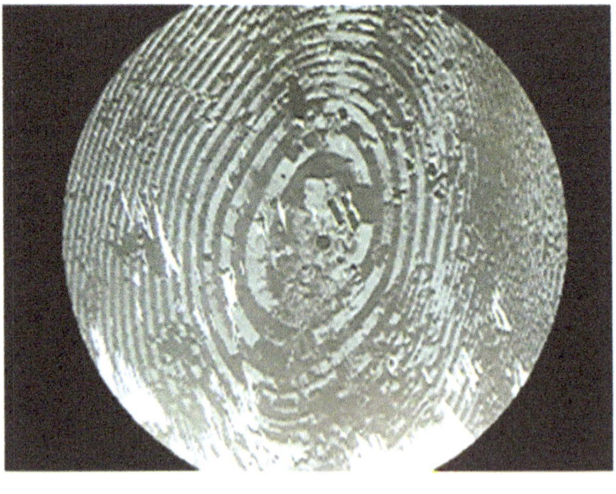

Fig. 2.50 Photomicrograph of the striped pattern of an equivalent-weight mixture of **12PY7OCB** and **12PY9OCB** contained in a homeotropically aligned cell with a cell gap of 5 μm in the induced SmCanti phase at 80 °C under crossed polarizers. The center was pressed to make it thinner

stripes moved forward while rotating over the bright background (see Video S3). The direction of rotation of the stripes at the top was the opposite to that of the pattern propagation. When the sample was rotated about 45° clockwise, the dark stripes almost kept the darkness, while the background texture partially changed (Fig. 2.52). We can say as follows: the black regions are close to optically isotropic while the background regions show the texture of the homeotropically aligned SmCanti phase. Such striped patterns have also been observed in a biaxial SmA phase of a binary system of a metallomesogen and 2,4,7-trinitrofluorenone [122], in that of a bent-core molecule [123] and in that of a nonsymmetric dimer [124]. We recently have given the following explanation for the pattern propagation [125]. A temperature gradient exists between the top and bottom of the sample. Therefore, Rayleigh–Bénard convection is thought to occur near the center of the pattern on cooling as shown in Fig. 2.53a. Figure 2.53b shows a model for the pattern formation and propagation in the left side of Fig. 2.53a. The thickness of the sample increases from the center of the pattern toward the periphery. The temperature gradient between the top and bottom of the sample increases with its thickness. We suppose that the smectic layers break to small smectic clusters at the boundary where the bright stripe changes to the dark stripe in the pattern progressing areas. The disordered nano clusters form the optically isotropic dark stripe. Then the smectic clusters reorganize to form the smectic layers at the boundary where the dark stripe changes to the bright stripe in the pattern's passing areas. The clusters rotate clockwise, moving the stripes to the left. Therefore, the black stripes emerging around the center propagate toward the side with the larger vertical temperature gradient, that is, toward the periphery. When the sample is heated, the smectic clusters rotate in the opposite direction, moving the stripes towards the lower-temperature gradient. The smectic phase has the interdigitated layer structure. There are two different interactions in the layer structure, i.e., the strong intra-layer dipole–dipole interaction between cyanobiphenyl units in antiparallel alignment and the weak inter-layer permeation of the tails of molecules in adjacent layers. The thermal convection partially breaks the weaker interlayer interactions, forming the smectic clusters with short correlation lengths.

2.4.2 Periodic Formation and Breakdown of Lamellar Aggregates in Water

We show a dissipative structure observed in a surfactant–solvent system. The surfactant **CB-B2ES** is a mixture of the two isomers possessing a 4-hexyloxy-4′-cyanobiphenyl (**6OCB**) unit and a butoxyethoxyethyl tail (Fig. 2.54) [126]. Figure 2.55 shows the time-dependent textures of an aqueous solution containing 30 wt% of **CB-B2ES** at 30 °C under crossed polarizers. Maltese cross textures can be seen in those photographs, therefore, the surfactant molecules are thought to form the lamellar aggregations in water. Interestingly, the maltese cross textures spontaneously emerged and disappeared in water. For example please see the maltese cross texture

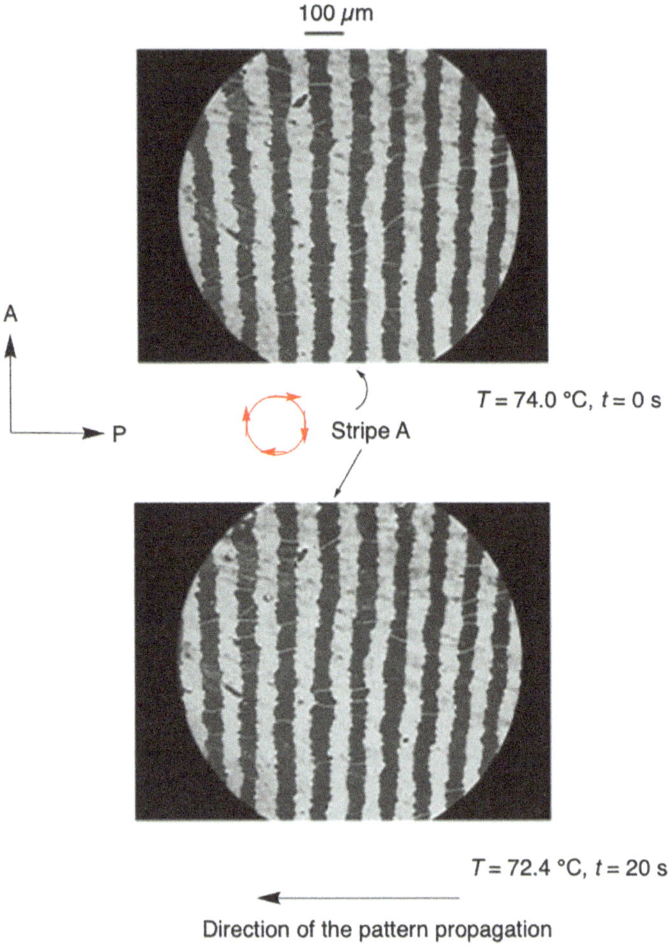

Fig. 2.51 Pattern propagation of the mixture in the periphery region on cooling under crossed polarizers. The cooling rate was 5 °C min^{-1}. Red arrows show the direction of the stripe rotation. Reprinted with permission from [121]. Copyright (2006) Taylor & Francis Ltd.

indicated by white arrow. The lifetime of a single occurrence of their lamellar aggregations was approximately ~10 s. That unstability is unusual phenomenon, because typical lyotropic lamellar LCs are stable over at least several hours [127–130].

The disappearance of the texture is due to the transformation occurring from the lamellar aggregation to an optically isotropic micellar aggregation. Sagisaka et al. proposed the following mechanism for the structural change as shown in Fig. 2.56 [126]. The hydrophilic/lipophilic balance of the ethoxylated tail fluctuates due to the change in the hydration of the tail at 30 °C. The enhanced oxyethylene hydration makes the tail to orient toward water and the hydrophilic head group away from the hydrophobic cyano moiety. This conformation decreases the CPR value of **CB-B2ES**.

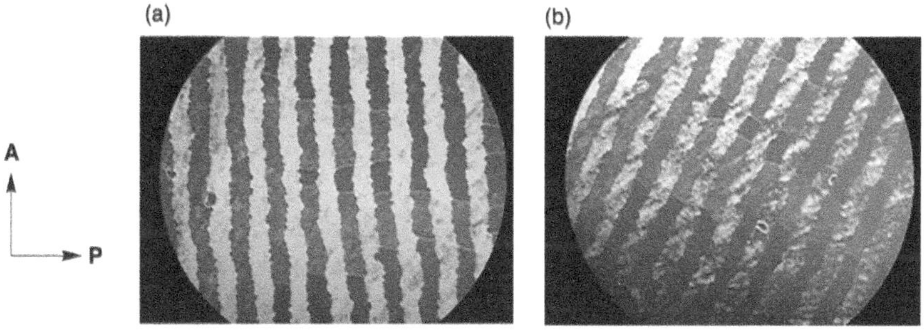

Fig. 2.52 **a** Optical texture of the pattern of the mixture in the periphery region and **b** that for which the sample was rotated about 45° clockwise

The lower CPR increases the curvature of the layer, driving the lamellar structure to the micellar aggregation. With respect to the reverse transformation, the antiparallel dipole–dipole interactions between cyanobiphenyl moieties also drives to form the bilayer structure in the lamellar aggregation. The reversible conformation change cooperatively occurs to give a rise to the periodic formation/breakdown of lamellar aggregates of **CB-B2ES** in water.

2.5 Concluding Remarks

We have demonstrated how the designed liquid-crystalline supermolecules drive the hierarchical phases such as cybotactic nematic, modulated smectic, and bicontinuous cubic phases. The LC supermolecues have a characteristic shape, inducing specific core–core interactions in the smectic liquid crystal phases. They also have an intramolecular dynamic order, which organizes cooperative motions in their aggregations. Therefore, these factors induce the smectic layer fluctuations, producing the frustrated phases. Furthermore, the striped pattern similar to those found in nature was formed in the interdigitated smectic phase composed of unsymmetric LC dimers. The pattern progressed with temperature changes. The pattern propagation is explainable in terms of the rotation of smectic clusters through Rayleigh–Bénard convection. This chapter shows that liquid-crystalline supermolecules are no longer perceived merely as molecules with peculiar structures but serve as design principles for hierarchical structures and dissipative structures.

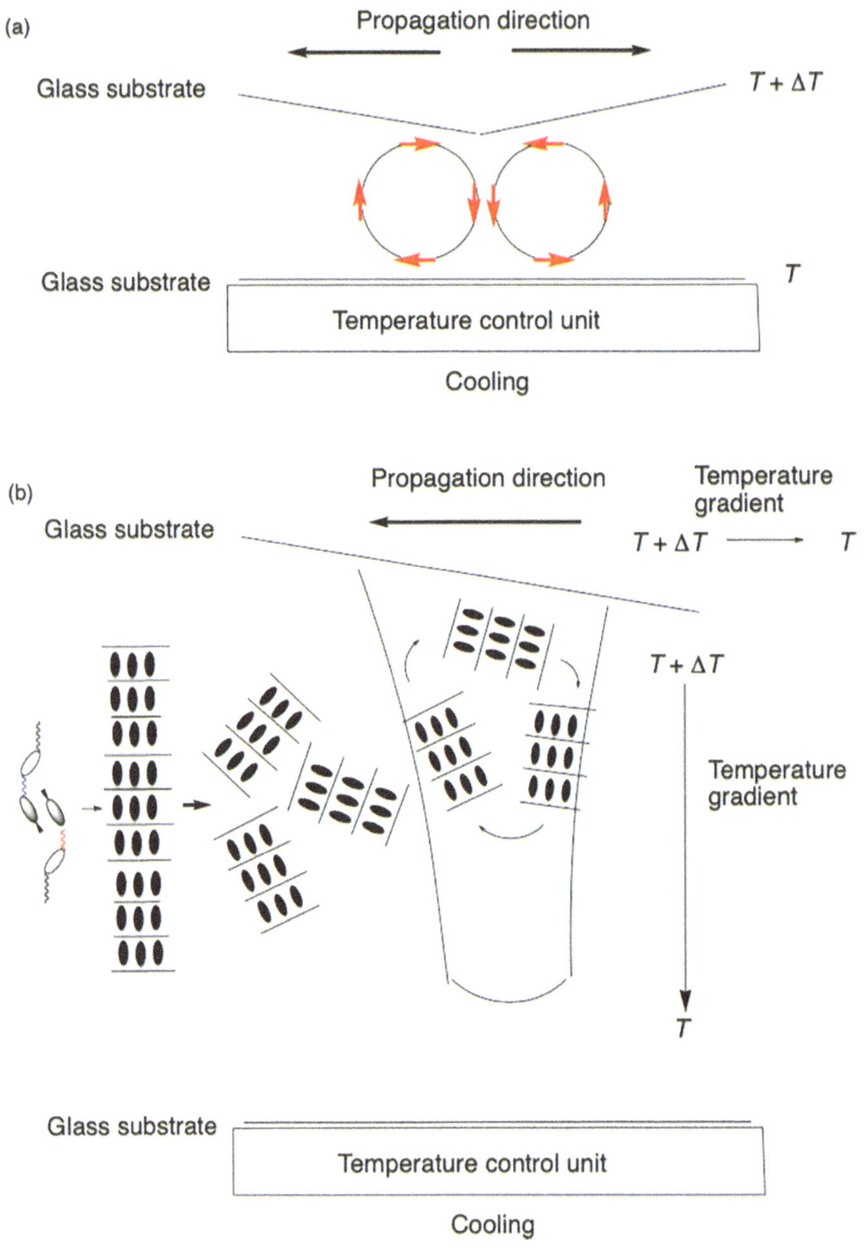

Fig. 2.53 **a** Rayleigh–Bénard convection (red arrows) near the center of the pattern on cooling. **b** Model for the pattern formation and propagation on cooling in the area of the left side of (**a**) [125]

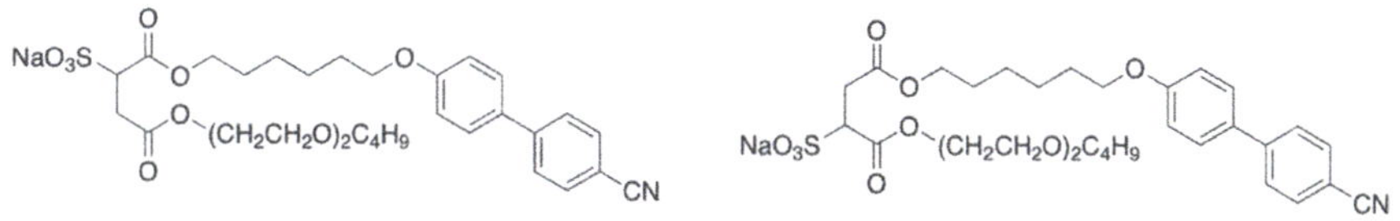

Fig. 2.54 The molecular structure of **CB-B2ES**. The surfactant has two isomers with different C-S locations

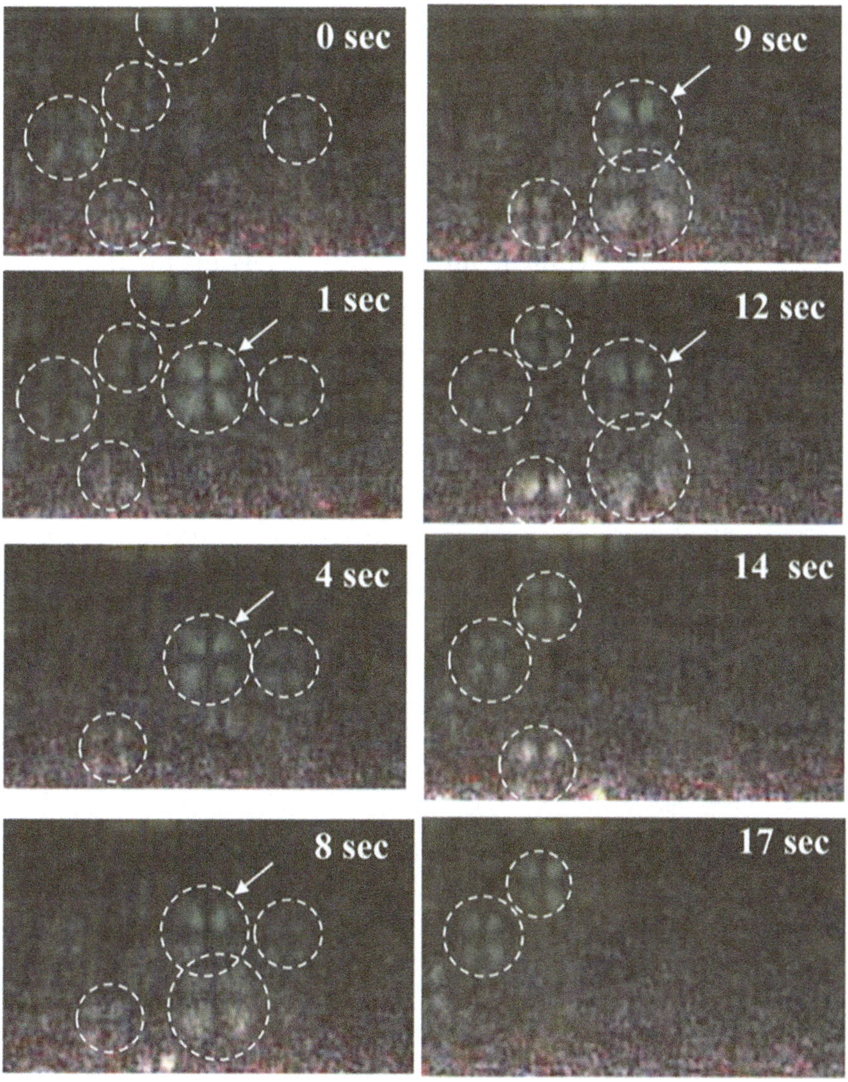

Fig. 2.55 Spontaneous appearance/disappearance of the maltase cross texture for the 30 wt% of the **CB-B2ES**/water mixture at 30 °C under crossed polarizers. Reprinted with permission from [126]. Copyright (2015) American Chemical Society

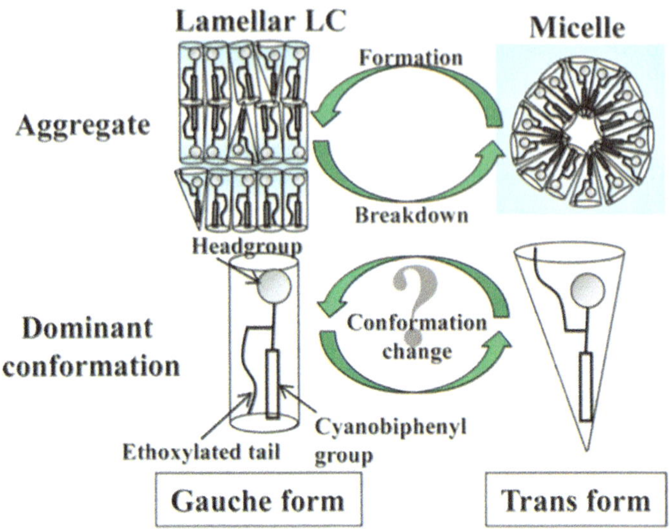

Fig. 2.56 *Gauche–trans* conformational changes for the **CB-B2ES** molecule inducing the periodical formation/breakdown of lamellar aggregates. Reprinted with permission from [126]. Copyright (2015) American Chemical Society

References

1. Cladis, P.E.: Fluctuations and liquid crystal phase transitions. In: Demus, D., Goodby, J.W., Gray, G.W., Spiess, H.W., Vill, V. (eds.) Handbook of Liquid Crystals, vol. 1, 1st edn, pp. 379–390. Wiley-VCH Verlag GmbH, Weinheim, Germany (1998)
2. Moses, T., Reeves, J., Pirondi, P.: Fluctuations near a phase transition in liquid crystals. Am. J. Phys. **75**, 220–225 (2007)
3. De Vries, A.: Evidence for the existence of more than one type of nematic phase. Mol. Cryst. Liq. Cryst. **10**, 31–35 (1970)
4. De Vries, A.: X-ray photographic studies of liquid crystals I. A cybotactic nematic phase. Mol. Cryst. Liq. Cryst. **10**, 219–236 (1970)
5. Chistyakov, I.G., Chaikowsky, W.M.: The structure of p-azoxybenzenes in magnetic fields. Mol. Cryst. Liq. Cryst. **7**, 269–277 (1969)
6. Samulski, E.T.: Meta-Cybotaxis and nematic biaxiality. Liq. Cryst. **37**, 669–678 (2010)
7. Leadbetter, A.J.: Structural classification of liquid crystals. In: Gray, G.W. (ed.) Thermotropic Liquid Crystals; Critical Reports on Applied Chemistry, vol. 22, pp. 1–27. Wiley, Chichester, UK (1987)
8. Nishiya, W., Takanishi, Y., Yamamoto, J., Yoshizawa, A.: Molecular design foe a cybotactic nematic phase. J. Mater. Chem. C **2**, 3677–3685 (2014)
9. Sigaud, G., Hardoin, F., Achard, M.F., Levelut, A.M.: A new type of smectic A phase with long range modulation in the layers. J. Phys. **42**, 107–111 (1981)
10. Prost, J., Barois, P.: Polymorphism in polar mesogens. II—Theoretical aspects. J. Chim. Phys. **80**, 65–80 (1983)
11. Seddon, J.M.: Structural studies of liquid crystals by X-ray diffraction. In: Demus, D., Goodby, J.W., Gray, G.W., Spiess, H.W., Vill, V. (eds.) Handbook of Liquid Crystals, vol. 1, 1st edn, pp. 635–679. Wiley-VCH Verlag GmbH, Weinheim, Germany (1998)

12. Imrie, C.T., Luckhurst, G.R.: Liquid crystal dimers and oligomers. In: Goodby, J.W., Collings, P.J., Kato, T., Tschierske, C., Gleeson, H.F., Raynes, P. (eds.) Handbook of Liquid Crystals, vol. 7, 2nd edn, pp. 137–210. Wiley-VCH Verlag GmbH: Weinheim, Germany (2014)
13. Sasaki, T., Takanishi, Y., Yamamoto, J., Yoshizawa, A.: A frustrated phase driven by competition among layer structures. Soft Matter **13**, 5194–5203 (2017)
14. Pociecha, D., Kardas, D., Gorecka, E., Szydlowska, J., Mieczkowski, J., Guillon, D.: Modulated and intercalated smectic phases formed by dimeric molecules. J. Mater. Chem. **13**, 34–37 (2003)
15. Ungar, G., Liu, F., Zeng, X.: Cubic and other 3D thermotropic liquid crystal phases and quasicrystals. In: Goodby, J.W., Collings, P.J., Kato, T., Tschierske, C., Gleeson, H.F., Raynes, P. (eds.) Handbook of Liquid Crystals, vol. 5, 2nd edn, pp. 363–436. Wiley-VCH Verlag GmbH, Weinheim, Germany (2014)
16. Kutsumizu, S.: Recent progress in the synthesis and structural clarification of thermotropic cubic phases. Isr. J. Chem. **52**, 844–853 (2012)
17. Levelut, A.-M., Clerc, M.: Structural investigation on 'smectic D' and related mesophases. Liq. Cryst. **24**, 105–115 (1998)
18. Dressel, C., Liu, F., Prehm, M., Zeng, X., Ungar, G., Tschierske, C.: Dynamic mirror-symmetry breaking in bicontinuous cubic phases. Angew. Chem. Int. Ed. **53**, 13115–13120 (2014)
19. Wang, Y., Yang, S.-G., Li, Y.-X., Cao, Y., Zeng, X.-B., Cseh, L., Ungar, G.: Supertwisted chiral gyroid mesophase in chiral rod-like compounds. Angew. Chem. Int. Ed. **63**, 202403156 (2024)
20. Ichikawa, T., Yoshio, M., Hamasaki, A., Taguchi, S., Liu, F., Zeng, X.-B., Ungar, G., Ohno, H., Kato, T.: Induction of thermotropic bicontinuous cubic phases in liquid-crystalline ammonium and phosphonium salts. J. Am. Chem. Soc. **134**, 2634–2643 (2012)
21. Kato, T., Yoshio, M., Ichikawa, T., Soberats, B., Ohno, H., Funahashi, M.: Transport of ions and electrons in nanostructured liquid crystals. Nat. Rev. Mater. **2**, 17001 (2017)
22. Prigogine, I.: Time, structure, and fluctuations. Science **201**, 777–785 (1978)
23. Prigogine, I.: Order through fluctuation: self-organization and social system. In: Jantsch, E. (ed.) Evolution and consciocness: human systems in transition, pp. 93–130. Addison-Wesley, Boulder, USA (1976)
24. Yoshizawa, A.: Uncoventional liquid crystal oligomers with a hierarchical structure. J. Mater. Chem. **18**, 2877–2889 (2008)
25. Saez, I.M., Goodby, J.W.: Supermolecular liquid crystals. J. Mater. Chem. **15**, 26–40 (2005)
26. Imrie, C., Henderson, P.A.: Liquid crystal dimers and higher oligomers: between monomers and polymers. Chem. Soc. Rev. **36**, 2096–2124 (2007)
27. Saez, I.M.: Supermolecular liquid crystals. In: Goodby, J.W., Collings, P.J., Kato, T., Tschierske, C., Gleeson, H.F., Raynes, P. (eds.) Handbook of Liquid Crystals, vol. 7, 2nd edn, pp. 211–258. Wiley-VCH Verlag GmbH, Weinheim, Germany (2014)
28. Cestari, M., Diez-Berart, S., Dunmur, D.A., Ferrarini, M.R., Jackson, D.J.B., Lopez, D.O., Luckhurst, G.R., Perez-Jubindo, M.A., Richardson, R.M., Salud, J.: Phase behavior and properties of the liquid-crystal dimer 1,7-bis(4-cyanobiphenyl-4-yl) heptane: a twist-bend nematic liquid crystal. Phys. Rev. E **84**, 031704 (2011)
29. Paterson, D.A., Abberley, J.P., Harrison, W.T.A., Storey, J.M.D., Imrie, C.T.: Cyanobiphenyl-based liquid crystal dimers and the twist-bend nematic phase. Liq. Cryst. **44**, 127–146 (2017)
30. Saez, I.M., Goodby, J.W.: Design and properties of "Janus-like" supermolecular liquid crystals. Chem. Comm. 1726–1727 (2003)
31. Saez, I.M., Goodby, J.W.: "Janus" supermolecular liquid crystals–giant molecules with hemispherical architectures. Chem. Eur. J. **9**, 4869–4877 (2003)
32. Yoshizawa, A.: Nanostructured assemblies of liquid-crystalline supermolecules: from display to medicine. Liq. Cryst. **46**, 1950–1972 (2019)
33. Yoshizawa, A., Kikuzaki, H., Fukumasa, M.: Microscopic organization of the molecules in smectic A and chiral (racemic) smectic C phases: dynamic molecular deformation effect on the SA to SC* (SC) transition. Liq. Cryst. **18**, 351–366 (1995)

34. Yoshizawa, A., Kogawa, Y., Kobayashi, K., Takanishi, Y., Yamamoto, J.: A binaphthyl derivative with a wide temperature range of a blue phase. J. Mater. Chem. **19**, 5759–5764 (2009)
35. Sasaki, H., Takanishi, Y., Yamamoto, J., Yoshizawa, A.: Achiral flexible liquid crystal trimers exhibiting chiral conglomerates. Soft Matter **12**, 3331–3339 (2016)
36. Vertogen, G., de Jeu, W.H.: Nematic liquid crystals. In: Thermotropic Liquid Crystals, Fundamentals. Springer Series in Chemical Physics, vol. 45. Springer, Berlin, Heidelberg (1988). https://doi.org/10.1007/978-3-642-83133-1_2
37. Mandle, R.J., Bevis, E., Goodby, J.W.: Phase structures of nematic liquid crystals. In: Goodby, J.W., Collings, P.J., Kato, T., Tschierske, C., Gleeson, H.F., Raynes, P. (eds.) Handbook of Liquid Crystals, vol. 3, 2nd edn, pp. 3–29. Wiley-VCH Verlag GmbH, Weinheim, Germany (2014)
38. Freiser, M.J.: Ordered states of a nematic liquid. Phys. Rev. Lett. **24**, 1041–1043 (1970)
39. Lehmann, M.; Görtz, V. Design of biaxial nematic mesogens. In: Goodby, J.W., Collings, P.J., Kato, T., Tschierske, C., Gleeson, H.F., Raynes, P. (eds.) Handbook of Liquid Crystals, vol. 3, 2nd edn, pp. 335–373. Wiley-VCH Verlag GmbH, Weinheim, Germany (2014)
40. Tschierske, C., Photinos, D.J.: Biaxial nematic phases. J. Mater. Chem. **20**, 4263–4294 (2010)
41. Bruce, D.W.: Towards the biaxial nematic phase through molecular design. Chem. Rec. **4**, 10–22 (2004)
42. Mandle, R.J.: A ten-year perspective on twist-bend nematic materials. Molecules **27**, 2689 (2022)
43. Mandle, R.J., Cowling, S.J., Goodby, J.W.: A nematic to nematic transformation exhibited by a rod-like liquid crystal. Phys. Chem. Chem. Phys. **19**, 11429–11435 (2017)
44. Mandle, R.J., Cowling, S.J., Goodby, J.W.: Rational design of rod-like liquid crystals exhibiting two nematic phases. Chem. Eur. J. **23**, 14554–14562 (2017)
45. Nishikawa, H., Shiroshita, K., Higuchi, H., Okumura, Y., Haseba, Y., Yamamoto, S., Sago, K., Kikuchi, H.: A fluid liquid-crystal material with highly polar order. Adv. Mater. **29**, 1702354 (2017)
46. Chen, X., Korblova, E., Glaser, M.A., Maclennan, J.E., Walba, D.M., Clark, N.A.: Polar in-plane surface orientation of a ferroelectric nematic liquid crystal: Polar monodomains and twisted state electro-optics. Proc. Natl. Acad. Sci. U.S.A. **118**, e2104092118 (2021)
47. Karcz, J., Herman, J., Rychłowicz, N., Kula, P., Górecka, E., Szydlowska, J., Majewski, P., Pociecha, D.: Spontaneous chiral symmetry breaking in polar fluid-heliconical ferroelectric nematic phase. Science **384**, 1096–1099 (2024)
48. Nasir, A.K., Rahman, M.: Cybotactic nematic liquid crystal—An overview. Liq. Cryst. **51**, 503–535 (2024)
49. Keith, C., Lehmann, A., Baumeister, U., Prehm, M., Tschierske, C.: Nematic phases of bent-core mesogens. Soft Matter **6**, 1703–1721 (2010)
50. Deb, R., Nath, R.K., Sarkar, D.D., Araoka, F., Ishikawa, K., Takezoe, H., Takanishi, Y.: Synthesis of four-ring unsymmetrical bent-core mesogens and cybotactic cluster formation in their nematic phase. J. Mol. Struct. **1317**, 139055 (2024)
51. Tadapartri, P., Hiremath, U.S., Yelamaggad, C.V., Krishnamurthy, K.S.: Permittivity, conductivity, elasticity, and viscosity measurements in the nematic phase of a bent-core liquid crystal. J. Phys. Chem. B **114**, 1745–1750 (2010)
52. Panarin, Y.P., Sreenilayam, S.P., Vij, J.K., Lehnmann, A., Tschierske, C.: Formation and development of nanometer-sized cybotactic clusters in bent-core nematic liquid crystalline compounds. Beilstein J. Nanotechnol. **9**, 1288–1296 (2018)
53. Monika, M., Roy, A., Prasad, V.: Smectic nanoclusters in the nematic mesophases of dimeric compounds composed of rod-like azo moieties with lateral substituents. New J. Chem. **41**, 11576–11583 (2017)
54. Vita, F., Sparnacci, K., Panzarasa, G., Placentino, I.F., Marino, S., Scaramuzza, N., Portale, G., Cola, E.D., Ferrero, C., Torgova, S.I., Galli, G., Laus, M., Francescangelo, O.: Evidence of cybotactic order in the nematic phase of a main-chain liquid crystal polymer with bent-core repeat unit. ACS Macro Lett. **3**, 91–95 (2014)

55. Yoshizawa, A., Yamaguchi, A.: Kinetically induced intermolecular association: unusual enthalpy changes in the nematic phase of a novel dimeric liquid-crystalline molecule. Chem. Commun. 2060–2061 (2002)
56. Vanakaras, A.G., Photinos, D.J.: Thermotropic biaxial nematic liquid crystals: spontaneous or field stabilized? J. Chem. Phys. **128**, 154512 (2008)
57. Lehmann, M.: Biaxial nematics from their prediction to the materials and the vicious circle of molecular design. Liq. Cryst. **38**, 1389–1405 (2011)
58. Taushanoff, S., Le, K.V., Williams, J., Twieg, R.J., Sadashiva, B.K., Takezoe, H., Jakli, A.: Stable amorphous blue phase of bent-core nematic liquid crystals doped with a chiral material. J. Mater. Chem. **20**, 5893–5898 (2010)
59. Francescangeli, O., Stanic, V., Torgova, S.I., Strigazzi, A., Scaramuzza, N., Ferrero, C., Dolbnya, I.P., Weiss, T.M., Beradi, R., Muccioli, L., et al.: Ferroelectric response and induced biaxiality in the nematic phase of bent-core mesogens. Adv. Funct. Mater. **19**, 2592–2600 (2009)
60. Vita, F., Adamo, F.C., Francescangeli, O.: Polar order in bent-core nematics: an overview. J. Mol. Liq. **267**, 564–573 (2018)
61. Nishikawa, H., Sano, K., Kurihara, S., Watanabe, G., Nisjonyanagi, A., Dhara, B., Araoka, F.: Nano-clustering mediates phase transitions in a diastereomerically-stabilized ferroelectric nematic system. Commun. Mater. **3**, 89 (2022)
62. Yoshizawa, A., Nishizawa, A., Takeuchi, K., Takanishi, Y., Yamamoto, J.: Interlayer interactions induced by amphiphilicities of a rod-like molecule produce frustrated structures in conventional calamitic phases. J. Phys. Chem. B **114**, 13304–13311 (2010)
63. Kimoto, Y., Nishizawa, A., Takanishi, Y., Yoshizawa, A., Yamamoto, J.: Layer modulated smectic-C phase in liquid crystals. Phys. Rev. E **89**, 042503 (2014)
64. Barois, P.: Phase transition theories. In: Demus, D., Goodby, J.W., Gray, G.W., Spiess, H.W., Vill, V. (eds.) Handbook of Liquid Crystals, vol. 1, 1st edn, pp. 281–309. Wiley-VCH Verlag GmbH, Weinheim, Germany (1998)
65. Lobko, T.A., Ostrovskii, B.I., Hasse, W.: X-ray studies of the smectic phase and transitions between reentrant nematic, smectic A_1, A_d and C phases in binary mixtures of polar mesogens. J. Phys. II **2**, 1195–1213 (1992)
66. Tinh, N.H., Destrade, A.M., Levelut, A.M., Malthete, J.: Biforked mesognes: a new type of thermotropic liquid crystals. J. Phys. **47**, 553–557 (1986)
67. Date, R.W., Luckhurst, G.R., Shuman, M., Seddon, J.M.: Novel modulated hexatic phases in symmetric liquid crystal dimers. J. Phys. II **5**, 587–605 (1995)
68. Faye, V., Babeau, A., Placin, F., Nguyen, H.T., Barois, P., Laux, V., Isaert, N.: SC*A and S[Ctilde]* phases in chiral non-symmetric dimesogens. Liq. Cryst. **21**, 485–503 (1996)
69. Walker, R., Pociecha, D., Storey, J.M., Gorecka, E., Imrie, C.T.: Remarkable smectic phase behaviour in odd-membered liquid crystal dimers: the CT6O.m series. J. Mater. Chem. C **9**, 5167–5173 (2021)
70. Pociecha, D., Vaupotic, N., Gorecka, E., Mieczkowski, J., Gomola, K.: 2-D density-modulated structures in asymmetric bent-core liquid crystals. J. Mater. Chem. **18**, 881–885 (2008)
71. Nguyen, H.T., Sigaud, G., Achard, M.F., Hardouin, F.: Rod-like mesogens with antipathetic fluorocarbon and hydrocarbon tails. Liq. Cryst. **10**, 389–396 (1991)
72. Davidson, P., Keller, P., Levelut, A.M.: Molecular organization in side chain liquid crystalline polymers. J. Phys. **46**, 939–946 (1985)
73. Watanabe, J., Nakata, Y., Shimizu, K.: Frustrated bilayer smectic phase in main-chain polymers with two different spacers. J. Phys. II **4**, 581–588 (1994)
74. Endres, B.W., Ebert, M., Wendorff, J.H., Reck, B., Ringsdorf, H.: Combined main chain/side chain polymers: a new class of liquid crystalline polymers with unusual structural thermotropic and dynamic properties. Liq. Cryst. **7**, 217–239 (1990)
75. Kashima, S., Takanishi, Y., Yamamoto, J., Yoshizawa, A.: Flexible taper-shaped liquid crystal trimer exhibiting a modulated smectic phase. Liq. Cryst. **41**, 1752–1761 (2014)
76. Takanishi, Y., Takezoe, H., Fukuda, A.: Visual observation of dispirations in liquid crystals. Phys. Rev. B **45**, 7684–7689 (1992)

77. Kashima, S.: Supermolecular liquid crystals exhibiting frustrated phases. Master's Thesis, Hirosaki University, Hirosaki, Japan (2014)
78. Rybalova, T.V., Bagryanskaya, I.Y.: C-F...π, F...H, and F...F intermolecular interactions and F-aggregation: Role in crystal engineering of fluoroorganic compounds. J. Struct. Chem. **50**, 741–753 (2009)
79. Brownsey, G.J., Leadbetter, A.J.: Incommensurate coexistent density fluctuations in liquid-crystal phases of cyano compounds. Phys. Rev. Lett. **44**, 1608–1611 (1980)
80. Leadbetter, J., Tucker, P.A., Gray, G.W., Tajbakhsh, A.R.: Unusual structures in some alkyl cyanoterphenyls. Liq. Cryst. **8**, 1–12 (1990)
81. Hardouin, F., Achard, M.F., Jin, J., Shin, J.-W., Yun, Y.-K.: Novel sequence with incommensurate SA phases in a new dimesogenic liquid crystal. J. Phys. II France **4**, 627–643 (1994)
82. Hardouin, F., Achard, M.F., Jin, J., Yun, Y.-K.: From incommensurability to commensurability in smectic phases for a series of dimesogenic liquid crystals. J. Phys. II France **5**, 927–935 (1995)
83. Yamaguchi, A., Nishiyama, I., Yamamoto, J., Yokoyama, H., Yoshizawa, A.: Unusual smectic phases organized by novel λ-shaped mesogenic molecules. J. Mater. Chem. **15**, 280–288 (2005)
84. Huang, Y., Gui, S.: Factors affecting the structure of lyotropic liquid crystals and the correlation between structure and drug diffusion. RSC Adv. **8**, 6978–6987 (2018)
85. Alfutimie, A., Curtis, R., Tiddy, G.J.T.: Lyotropic surfactant liquid crystals: micellar systems. In: Goodby, J.W., Collings, P.J., Kato, T., Tschierske, C., Gleeson, H.F., Raynes, P. (eds.) Handbook of Liquid Crystals, vol. 6, 2nd edn, pp. 377–420. Wiley-VCH Verlag GmbH, Weinheim, Germany (2014)
86. Gray, G.W., Jones, B., Marson, F.: 71. Mesomorphism and chemical constitution. Part VIII. The effect of 3-substituents on the mesomorphism of the 4 -n-alkoxydiphenyl-4-carboxylic acid and their alkyl esters. J. Chem. Soc. 393–401 (1954)
87. Diel, S., Göring, P.: Thermotropic cubic phases. In: Demus, D., Goodby, J.W., Gray, G.W., Spiess, H.W., Vill, V. (eds.) Handbook of Liquid Crystals, vol. 2B, 1st edn, pp. 887–900. Wiley-VCH Verlag GmbH, Weinheim, Germany (1998)
88. Rowe, K.E., Bruce, D.W.: The synthesis and mesomorphism of di-, tetra- and hexa-catenar liquid crystals based on 2,2-bipyridine. J. Mater. Chem. **8**, 331–341 (1998)
89. Borisch, K., Diel, S., Göring, P., Müller, H., Tschierske, C.: Amphiphilic N-benzoyl-1-amino-1-deoxy-D-glucitol derivatives forming thermotropic lamellar, columnar and different types of cubic mesophases. Liq. Cryst. **22**, 427–443 (1997)
90. Tschierske, C., Ungar, G.: Mirror symmetry braking by chirality synchronization in liquids and liquid crystals of achiral molecules. ChemPhysChem **17**, 9–26 (2016)
91. Vaupotic, N., Salamonczyk, M., Matraszek, J., Vogrin, M., Pociecha, D., Górecka, E.: New structural model of a chiral cubic liquid crystalline phase. Phys. Chem. Chem. Phys. **22**, 12814–12820 (2020)
92. Reppe, T., Poppe, S., Tschierske, C.: Controlling mirror symmetry breaking and network formation in liquid crystalline cubic, isotropic liquid and crystalline phases of benzil-based polycatenars. Chem. Eur. J. **26**, 16066–16079 (2020)
93. Oka, T., Yamamura, Y., Kutsumizu, S., Saito, K.: Aggregation structure of chiral cubic liquid crystals revealed by X-ray diffraction utilizing a new algorithm. Soft Matter **19**, 1194–1201 (2023)
94. Rancon, Y., Charvolin, J.: Fluctuations and phase transformations in a lyotropic liquid crystal. J. Phys. Chem. **92**, 6339–6344 (1988)
95. Yamaguchi, A., Maeda, Y., Yokoyama, H., Yoshizawa, A.: Self-assembly of amphiphilic liquid-crystalline oligomers possessing a semiperfluorinated alkyl chain. Chem. Mater. **18**, 5704–5710 (2006)
96. Safinya, C.R., Roux, D., Smith, G.S., Sinha, S.K., Dimon, P., Clark, N.A., Bellocq, A.M.: Steric interactions in a model multimembrane system: a synchrotron X-ray study. Phys. Rev. Lett. **57**, 2718–2721 (1986)

97. Hjduk, D.A., Takenouchi, H., Hilmyer, M.A., Bates, F.S., Vigild, M.E., Almdal, K.: Stability of the perforated layer (PL) phase in deblock copolymer melts. Macromolecules **30**, 3788–3795 (1997)
98. Qi, S., Wang, Z.-G.: Kinetics of phase transitions in weakly segregated block copolymers: pseudostable and transient states. Phys. Rev. E **55**, 1682–1697 (1997)
99. Laradji, M., Shi, A.-C., Noolandi, J., Desai, R.C.: Stability of ordered phases in deblock copolymer melts. Macromolecules **30**, 3242–3255 (1997)
100. Imai, M., Kawaguchi, A., Saeki, A., Nakaya, K., Kato, T., Ito, K., Amemiya, Y.: Fluctuations of a lamella r gyroid transition in a nonionic surfactant system. Phys. Rev. E **62**, 6865–6874 (2000)
101. Takeuchi, K., Takanishi, Y., Yamamoto, J., Yoshizawa, A.: Amplification taper-shaped oligomer exhibiting a monolayer smectic A to columnar phase transition. Liq. Cryst. **37**, 507–515 (2010)
102. Yamaguchi, A., Uehara, N., Yamamoto, J., Yoshizawa, A.: Lamellar to lamellar phase transition driven by conformation change of an amphiphilic liquid crystal oligomer. Chem. Mater. **17**, 6445–6450 (2007)
103. Boussinesq, M.J.: Theorie de l'intumescene liquid appellee onde solitaire ou de translation, se propagent dans un canal rectangulaire. Comptes Rendus Acad. Sci. **72**, 755–759 (1871)
104. Oberbeck, A.: Über die wärmeletiung der flüssigkeiten bei berücksichtigung der strömungen infolge von temperaturdifferenzen. Ann. Phys. Chem. **7**, 271–292 (1879)
105. Getling, A.V.: Rayleigh-Bénard Convection: structures and Dynamics. World Scientific, Singapore (1998)
106. Belousov, B.P.: A periodic reaction and its mechanism. In: Collection of Short Papers on Radiation Medicine for 1958; Meditsina Publishers, Moscow, Russia (1959)
107. Zhabotinsky, A.M.: Periodical oxidation of malonic acid in solution (a study of the Belousov reaction kinetics). Biofizika **9**, 306–311 (1964)
108. Cassani, A., Monteverde, A., Piumetti, M.: Belousov-Zhabotinsky type reactions: the nonlinear behavior of chemical systems. J. Math. Chem. **59**, 792–826 (2021)
109. Buka, A., Kramer, L. (eds.): Pattern Formation in Liquid Crystals. Springer, New York, NY, USA (1996)
110. Wang, Z., Servio, P., Rey, A.D.: Structure and pattern formation in biological liquid crystals: Insights from theory and simulation self-assembly and self-organization. Front. Soft Matter **2**, 904069 (2022)
111. Morris, S.W., De Bruyn, J.R., May, A.D.: Electroconvection and pattern formation in a suspended smectic film. Phys. Rev. Lett. **65**, 2378–2381 (1990)
112. Migler, K.B., Meyer, R.B.: Solitons and pattern formation in liquid crystals in a rotating magnetic field. Phys. Rev. Lett. **66**, 1485–1488 (1991)
113. Zhang, B., Kitzerow, H.: Pattern formation in a nematic liquid crystal mixture with negative anisotropy of the electric conductivity–a long-known system with "inverse" light scattering revisited. J. Phys. Chem. B **120**, 6865–6871 (2016)
114. Sing, Z., Li, Z., Shang, X., Li, C., Ma, L., Lu, Y., Li, B.: Electrically switchable structural patterns and diffractions in a dual frequency nematic liquid crystal. Chinese Opt. Lett. **21**, 010501 (2023)
115. Eber, N., Salamon, P., Buka, A.: Electrically induced patterns in nematics and how to avoid them. Liq. Cryst. Rev. **4**, 101–134 (2016)
116. Carr, E.F.: Influence of an electric field on the dielectric loss of the liquid crystal p-azoxyanisole. J. Chem. Phys. **39**, 1979–1983 (1963)
117. Carr, E.F.: Influence of electric fields on the molecular alignment in the liquid crystal p-(anisalamino)-phenyl acetate. Mol. Cryst. Liq. Cryst. **7**, 253–268 (1969)
118. Helfrich, W.: Conduction-induced alignment of nematic liquid crystals: basic model and stability considerations. J. Chem. Phys. **51**, 4092–4105 (1969)
119. Clerc, M.G., Gonzalez-Cortes, G., Morel, M.J., Hidalgo, P.I., Vergara, J.: Finger front propagation in smectic-A Fréedericksz transition. Phys. Rev. E **105**, 054701 (2022)

120. Tang, X., Atzin, N., Mozaffari, A., Das, S., Abbott, N.L., De Pablo, J.J.: Generation and propagation of flexoelectricity-induced solitons in nematic liquid crystals. ACS Nano **18**, 10768–10775 (2024)
121. Yoshizawa, A., Kurauchi, M., Kohama, Y., Dewa, H., Yamamoto, K., Nishiyama, I., Yamamoto, T., Yamamoto, J., Yokoyama, H.: Synthesis and physical properties of α-(4-cyanobiphenyl-4-yloxy)-ω-[4-(5-alkylpyrimidine-2-yl)phenyloxy]alkanes. Liq. Cryst. **33**, 611–619 (2006)
122. Hegmann, T., Kain, J., Diel, S., Pelzl, G., Tschierske, C.: Evidence for the existence of the McMillan phase in a binary system of a metallomesogen and 2,4,7-trinitrofluorenone. Angew. Chem. Int. Ed. **40**, 887–890 (2001)
123. Sadashiva, B.K., Reddy, R.A., Pratibha, R., Madhusudana, N.V.: Biaxial smectic A phase in homologous series of compounds composed of highly polar unsymmetric substituted core molecules. J. Mater. Chem. **12**, 943–950 (2002)
124. Yelamaggad, C.V., Shashikala, I.S., Rao, D.S.S., Nair, G.G., Prasad, S.K.: The biaxial smectic (SmAb) phase in nonsymmetric liquid crystal dimers comprising two rodlike anisometric segments: an unusual behavior. J. Mater. Chem. **16**, 4099–4102 (2006)
125. Yoshizawa, A.: Liquid-crystalline supermolecules inducing layer fluctuations: from hierarchical to dissipative structures. Crystals **14**, 681 (2024)
126. Sagisaka, M., Fujita, Y., Nakanishi, Y., Takahashi, H., Tsuyoshi, N., James, C., Yoshizawa, A., Mohamed, A., Guittard, F., Eastoe, J.: Periodic formation/breakdown of lamellar aggregates with anionic cyanobiphenyl surfactants. Langmuir **31**, 13040–13047 (2015)
127. Kunitake, T., Okahata, Y.: A totally synthetic bilayer membrane. J. Am. Chem. Soc. **99**, 3860–3861 (1977)
128. Okahata, Y., Kunitake, T.: Formation of stable monolayer membranes and related structures in dilute aqueous solution from two-headed ammonium ampliphiles. J. Am. Chem. Soc. **101**, 5231–5234 (1979)
129. Fendler, J. H.: Surfactant vesicles as membrane mimetic agents: characterization and utilization. Acc. Chem. Res. **13**, 7–13 (1980)
130. Zana, R. (ed.): Dynamics of Surfactant Self-Assemblies: Micelles, Microemulsions, Vesicles, and Lyotropic Phases. CRC Press, Boca Raton, FL, USA (2005)

Chapter 3
Ferroelectric Liquid Crystals

Abstract Ferroelectricity in liquid crystals has attracted significant attention in the field of soft matter. In the first half of this chapter, the evolution of the design concept for ferroelectric smectic liquid crystals is the primary focus. The progression of the subject is event in the transition from the consideration of macroscopic phase symmetry to the design of intermolecular interactions. The dynamic behavior of molecules in the smectic phases is elucidated by ^{13}C NMR studies of ferroelectric liquid crystals. The development of liquid crystal materials for display purpose is described, and de Vries-like materials that can provide defect-free alignment are reconsidered. It has been demonstrated that certain liquid-crystalline supermolecules are capable of exhibiting polar switching in the uniaxial SmA phase. In the latter part of this chapter, the research status of recently discovered ferroelectric nematic liquid crystals is introduced. The fluidity exhibited by the nematic FLCs is comparable to that of conventional nematic liquid crystals, whilst simultaneously demonstrating an exceptionally pronounced spontaneous polarization of approximately 5 μC cm^{-2}. The polar orientational order of molecules in the N_F phase has been demonstrated to trigger chirality in the fluid state, despite of the absence of intrinsic chirality. The ferroelectric molecular ordering in the nematic phase coupled with fluidity, is of fundamental significance and has great potential to revolutionize liquid crystal technologies.

Keywords Ferroelectric smectic liquid crystals · De Vries smectic A · ^{13}C NMR · Display · Antiferroelectric liquid crystals · Ferroelectric nematic liquid crystals

3.1 Introduction

Ferroelectricity in liquid crystals was first discovered in the chiral smectic C (SmC*) phase of the rod-like chiral compound **DOBAMBC** in 1975 [1]. This ferroelectricity arose from stabilization at the substrate surface and did not manifest as bulk ferroelectricity. The surface-stabilized ferroelectric liquid crystals (SSFLCs) exhibit

A. Yoshizawa, *Liquid-Crystalline Supermolecules*, The Materials Research Society Series, https://doi.org/10.1007/978-981-92-2524-8_3

spontaneous polarization, enabling ultrafast electro-optical switching with a response time of the μs order [2]. These discoveries brought ferroelectric liquid crystals into the spotlight, making them a major trend in liquid crystal science and technology [3–8]. Figure 3.1a shows polarization (*P*) as a function of an applied electric field (*E*) in the ferroelectric SmC* phase. Even when the electric field is removed, the polarization does not become zero. Two stats with spontaneous polarization exist stably (also see Fig. 3.1b). Liquid-crystalline molecules switch between these two states depending on the direction of the electric field. Each state is maintained even after the electric field is removed. Therefore, the high-speed bistable switching is achieved. Figure 3.1c shows the peak of the polarization reversal current that occurs during switching between the two states. Ferroelectric liquid crystal displays (FLCDs) do not require a thin film transistor (TFT) in each pixel due to the memory effect. The dark state is obtained when one of the polarizers is set parallel to the director, and the brightest state is obtained when the director is tilted with 45 degrees from each of the crossed polarizers (see Fig. 3.2). The in-plane switching provides a wide viewing angle. Although the bistable in-plane switching with a response time of the μs order is very attractive for display applications, the FLCDs were unable to enter the large-screen LCD display market. The reason is that the significant reduction in manufacturing costs of competing TFT-LCDs has made it possible to produce larger screens. Furthermore, the FLCDs offered good black-and-white contrast but struggled to produce gray levels. Now, FLCDs are used in reflective micro displays based on liquid crystal-on-silicon (LCOS) technology for applications in 3D head-mounted displays and projectors [9, 10]. Figure 3.3a shows an illustration of the LCOS structure and Fig. 3.3b shows that of the LCOS projector.

Chandani et al. discovered the existence of an antiferroelectric phase during their research on the ferroelectric liquid crystal **MHPOBC** in 1989 [11]. As show in Fig. 3.4a, the *P*–*E* curve in the antiferroelectric phase shows a double-hysteresis. When the electric field is zero, the polarization is also zero. Above a certain threshold, the antiferroelectric state changes to a ferroelectric state. Figure 3.4b shows the electric field-induced transition between the antiferroelectric chiral smectic C (SmC^*_A) phase and the ferroelectric SmC* phase. In the SmC^*_A phase, polarization occurs within each layer, but the molecular orientation reverses in adjacent layers, resulting in no overall spontaneous polarization. Unlike a ferroelectric phase, two peaks of the polarization reversal current appear corresponding to each transition (Fig. 3.5c). A lot of research on antiferroelectric liquid crystal materials have been developed, and prototype displays have also been created [12–15]. Interestingly, Inui et al. reported materials exhibiting a continuous antiferroelectric–ferroelectric phase transition without a threshold [16]. This V-shaped switching is interesting in terms of its ability to accelerate response speed and reduce operating voltage [14].

It had been believed that molecular chirality is essential for the manifestation of ferro- and antiferro-electricity. Niori et al. discovered ferroelectric switching in a tilted smectic B2 phase of the achiral bent-core molecule **P-*n*–O-PIMB** in 1996 [17]. This discovery has opened up new avenues for research into polar order and chiral structures in liquid crystals [18–23]. Reddy et al. reported the emergence of ferroelectricity in the uniaxial SmA phase of an achiral bent-core liquid crystal

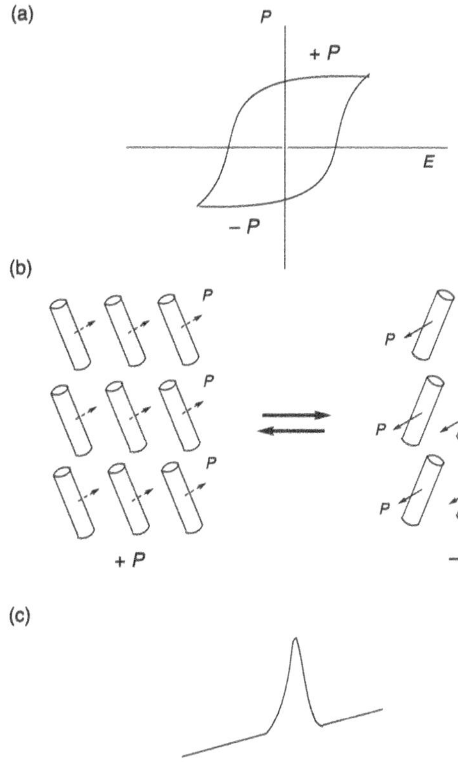

Fig. 3.1 **a** *P*-*E* curve in ferroelectrics. **b** Ferroelectric switching in the SmC* phase. **c** Electrical response in the switching current when applying a triangular wave voltage

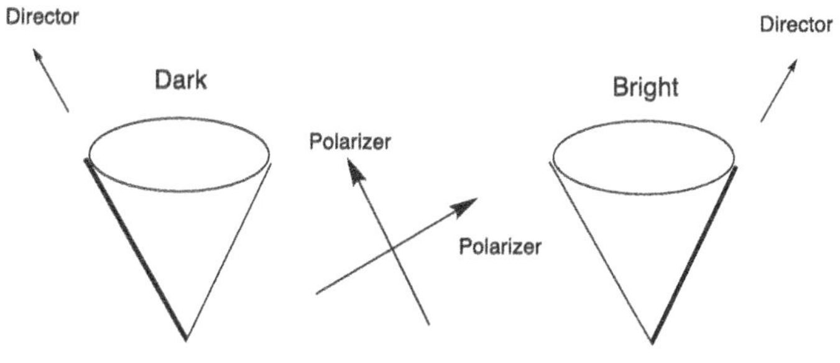

Fig. 3.2 Dark and bright states of FLCDs. The director is a unit vector showing the average direction of the molecular long axis

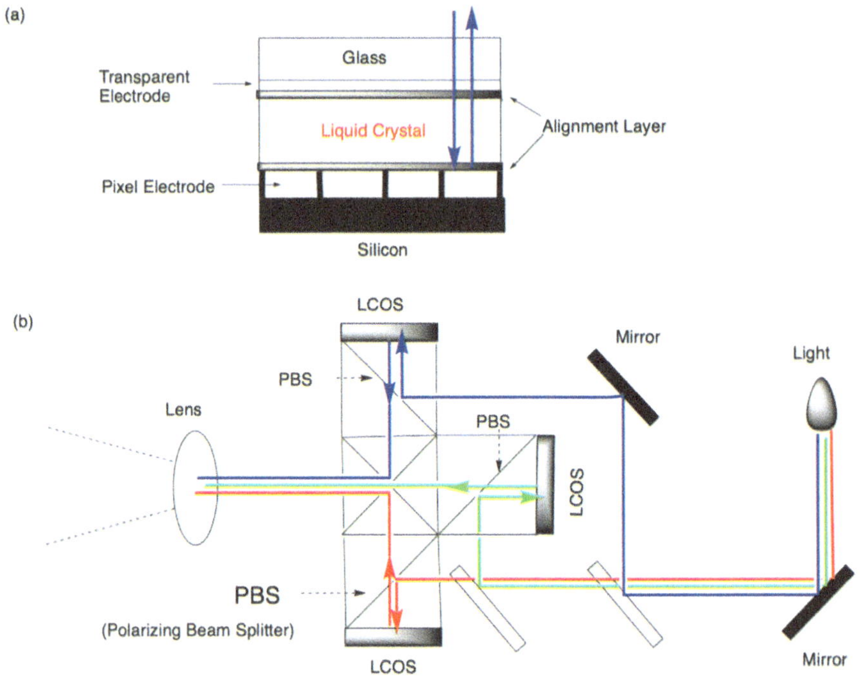

Fig. 3.3 **a** Illustration of the LCOS structure and **b** that of the LCOS projector

[24]. In 2017, two research groups independently reported that two achiral rod-like molecules, **RM734** [25, 26] and **DIO** [27], exhibit a nematic–nematic phase transition. It was revealed that the low-temperature phase exhibits nematic-like fluidity and ferroelectric order [27, 28]. That liquid crystal phase was named the ferroelectric nematic (N_F) phase. Figure 3.5 shows an image of the molecular arrangement in the N_F phase. The N_F phase exhibits a huge spontaneous polarization along the molecular long axis. The ferroelectric molecular ordering in the nematic phase coupled with fluidity has not only fundamental significance but also great potential to revolute liquid crystal technologies [29]. **DOMAMBC**, **MHPOBC**, **P-8-O-PIMB**, **RM734**, and **DIO** (Fig. 3.6) changed the liquid crystal world.

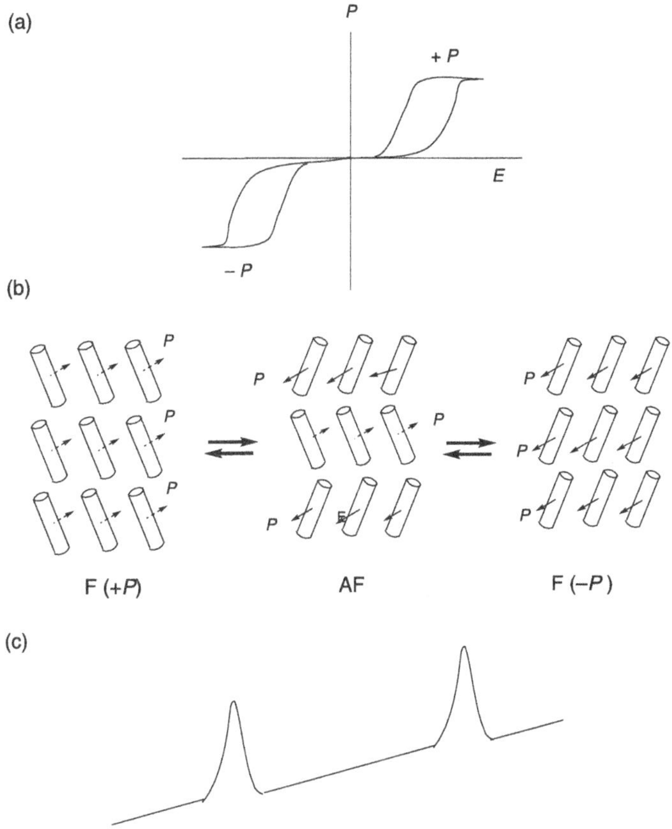

Fig. 3.4 **a** *P-E* curve in antiferroelectric. **b** Electric field-induced transition between Sm C^*_A and SmC* phases. **c** Electrical response in the switching current when applying a triangular wave voltage

3.2 Ferroelectricity in Liquid Crystals

Meyer predicted the ferroelectricity in the SmC* phase based on Neumann's principle. In a nematic phase, molecules exhibit fluidity similar to isotropic liquids, but they spontaneously align such that their molecular long axes are parallel to one another. On the other hand, smectic A and C phases have a layered structure, which results in increased viscosity. However, the order inside the layer is liquid-like in these phases. In the SmA phase, though locally tilted, the average direction of the molecules is parallel to the layer normal. In the SmC phase, the molecules are tilted with respect to the layer normal, and the tilt direction is the same in adjacent layers. In the SmC* phase, the tilt direction twists between adjacent layers, forming a helix with the layer normal direction as its helix axis. Figure 3.7 shows the phase symmetry of the N, SmA, and SmC phases. According to Lagerwall [5], let us apply the Neumann's principle to the liquid crystal phases formed by rod-like molecules. It is described as

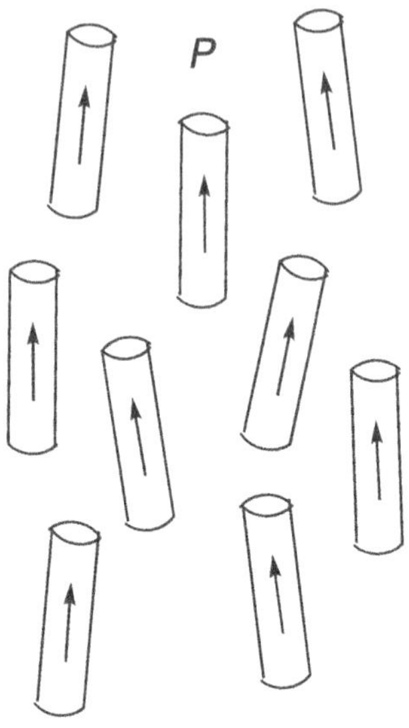

Fig. 3.5 Schematic image of the N_F phase. The arrow shows a dipole moment

follows: "the symmetry elements of an intrinsic property must include the symmetry elements in the medium". The intrinsic property and the medium, here, are polarization and a liquid crystal phase, respectively. We assume that there is no distinction between the head and tail of the constituent molecule. In the nematic phase or smectic A phase, each medium remains symmetric even when rotated 180° rotation about the z-axis. Therefore, by the Neumann's principle, its corresponding polarity elements also satisfy $P\mathrm{x} = -P\mathrm{x}$ and $Py = -Py$. It is also symmetrical by 180° rotation around the y-axis, corresponding to $Pz = -Pz$. Therefore, $P\mathrm{x} = P\mathrm{y} = P\mathrm{z} = 0$, and no macroscopic polarization occurs in the N phase or SmA phase. With respect to a smectic C phase, the medium remains symmetric when rotated 180° rotation around the y-axis. It has mirror symmetry about the x–z plane (Fig. 3.8a). The SmC phase also exhibits no macroscopic polarization. If the molecule is chiral, it does not overlap its mirror image (Fig. 3.8b). The mirror symmetry does not exist. Therefore, $Py \neq -Py$. Polarization occurs in the y-axis direction of each layer. However, in the SmC* phase, a helical structure forms, cancelling the local polarization in each layer, resulting in zero macroscopic polarization in the bulk (Fig. 3.9).

Clark and Lagerwall demonstrated the macroscopic polarization in the SmC* phase using thin samples approximately 1 μm thick [2]. The *P*–*E* curve in the SmC* phase exhibited a hysteresis loop characteristic of ferroelectricity, as show in Fig. 3.1a. Their design concept is as follows: "When the smectic layers are formed perpendicular to the confining glass plates, there are no boundary conditions that are compatible with the helical arrangement. As the distance between the surfaces gradually shortens, the competition between the helical structure and the surface order ultimately cause the helical structure to unravel elastically via the surface forces. Below a certain sample thickness, the helix cannot be formed. The orientation of

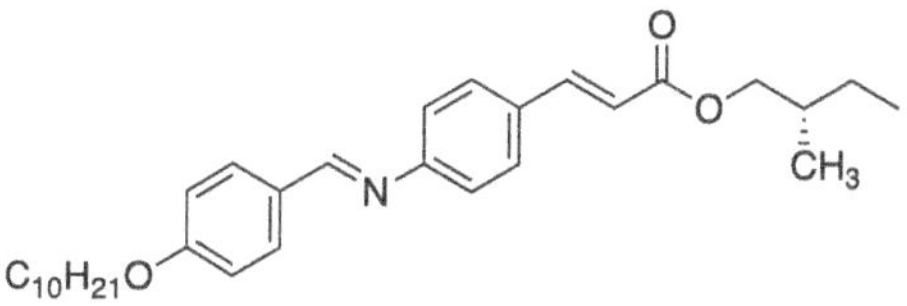

DOBAMBC

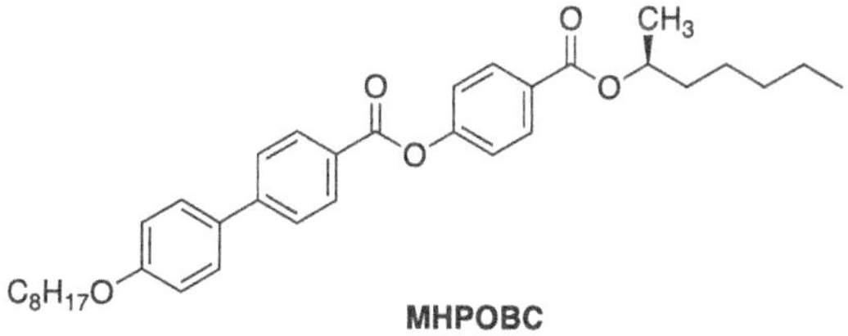

MHPOBC

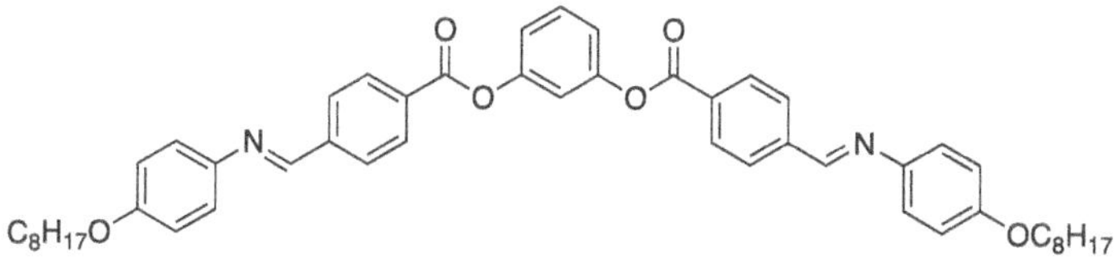

P-8-O-PIMB

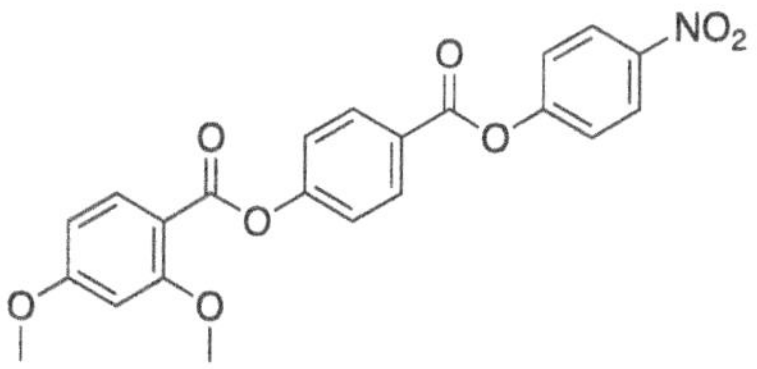

RM734

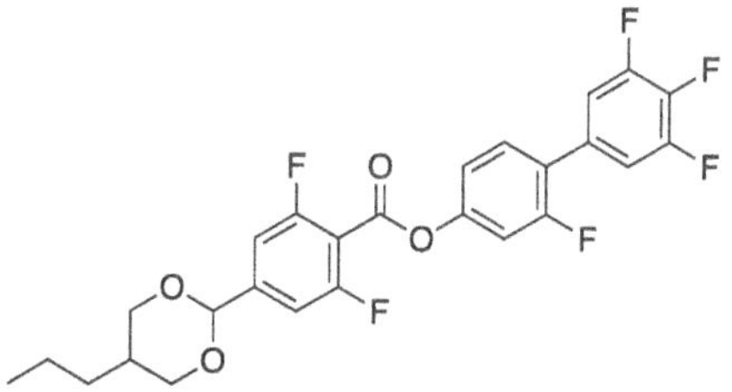

DIO

Fig. 3.6 Molecular structures of **DOBAMBC**, **MHPOBC**, **P-8-O-PIMB**, **RM734**, and **DIO**

that director is expected to be permitted only at the positions where the surface intersects the smectic cone. This is because the intrinsic cone constraint and the surface constraint are simultaneously satisfied (Fig. 3.10a) [5]". The surface stabilization realizes a ferroelectric liquid crystal. The angle between the layer normal and the optical axis (*n* director) is defined as tilt angle θ (Fig. 3.10b). The magnitude of polarization increases with the increase in the tilt angle. The sense of polarization is defined as shown in Fig. 3.11. In a given compound, the sense of the helix and that of the spontaneous polarization are not necessarily related. The origin of spontaneous polarization is not yet fully understood. Blinov and Beresnev noted that it is the directional bias for a lateral dipole of a molecule [30]. There is a hypothesis that

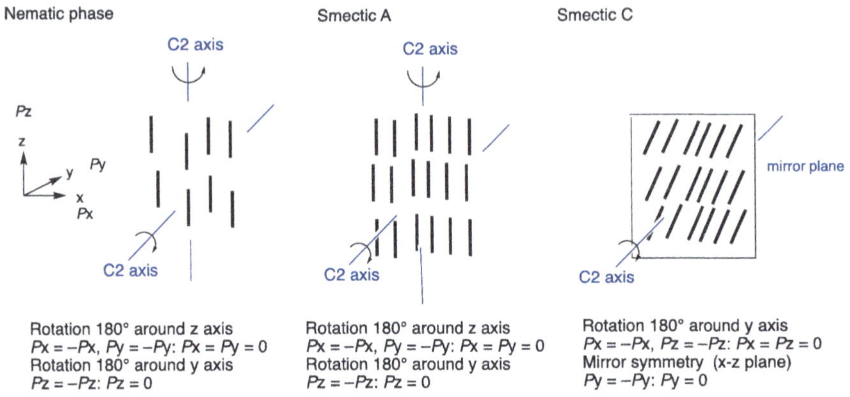

Fig. 3.7 Phase symmetry of N, SmA, and SmC phases

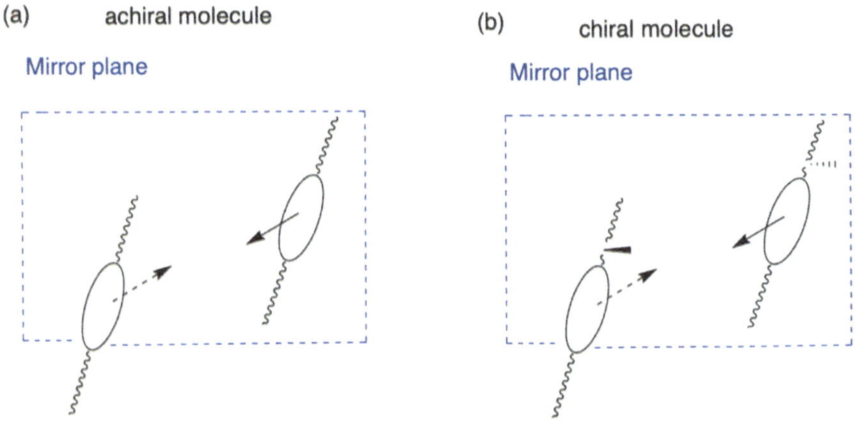

Fig. 3.8 Relationship between a molecule and its mirror image. **a** Achiral molecule **b** Chiral molecule

the non-uniform rotation around the molecular long axis causes the time-averages localization of the molecular dipole moment. Two opposite experimental results have been reported for the rotation of the molecules around their long axis. Lalanne et al. detected the slowing down of molecular rotation at the SmA–SmC* transition by optical four-wave mixing experiments [31]. On the other hand, Kremer et al. reported constant molecular rotation at the SmA–SmC* transition by using dielectric measurements [32]. Largerwall suggested that a polar bias in the SmC* phase gives a different probability for the dipole to be directed at one side out of the tilt plane rather than in the opposite direction [5].

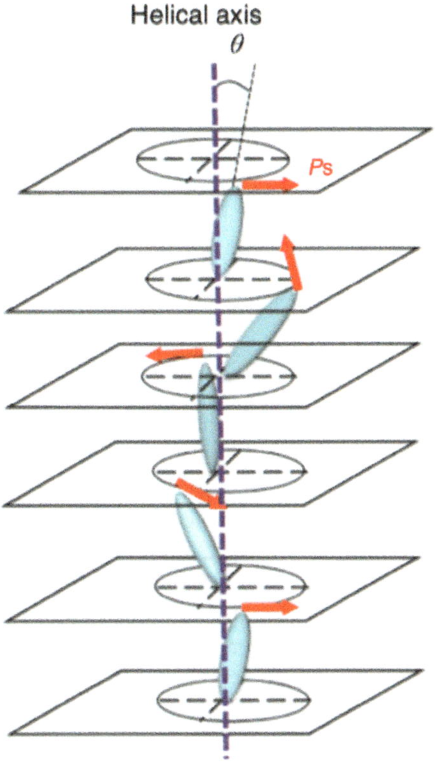

Fig. 3.9 Helical structure of a chiral smectic C phase

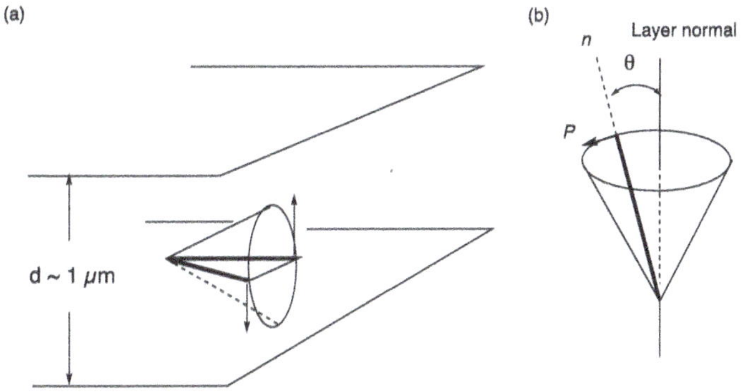

Fig. 3.10 **a** Surface-stabilized bistable states. **b** The definition of tilt angle

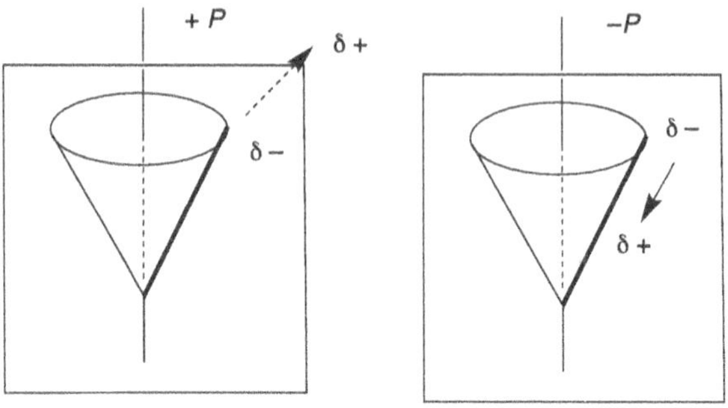

Fig. 3.11 Sense of polarization

Next, we describe the design approaches to date for realizing ferroelectricity in a nematic phase. Flexoelectricity represents a potential mechanism for the emergence of ferroelectricity in the nematic phase. It is a property of dielectric materials where electric polarization is generated by a non-uniform mechanical deformation. Three director deformations, i.e., splay, twist, and bend, are permitted in the N phase. In the splay deformation, as shown in Fig. 3.12a, the polarization may occur along the molecular long axis. On the other hand, it may appear along the molecular short axis in the bend deformation (Fig. 3.12b). There are also attempts to create ferroelectric nematic liquid crystals by introducing suitable dipole–dipole interactions. Zannoni et al. developed a simple candidate structure that is a taper-shaped molecule possessing two dipole units with opposite directions and differing strengths for the computer simulation of the ferroelectric nematic phase (Fig. 3.13) [33]. These molecules can be distinguished by their head and tail. The computer simulation revealed that a system of tapered particles can exhibit a phase sequence of isotropic liquid–apolar nematic–polar nematic phase–polar smectic. The polar SmA phase resulting from this design will be discussed later. However, it cannot explain the ferroelectricity of **RM734** or **DIO**.

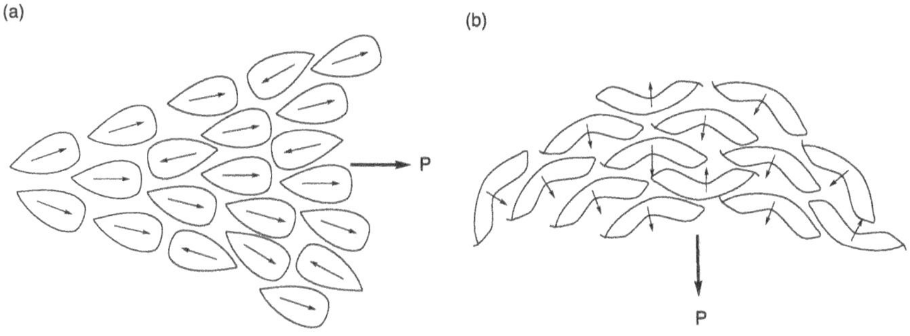

Fig. 3.12 **a** Flexoelectric polarization in the splay deformation and **b** that in the bend deformation

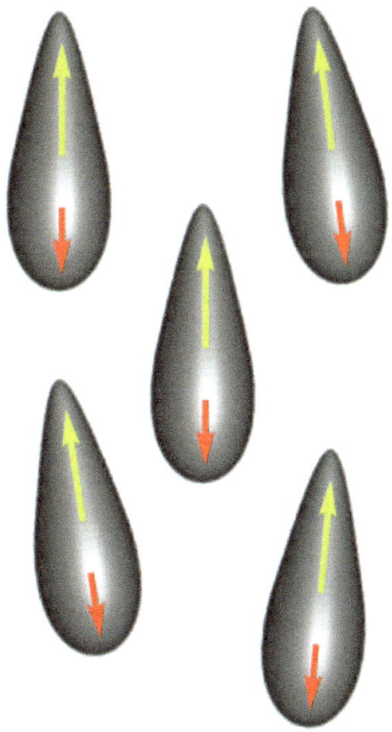

Fig. 3.13 Schematic illustration of molecular organization of tapered molecules in a nematic phase with polar order

3.3 ^{13}C NMR Studies on Smectic Liquid Crystals

In order to understand how the liquid-crystalline molecules organize in the smectic phases. we observed solid-state ^{13}C nuclear magnetic resonance (NMR) spectroscopy of a ferroelectric liquid crystal **MONBIC** [34]. Figure 3.14 shows the molecular structure and phase-transition temperatures. **MONBIC** exhibited a phase sequence of Iso–SmA–SmC*–Cry and showed a large spontaneous polarization value at about 100 nC cm^{-2} in the SmC* phase at 10 K below the SmA–SmC* transition.

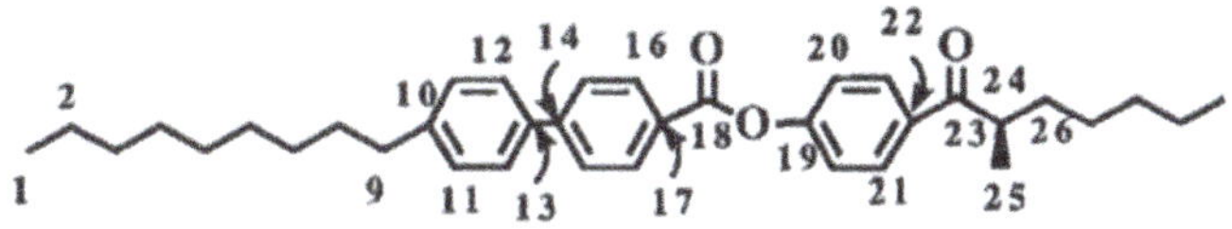

MONBIC: Cry 79 °C SmC* 101 °C SmA 118 °C Iso

Fig. 3.14 Molecular structure and phase-transition temperatures of **MONBIC**. The carbon atoms of **MONBIC** are numbered as shown in the figure

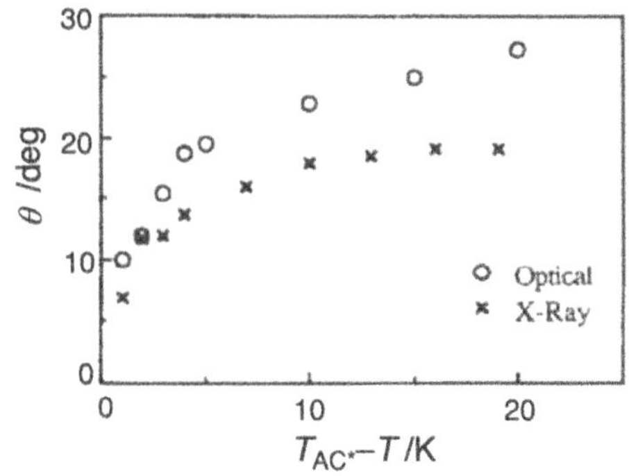

Fig. 3.15 The temperature dependence of the tilt angle, as determined by POM and XRD measurements. T_{AC^*} is the SmA–SmC* phase-transition temperature. Reprinted with permission from [34]. Copyright (1995) Taylor & Francis Ltd.

The molecular shape of **MONBIC** in the SmC* phase was investigated by comparing the tilt angles determined by X-ray diffraction (XRD) measurements with those determined by polarized optical microscopy (POM). The layer spacings of the SmA and SmC* phases were determined, and the tilt angles were calculated using Eq. 3.1.

$$\theta = \cos^{-1}(l_C/l_A) \tag{3.1}$$

where l_A is a layer spacing in the SmA phase, and l_C is that in the SmC phase. Figure 3.15 shows the temperature dependence of the tilt angle determined by polarized optical microscopy (POM) or X-ray diffraction (XRD) measurements. Both increase as the temperature decreases, but a difference can be seen between them. The tilt angle determined by POM observation reflects the tilt of the core part with the delocalized electrons of the molecule. The tilt angel determined by XRD reflects the average tilt of the molecule. The former shows larger tilt angles than the latter, indicates that the constituted molecules form a zigzag shape in the SmC* phase.

Figure 3.16 shows the ^{13}C NMR spectra of **MONBIC** in the isotropic liquid, the SmA phase, and the SmC* phase upon cooling from the isotropic liquid without sample-spinning. Liquid-crystalline molecules with phenyl rings exhibiting positive magnetic anisotropy align their molecular long axes along the magnetic field direction upon phase transition from isotropic liquid to liquid crystal. The ^{13}C chemical shifts of the keto carbon (C23), the ester carbon (C18), and the aromatic carbons (C17, C16, and C20) increase markedly, but those of the aliphatic carbons (C24, C9, and C1) decrease from the isotropic liquid to the SmA phase. These changes in the chemical shifts indicate that both of the core and the tails of the molecule tend to orient parallel to the magnetic field in the SmA phase. Upon cooling to the SmC* phase, those of the keto, ester, and aromatic carbons decrease on the contrary; furthermore, the marked line-broadening can be seen for those carbons. These indicate that the angle between the core part and the magnetic field has increased, and that the orientation of the core part has become disordered. On the other hand, those of the aliphatic carbons continuously decrease. Figure 3.17 shows a schematic image of the molecular orientation of **MONBIC** in the SmA and SmC* phases. On cooling, the tilt angle of the core part decreases in the SmA phase, but that increases in the SmC* phase. On the other hand, that of the tail continuously decreases in the SmA and SmC* phases. The line-broadening for the core carbons indicates that the core parts form the helicoidal structure in the SmC* phase.

Next, we obtained information on the dynamic behavior of each molecule in the smectic phases. We observed ^{13}C spin–lattice relaxation times in the laboratory frame (T_1) of **MONBIC** in the Iso, SmA, and SmC* phases. Molecular motions with a relaxation time in the range from 10^{-12} to 10^{-8} s are detected in the ^{13}C T_1 measurements. Figure 3.18a shows the ^{13}C T_1 values for the tail carbon (C2), the

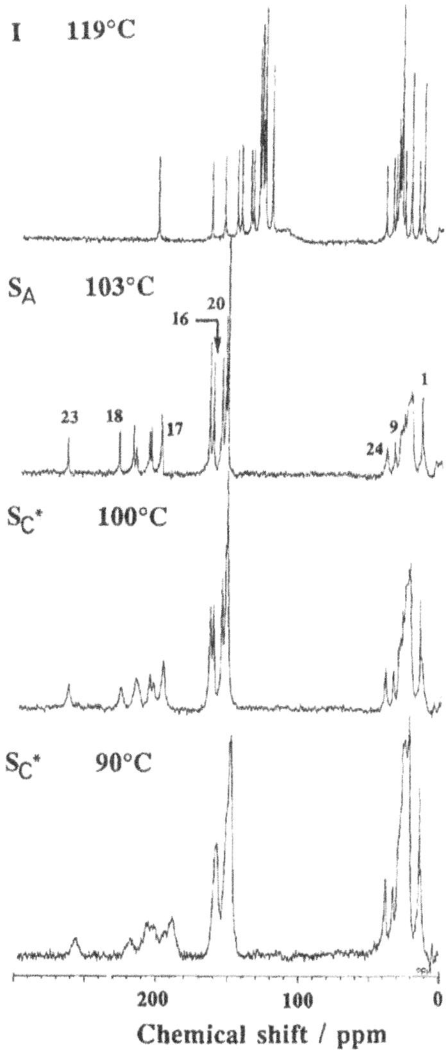

Fig. 3.16 ^{13}C NMR spectra of **MONBIC** in the isotropic liquid, SmA, and SmC* without sample spinning. The numbers are carbon numbers indicated in Fig. 3.14. Reprinted with permission from [34]. Copyright (1995) Taylor & Francis Ltd.

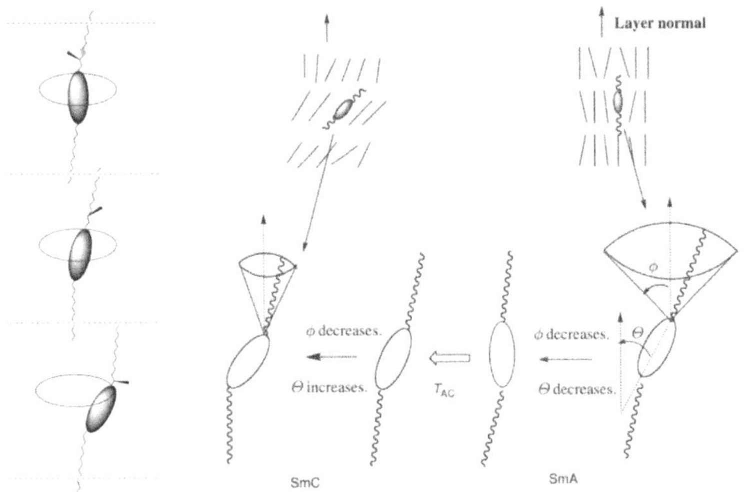

Fig. 3.17 Microscopic behavior of **MONBIC** in the SmA and SmC* phases

asymmetric carbon (C24), and the core carbon (C12) as a function of temperature. The T_1 values of those carbons continuously decrease, indicating that the molecular rotation around the long axis continuously slows down in the isotropic liquid, the SmA phase, and the SmC* phase. This result refutes the hypothesis that hindered molecular rotation around the long axis is the origin of spontaneous polarization in the SmC* phase. Figure 3.18b shows ^{13}C spin–lattice relaxation times in the rotating frame ($T_{1\rho}$) of an aromatic carbon (C21) in the SmA and SmC* phases. Molecular motions with a relaxation time in the range 10^{-6} to 10^{-4} s are effective in ^{13}C $T_{1\rho}$ values. ^{13}C $T_{1\rho}$ values of the core carbon (C21) are independent of temperature in each phase and show the discontinuous decrease at the SmA–SmC* transition. On the other hand, those of the tail carbons were not detected. Therefore, we can say that the cooperative molecular motion for the core part contributes to the layer ordering in each smectic phase, and that the motion mode in the SmC* phase differs from that in the SmA phase. The collective motion in the SmA phase corresponds to the tilt fluctuations, and that in the SmC* phase corresponds to the molecular fluctuations around the cone.

The question arises what happens at the SmA–SmC* phase transition. Figure 3.19 shows the temperature dependence of some protonated aromatic carbons. ^{13}C T_1 values of C12 continuously decrease at the SmA–SmC* transition, whereas ^{13}C T_1 values of C20 and C21 exhibit marked jump. The fluctuation of those values of C20 and C21 is attributed to the change of the angle between C–H vector and the long axis. The angle between the phenyl ring of C20 and the molecular long axis undergoes reversible changes neat the SmA–SmC* phase transition. Furthermore, ^{13}C T_1 values of unprotonated core and carbonyl carbons (C18, C19) also showed a significant jump at the transition [39]. The jumps of ^{3}C T_1 values of the unprotonated carbons is due to the increase in the internuclear C–H distance between the unprotonated caron and

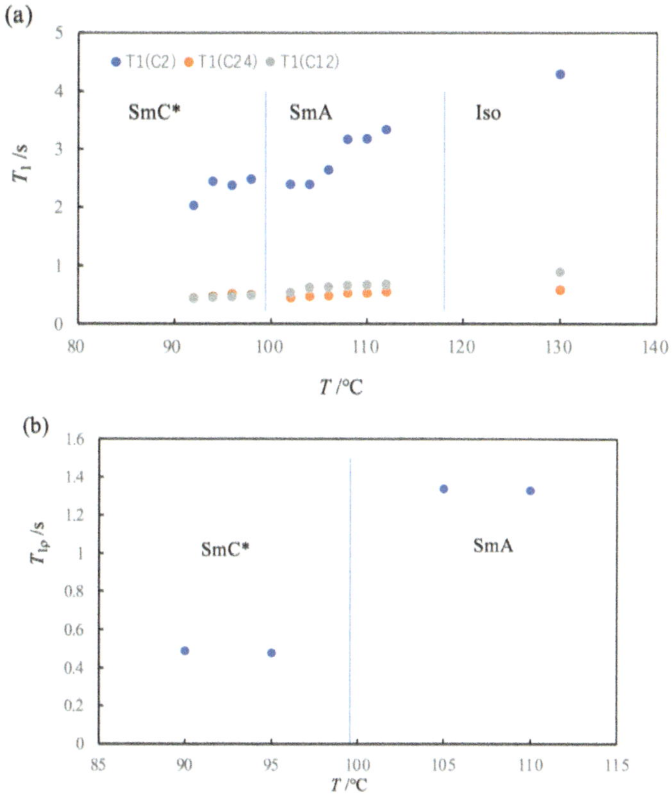

Fig. 3.18 **a** Temperature dependence of ^{13}C T_1 values of C2, C12, and C24 of **MONBIC** in the Iso, SmA, and SmC*. **b** That of ^{13}C $T_{1\rho}$ values of C21 of **MONBIC** in the SmA and SmC*. The carbon numbers (C2, C12, and C24) are indicated in Fig. 3.14

its intramolecular neighboring proton. We can say that the molecular deformation, such as a twist or bend for the core part, occurs near the SmA–SmC* transition (Fig. 3.20).

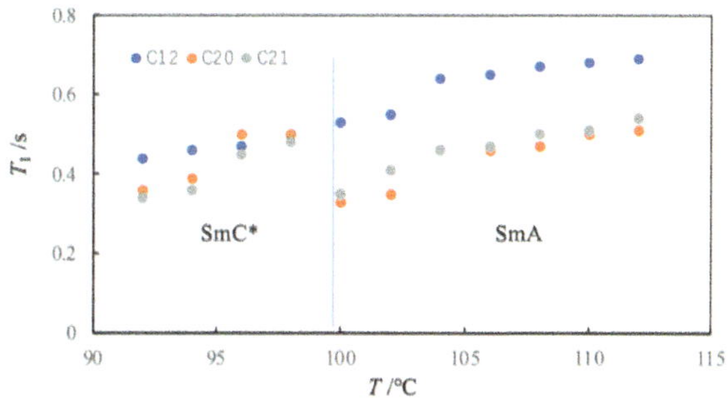

Fig. 3.19 Temperature dependence of ^{13}C T_1 values of C12, C20, and C21 of **MONBIC** in the SmA and SmC* phases. The carbon numbers (C12, C20, and C21) are indicated in Fig. 3.14

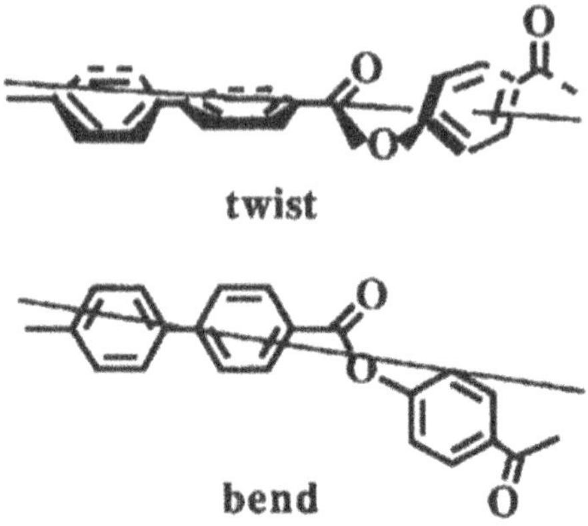

Fig. 3.20 Molecular deformation for the core part. The estimated long axis is shown. Reprinted with permission from [34]. Copyright (1995) Taylor & Francis Ltd.

The NMR studies reveal the dynamic behavior of liquid-crystalline molecules in the smectic phases: (1) the core parts perform cooperative motions; (2) interlayer permeation of the tails induces the correlation between cores in adjacent layers; and (3) the conformation change of the core part occurs near the SmA–SmC*(or SmC) transition. Figure 3.21 shows a schematic image of the molecular organization model in the smectic phases. This model is useful for the molecular design of smectic liquid crystals. The cooperative motions of core parts affect the electro-optical switching in the smectic phases. Introducing a polar group into the core unit improves the *P*s-driven switching speed. Enhancing the correlation between cores in each layer leads to the formation of nanoclusters (see Chap. 2). The interlayer correlation relates to the molecular tilt direction or twist in adjacent layers.

Next, we introduce the ^{13}C NMR studies of an antiferroelectric liquid crystal [35]. The phase transition temperatures of the liquid crystal (**MOPBIC**) on cooling were as follows: Iso 157 °C SmA 138 °C SmC* 81 °C SmC_A* 64 °C Sm4. The NMR spectra were obtained by cross-polarization (CP) and magic angle spinning (MAS)

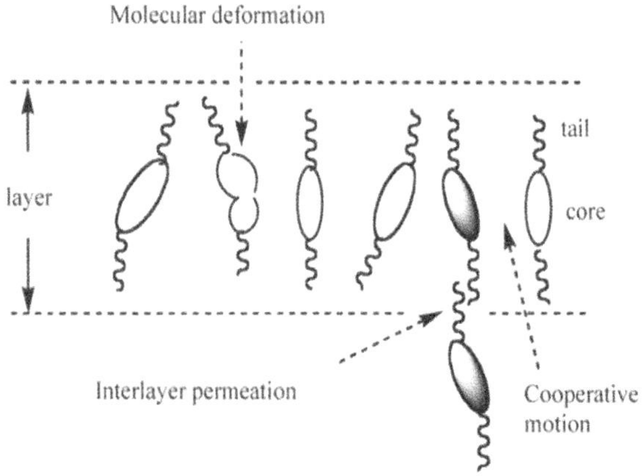

Fig. 3.21 Model for molecular organization near the SmA–SmC* transition

techniques at a spinning speed of 4.5 kHz on cooling. Figure 3.22 shows the ^{13}C NMR spectra of **MOPBIC** in the SmC*, SmC_A*, and Sm4 phases. In this measurement, the liquid-crystalline molecules are not aligned in the direction of the magnetic field. Upon transition from the SmC* to the SmC_A* phase, the line widths of the core carbons significantly become broader, but the line broadening is not observed for the tail carbons (Fig. 3.23a). They narrow again upon transition from the SmC_A* phase to the Sm4 phase. The line broadening of the core carbons indicates that the molecular motion with a relaxation time of 10^{-6} s exists for the core part in the SmC_A*. The peak intensity of both terminal carbons (C1 and C30) in the tails increases at the SmC–SmC_A* phase transition (Fig. 3.23b). The increase in the peak intensity results from the increase in the cross-polarization efficiency, which reflects the decrease of the mobility of the terminal part. The increased interlayer interactions due to the tail penetration induces the characteristic motion for the core part, causing the molecular tilt to alternate between adjacent layers. Nishiyama et al. reported that compounds bearing a terminal branching unit exhibit an anticlinic SmC phase [36, 37]. These experimental results also support the notion that interactions between terminal alkyl chains in adjacent layers stabilize the anticlinic molecular arrangement.

3.4 Materials for Ferroelectric Smectic Liquid Crystal Displays

3.4.1 FLCs Exhibiting Large Spontaneous Polarization

Practical FLC mixtures consist of a host smectic C liquid crystal doped with a small amount of a chiral compound with large spontaneous polarization. Therefore, it is important to obtain compounds with large spontaneous polarization, and the correlation between molecular structure and spontaneous polarization has been investigated [38–41]. Figure 3.24 shows the molecular structures of some chiral compounds exhibiting large spontaneous polarization.

Here we present two examples illustrating the effect of molecular structure on spontaneous polarization. The compounds are ferroelectric liquid crystals possessing a 2-methyalkanoyl group or a 2-fluro-2-methyalkanolyoxy group developed in Nippon Mining Co. Ltd. (later Japan Energy Corporation). Those chiral groups were derived from (*R*)-(+)-1,2-epoxyalkane. One of the compounds is **MOPBIC** with a (*S*)-2methyloctanoly group, and it showed both ferroelectric and antiferroelectric phases as described above [35, 45]. The spontaneous value in the SmC* phase at T_{AC*} −10 °C was 112 nC cm^{-2}. Let us estimate the degree of alignment of dipoles in the SmC* phase of **MOPBIC**. If they are perfectly aligned in the same direction, the magnitude of normalized polarization $P_0 (= P_S / \sin\theta)$ can be approximated by Eq. (3.2),

$$P_0 \sim N_A \cdot \left(V \cdot \frac{d}{M}\right)\left(\frac{1}{V}\right) \cdot \mu_\perp = N_A \cdot d \cdot \mu_\perp / M \qquad (3.2)$$

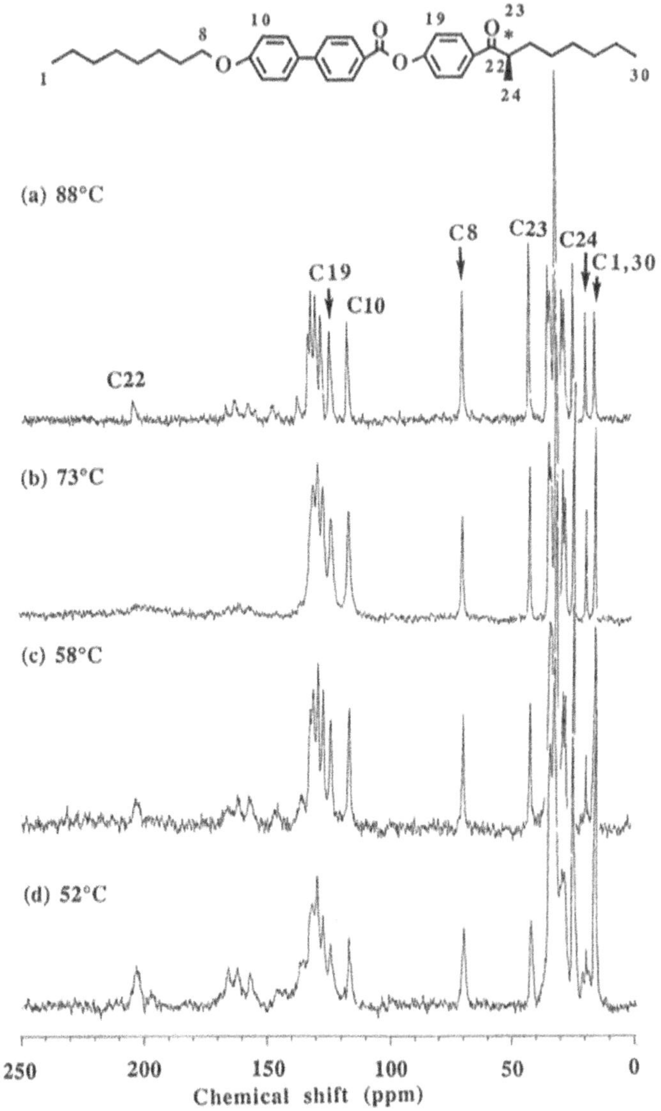

Fig. 3.22 Temperature-dependent CP/MAS ^{13}C NMR spectra of **MOPBIC** on cooling. Reprinted with permission from [35]. Copyright (1992) The Japan Society of Applied Physics

where N_A is Avogadro constant, d is density, M is a molecular weight, $\mu_\perp$ is a dipole moment perpendicular to the long axis, and V is a unit volume. Here, we assume that the dipole moment of the functional group adjacent to the chiral carbon contributes to the magnitude of Ps in the SmC* phase [39]. The perpendicular dipole moment ($\mu_\perp$) of the keto group is calculated to be 1.99 D. Assuming a density of 0.8 g/cm^3

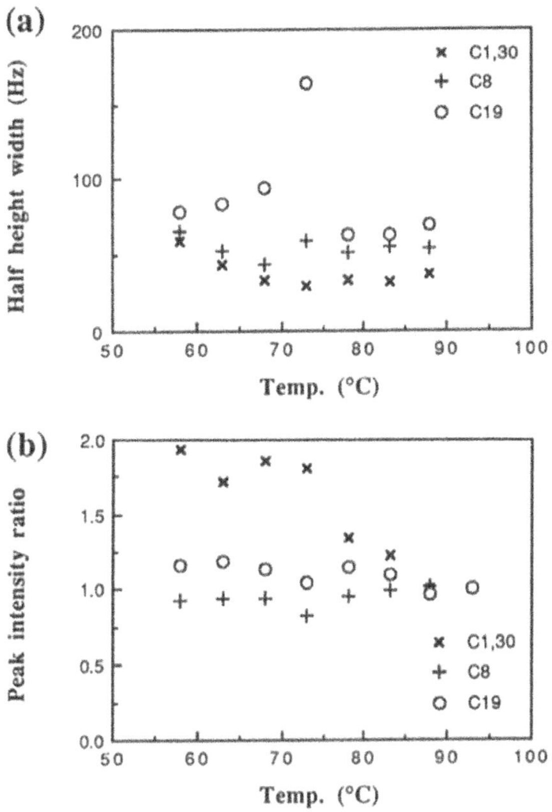

Fig. 3.23 The temperature dependence of **a** half-height widths and **b** relative peak intensity ratios for core and chain carbons of **MOPBIC** on cooling. The relative intensity of each peak was calculated by dividing the peak intensity at each temperature by that at 93 °C. Reprinted with permission from [35]. Copyright (1992) The Japan Society of Applied Physics

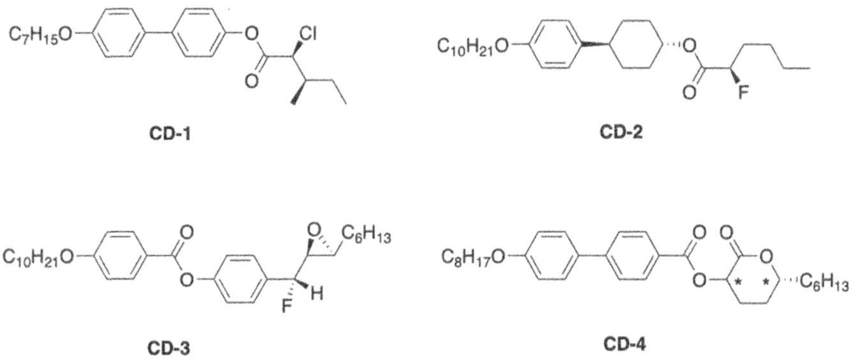

Fig. 3.24 Molecular structure of chiral dopants, **CD-1** [42], **CD-2** [43], **CD-3** [38], and **CD-4** [44]. The carbon indicated by asterisk is an asymmetric carbon

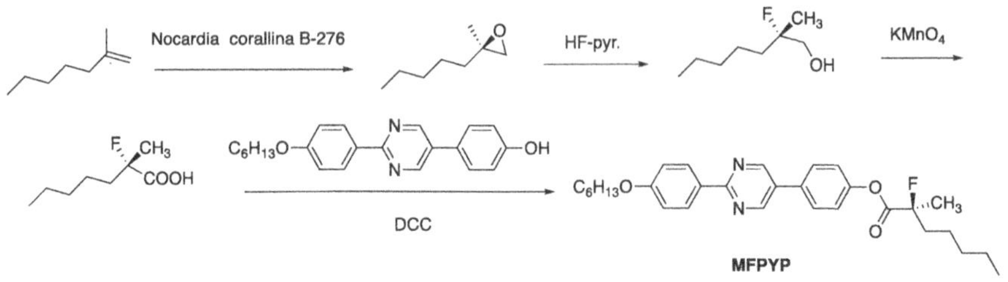

Fig. 3.25 Synthetic scheme of **MFPYP**

for **MOPBIC** in the SmC* phase, the magnitude of P_0 is estimated to be about 600 nC cm^{-2} at the first approximation. The P_0 value calculated from the measured *P*s value of 112 nC cm^{-2} at T_{AC*} −10 °C is 313 nC cm^{-2}. Approximately 50% of the dipoles align at 10 °C below the SmA–SmC* phase transition.

Another compound is **MFPYP** with a 2-fluoro-2-methylheptanoyloxy group [46]. Figure 3.25 shows the synthetic scheme. 2-Methylheptene was converted to (*R*)-2-methyl-1,2-epoxyheptane by using Norcardia corallina B-276. The epoxide was treated with hydrogen fluorine to give (*S*)-2-fluoro-2-methylheptanol. Then it was oxidized with potassium manganate to the corresponding carboxylic acid. The optical purity was 80%ee. (*S*)-2-fluoro-2-methylheptanoic acid was treated with 2-(4-hexylpoxyphenyl)-5-(4-hydroxyphenyl)pyrimidine in the presence of *N*,*N*'-dicyclohexylcarbodiimide to give the target compound **MFPYP**. The compound showed a phase sequence of Iso 130 °C SmC* 88 °C Cry on cooling. The magnitude of *P*s at T_{AC*} −2 °C was 250 nC cm^{-2}. The maximum value was 438 nC cm^{-2}. This compound exhibits good miscibility with host smectic liquid crystal mixtures composed of phenylpyrimidine derivatives. for developing practical FLC materials. The **MFPYP** is a useful chiral dopant for developing practical FLC mixtures [46].

Not only the polar group but also the core structure plays an important role in the manifestation of spontaneous polarization. Replacing 2,4-diphenylpyrimidine in **MFPYP** with pheny biphenylcarboxylate resulted in a significant reduction in its spontaneous polarization (Fig. 3.26).

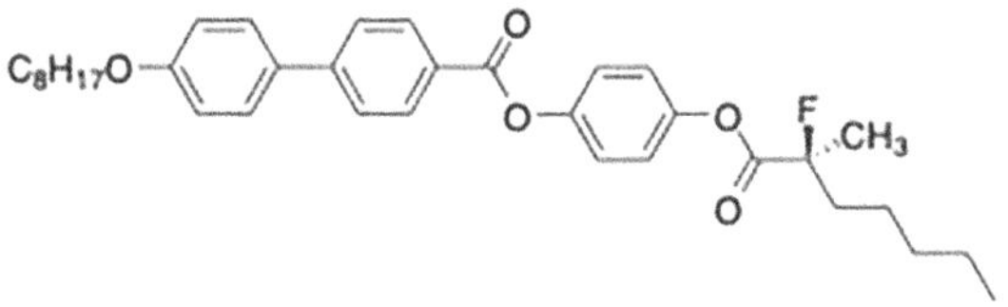

Iso 156 °C SmA 136 °C SmC* 104 °C Cry *Ps* = 33 nC cm^{-2} at T_{AC*}−2 °C: *Ps* max = 146 nC cm^{-2}

Fig. 3.26 Effects of core structure on the spontaneous polarization

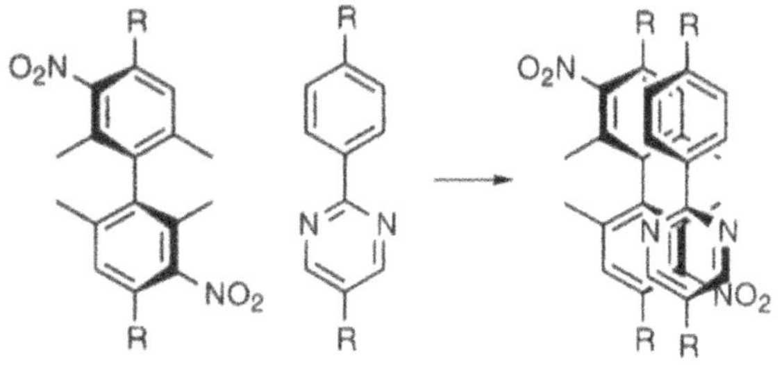

Fig. 3.27 Model for core–core chirality transfer via chiral conformational interactions. Reprinted with permission from [40]. Copyright (2001) American Chemical Society

How is the chirality induced in achiral smectic liquid crystal phases by chiral dopants? Lemuieux explained the chirality induction in terms of host–guest molecular recognition using a model for core–core chirality transfer via chiral conformational interactions [40, 47]. The dopant with an axial chirality behaves as an active guest and transfers its chirality to the surrounding host molecules through core–core interactions (Fig. 3.27). This chirality transfer via core–core interactions requires a good structure match. According to Lemieux [40], the best structural match is phenylpyrimidine, while the worst is phenyl benzoate.

3.4.2 Host SmC Materials

Achiral smectic liquid crystals for the ferroelectric host mixture have been developed [48, 49]. For manufacturing FLCDs, the FLC materials have to satisfy many important parameters, such as phase sequence, SmC* temperature range, spontaneous polarization, tilt angle, birefringence, dielectric anisotropy, rotational viscosity, elastic constant, N* pitch, and SmC* pitch. Their temperature dependence is also critical. Numerous achiral host SmC liquid crystals have been developed to adjust the aforementioned physical properties. Response time (τ) is expressed by Eq. (3.3),

$$\tau \sim \eta/(Ps \cdot E) \tag{3.3}$$

where Ps is spontaneous polarization, E is an electric field, and η is rotational viscosity. As is evident from Eq. (3.3), to reduce this response time, it is necessary to decrease the rotational viscosity and increase the spontaneous polarization. To increase the spontaneous polarization, a large dipole must be introduced into the chiral dopant. As a result, the rotational viscosity of the FLC mixture increases due to strong dipole–dipole interactions. Therefore, low-viscosity host SmC liquid crystals are required. Furthermore, FLCs must exhibit a phase sequence of Iso–N–SmA–SmC to obtain a favorable molecular alignment. Phenylpyrimidine derivatives have been used for the preparation of host SmC mixtures. They exhibit a low-viscosity

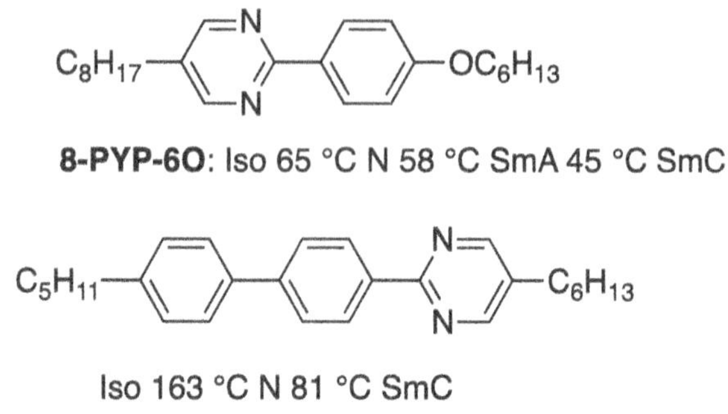

Fig. 3.28 Molecular structures of the 2-ring and 3-ring phenylpyrimidine derivatives

SmC phase without inhibiting the formation of the nematic phase. Figure 3.28 shows typical 2-ring and 3-ring phenylpyrimidine LCs.

We introduce compounds that meet the above requirements–namely, the phase sequence and low viscosity–and broaden the temperature range of the SmC phase. According to the scheme shown in Fig. 3.29, we synthesized 6-alkyl-2-(4-alkyloxyphenyl)quinoline (**PQm,n**) [50]. 4-Alkylaniline was converted to 6-alkylquinoline by the Skraup method [51]. It was treated with 4-methoxyphenyl lithium to give 6-alkyl-2-(methoxyphenyl)quinoline. The compound was deprotected with boron tribromide and then reacted with alkyliodide to give the target compound **PQm,n**.

The phase transition temperatures of **PQ8,6** were Iso 117 °C N 113 °C SmA 104 °C SmC 69 °C Cry. The compound exhibited an N phase and a broad SmC phase. We investigated the compatibility between phenylquinoline and phenylpydimidine derivatives. **PQ8,6** and 2-ring phenylpyrimidie derivative **8-PYP-6O** showed continuous miscibility across the full composition range for the N, SmA, and SmC phases of both compounds. Next, the viscosity of **PQ8,6** was examined. We prepared two FLC mixtures, i.e., mixture A (**PQ8,6**: 54 wt%/**8-PYP-6O**: 36%/chiral dopant: 10 wt%)

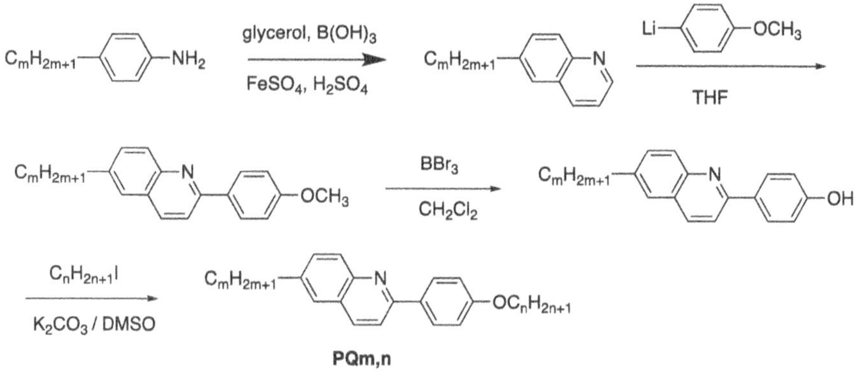

Fig. 3.29 Synthetic scheme of **PQm,n**

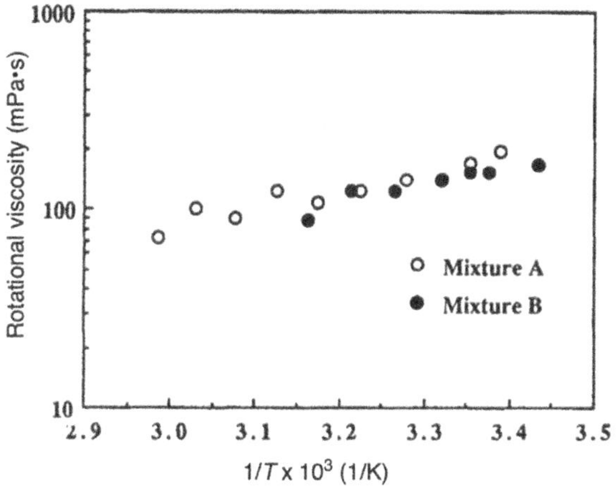

Fig. 3.30 Rotational viscosity for mixtures A and B as a function of reciprocal temperature. The rotational viscosity was obtained from the following equation, $\eta = \tau \cdot Ps \cdot E/(1.75 \sin^2 \theta)$. Reprinted with permission from [50]. Copyright (1993) Taylor & Francis Ltd.

and mixture B (**8-PYP-6O**: 90%/chiral dopant: 10 wt%). The chiral dopant used for the ferroelectric mixtures was (*S*)-4-(2-methyloctanoyl)-4'-biphenyl 3-chloro-4-octyloxybenzoate. Both mixtures showed a phase sequence of Iso–N*–SmA–SmC*. The SmA–SmC* transition temperature for mixture A was higher by 19 K than that for mixture B. Figure 3.30 shows the rotational viscosity, evaluated by switching studies, for mixtures A and B as a function of reciprocal temperature. Temperature-dependence of the rotational viscosity for mixture A is the same as that for mixture B. Compound **PQ8,6** widens the SmC temperature range of the 2-ring phenylpyrimidine derivative without increasing the viscosity. The phenylquinoline derivatives are effective for producing FLC materials.

A challenging issue in developing FLCDs was layer deformation originating from the SmC phase structure. The layer thickness decreases with decreasing temperature due to molecular tilting, which deforms the smectic layer structure from "a bookshelf structure" (Fig. 3.31a) to "a chevron structure" (Fig. 3.31b) [52–54]. The chevron structure reduces the effective switching angle, thereby decreasing the transmittance. If the rubbing direction is the same at both surfaces, there are two types of chevron structures depending on the orientation of the layer deformation, called as C1 and C2. In boundaries between the two different chevron structures, the smectic layers cannot fill the space, resulting in the formation of defects (Fig. 3.31c).

To solve the problems, extensive research had been conducted on pretilt control, electric field application, and material development [5]. We introduce the development of liquid crystals to achieve a defect-free alignment. Mochizuki et al. designed naphthalene derivatives stabilizing a bookshelf layer structure and demonstrated a defect-free prototype FLC display using the compounds [55, 56]. Janulis et al. reported a fluorinated phenylpyrimidine derivative exhibiting a weak negative layer expansion coefficient [57]. The layer spacing of the fluorinated compound increases continuously through the SmA–SmC phase transition. Rieker and Janulis proposed a model in which steric interactions drive the antiparallel alignment of nearest neighbors in the SmA and SmC phases [58].

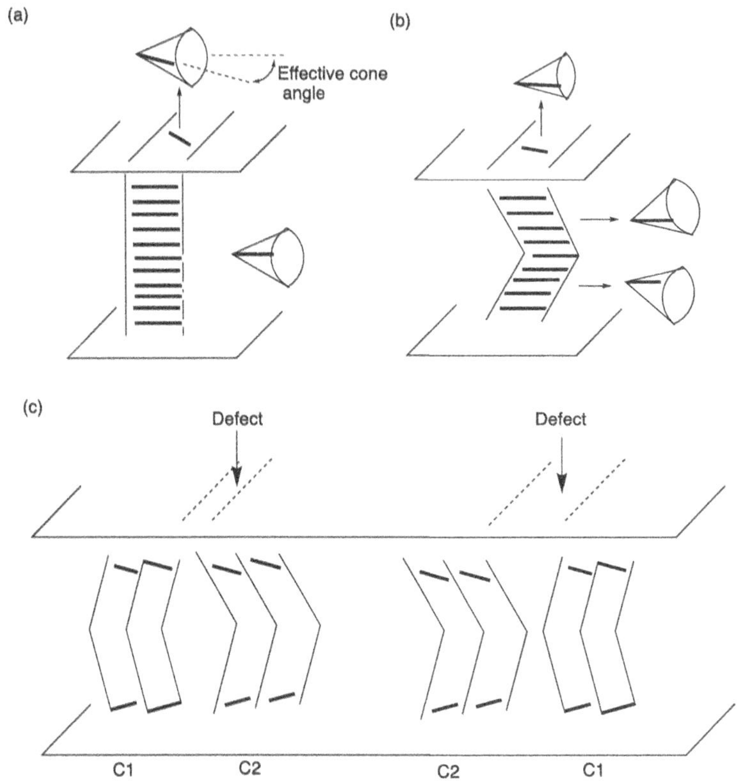

Fig. 3.31 **a** Bookshelf-layer structure. **b** Chevron-layer structure. **c** Defects in boundaries between the two different chevron-layer structures

Figure 3.32 shows the molecular structures of those developed compounds.

Compounds exhibiting small layer shrinkage during the SmA–SmC phase transition are designated as "de Vries-like" liquid–crystal materials. They are characterized by a maximum layer contraction of 1% upon transition from the SmA phase to the SmC phase [59]. In the SmA phase, if molecules within the layer are tilted at the

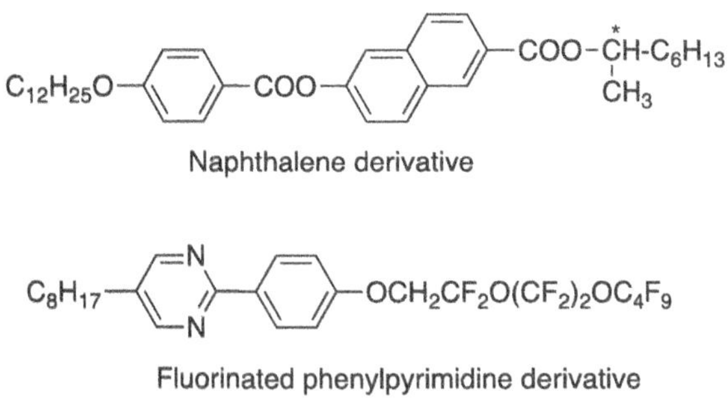

Fig. 3.32 Molecular structures of materials stabilizing a bookshelf-layer structure in the SmC phase. The carbon indicated by asterisk is an asymmetric carbon

same angle relative to the layer normal and their tilt direction is random, the director is parallel to the layer normal, and no layer contraction occurs during the SmA–SmC phase transition. De Vries proposed two diffuse cone models that describe the SmA phase in which mesogens have a tilted orientation and a random azimuthal distribution [60–62]. In de Vries I, the molecules are tilted in the same direction in each layer, while the azimuthal angle is disordered from layer to layer (Fig. 3.33a). In de Vries II, the molecules are tilted, but the azimuthal angle is disordered even in a single layer (Fig. 3.33b). According to these models, during the transition from the SmA phase to the SmC phase, the director tilts relative to the layer normal, but the layer spacing remains unchanged. The chiral version of this phase, termed de Vries SmA*, has attracted particular attention. Because it is expected that the occurrence of zigzag defects due to the layer shrinkage will be suppressed in the SmC* phase. Figure 3.34 shows the molecular structures of "de Vries-like" liquid–crystal materials developed to date. Most of reported de Vries compounds are amphiphilic rod-like molecules with polysiloxane or polysilane as a terminal chain [63, 64]. Recently, Kaur et al. reported unsymmetric bent-core molecules possessing de Vries-like properties [65]. The development of compounds exhibiting de Vries-like characteristics for improving FLCDs has to satisfy the following requirements: 1) good miscibility with a host SmC material and 2) no influence on the other physical properties of the FLC mixture.

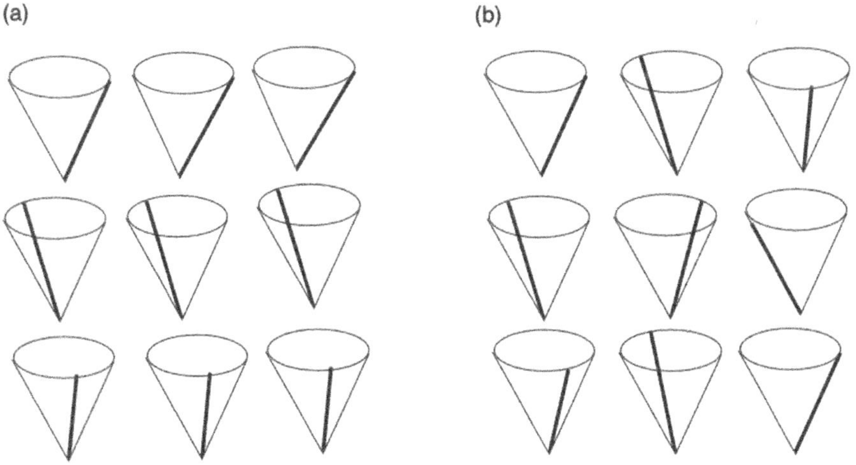

Fig. 3.33 **a** Model for de Vries I and **b** that for de Vries II

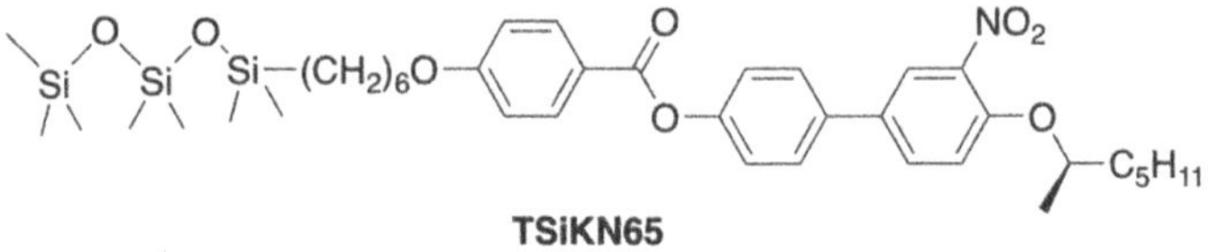

TSiKN65

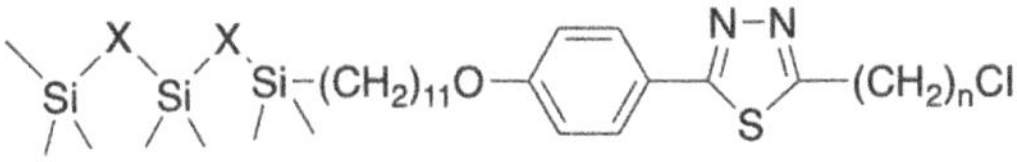

QL5-*n*: X = O
QL6-*n*: X = CH_2

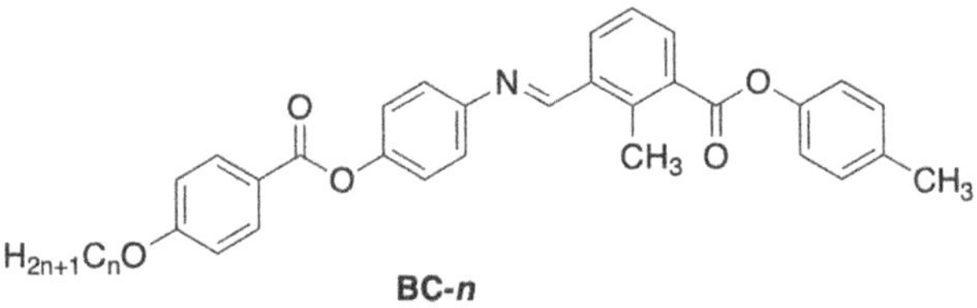

BC-*n*

Fig. 3.34 Typical molecules possessing de Vries-like properties, **TSiKN65** [63], **QL5-*n*** [64], **QL6-*n*** [64] and **BC-*n*** [65]

3.5 Electro-Optical Switching in the de Vries Smectic A Phase

The amphiphilicity plays an important role in the formation of the de Vries SmA phase. An amphiphilic compound (**SFDPB**) possessing a semiperfluorinated chain and a 2,3-difluoro-1,4-diphenylbenzene core exhibited the SmA–SmC phase transition without layer contraction [66]. Figure 3.35 shows the molecular structure and phase transition temperatures. Figure 3.36 shows the layer spacings of **SFDPB** in the SmA and SmC phases as a function of the reduced temperature, $T–T_{AC}$. The layer spacing in the SmA phase slightly increases with decreasing temperature, and this temperature dependence remains unchanged during the SmA–SmC phase transition. This no layer contraction at the SmA–SmC transition supports that the SmA phase is the de Vries SmA phase. Compound **SFDPB** was mixed with 20 wt% of a ferroelectric liquid crystal **FLC-1** (see Fig. 3.35), and the mixture showed a phase sequence of Iso 159 °C Iso + chiral smectic A (SmA*) 145 °C SmA* 66 °C Cry. It exhibited an electro-optical switching over a wide temperature rang in the SmA* phase. The response times in the SmA* at 91 °C and 71 °C with an AC field of ± 20 V cm^{-1} at a frequency of 10 Hz were 10 μs and 14 μs, respectively.

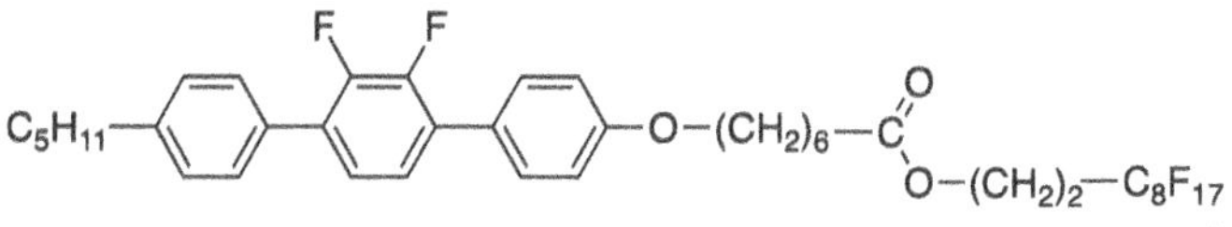

SFDPB

Iso 175 °C SmA 112 °C SmC 69 °C Cry

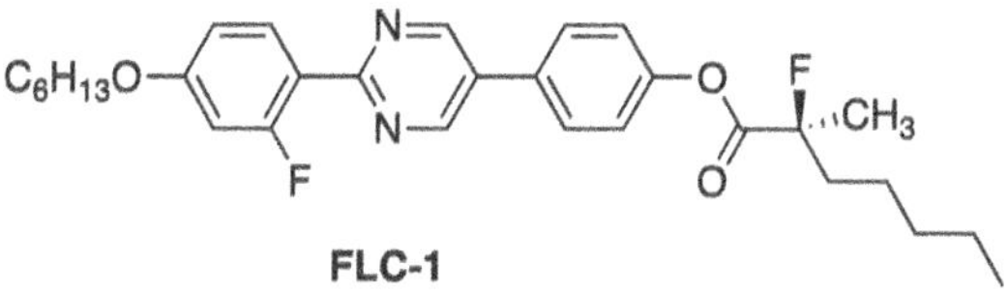

FLC-1

Iso 90 °C SmC* 76 °C Cry; P_s = 251 nC cm^{-2} at 80 °C

Fig. 3.35 Molecular structures and physical properties of the amphiphilic compound (**SFDPB**) and the ferroelectric liquid crystal (**FLC-1**)

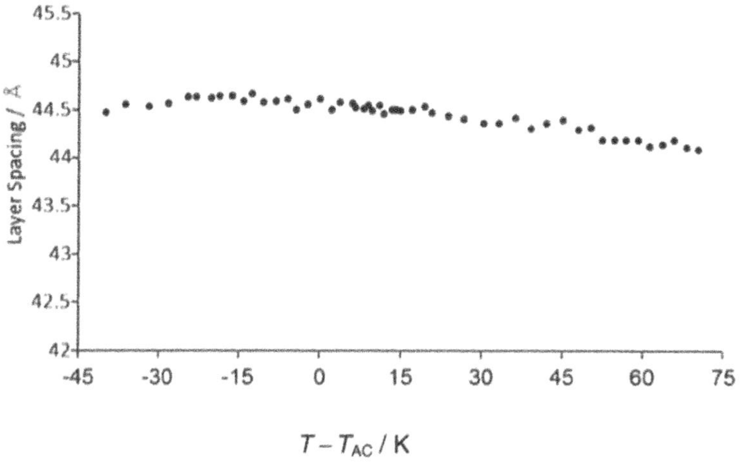

Fig. 3.36 Layer spacing of **SFDPB** in the SmA and SmC phases as a function of reduced temperature. Reprinted with permission from [66]. Copyright (2011) The Japan Society of Applied Physics

An electro-optical switching in the de Vries SmA* phase has been observed [59, 63, 67–72]. This switching phenomenon arises from the director tilting relative to the layer normal due to the applied electric field. Garoff and Meyer discovered that when an electric field is applied to the layers of an SmA phase, the director tilts with respect to the layer normal (Fig. 3.37) [73]. This phenomenon is called the electroclinic effect. The electric-induced tilt angle is described by Eq. (3.4) [74],

Fig. 3.37 Electroclinic switching in an SmA* phase

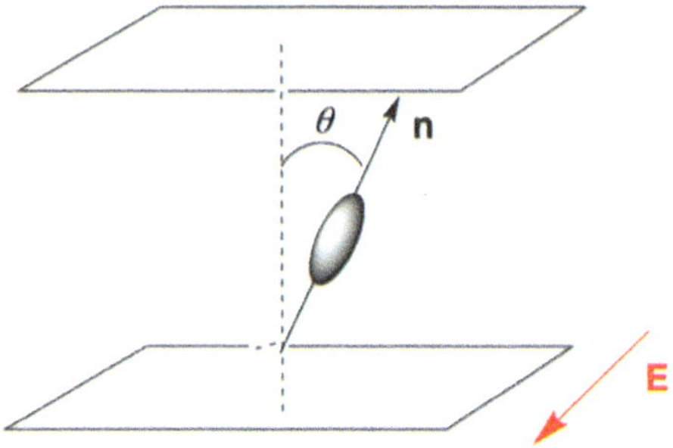

$$\theta = cE/\{\alpha(T - T_c)\} \tag{3.4}$$

where c is the electroclinic coupling constant between θ and Ps, α is a nonchiral parameter known as the tilt-susceptibility coefficient, and T_C is the Curie point corresponding to the second-order transition from the SmC* phase to the SmA* phase. As the temperature approaches to T_C, the tilt angle has a large value. The induced tilt angle is proportional to the strength of the applied electric field. Therefore, the transmittance can be controlled by the strength of the applied electric field, enabling continuous adjustment of gray levels.

The electroclinic effect in FLCs is observed in the SmA* phase in the vicinity of the SmA*–SmC* phase transition. However, in the de Vries SmA phase, the electro-optic response is also observed far from the phase transition. Karpernaum et al. reported a large electroclinic effect in the whole SmA* phase of a ferroelectric liquid crystal possessing substantial the de Vries character [70]. Prasad et al. observed a double-peak polarization current in the de Vries SmA* phase of amphiphilic organosiloxane derivatives, and proposed an antiferroelectric-like ordering for its phase structure [71]. Ghosh et al. supported the antiferroelectric-like model based on their observation of the double-peak current in the de Vries SmA* phase of a fluorinated antiferroelectric liquid crystal [72].

We discuss how the molecules organize in the SmA phase of **SFDPB**. The layer spacing at $T–T_{AC}$ = –40 K was 44.47 Å. The extend molecular length of the amphiphilic compound is estimated to be 42 Å. The SmA phase might have a partially bilayered structure in which the molecules are tilted with respect to the layer normal. The electro-optical switching in the SmA* phase of the mixture with **FLC-1** suggests that the SmA* phase has an SmC-like ordering in each layer. The repulsion between the fluorinated chains in the interlayer regions decreases the correlation of the azimuthal distribution between adjacent layers. The SmA phase of **SFDPB** can be explained by de Vries model I as shown in Fig. 3.38.

Figure 3.39 shows an illustration of the electro-optical switching in the de Vries SmA* phase of **SFDPB** mixed with **FLC-1**. When $E = 0$, the molecules tilt in the same direction in each layer, while the azimuthal angle is disordered from layer to layer (de Vries I). Therefore, the director is parallel to the layer normal. When

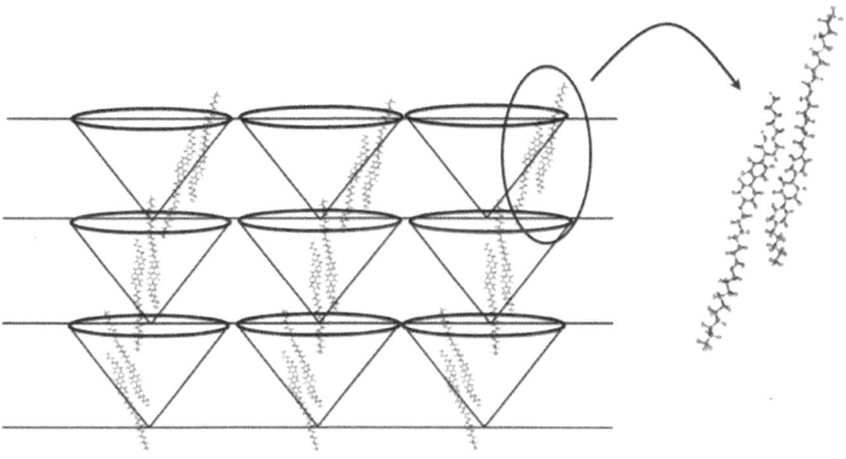

Fig. 3.38 Model for the molecular organization in the SmA phase of **SFDPB**. Reprinted with permission from [69]. Copyright (2011) The Japan Society of Applied Physics

applying an electric field, all the molecules reoriented along the same direction. The dark and bright states can be achieved.

A liquid-crystalline supermolecule exhibited an optically uniaxial smectic phase with a double-peak polarization [75]. The chiral trimer **MOHBC** (Fig. 3.40) showed two SmA* phases, i.e., the high-temperature SmA* (SmA*-H) phase and the low-temperature SmA* (SmA*-L) phase. The electro-optic response was observed in the SmA*-L phase. The phase sequence during cooling of **MOHBC** was Iso–SmA*-H–SmA*-L–SmC*–Cry. Figure 3.41 presents optical textures of **MOHBC** in the SmA*-L phase with or without an alternating electric field applied. The extinction direction was rotated counterclockwise with $+E$, but it was rotated clockwise with $-E$. Applying an electric field causes a distinct change in interference color from yellow to red. The increase in birefringence reveals that the orientation of the molecules without E is more disordered than that with E, suggesting that the SmA*-L phase has a de Vries-like character. Figure 3.42 shows the temperature dependence of electrical response in switching current under triangular waves. No switching current was detected in the SmA*-H phase (Fig. 3.42a). The double-peak polarization current per half-period of the switching voltage appeared at $T - T_{IA*} = -4.5$K in the SmA*-L phase, and it became lager with the decreasing temperature (Fig. 3.42b, c). After further cooling, in addition to the double peaks, a single sharp peak characteristic of ferroelectric switching appeared in the SmA*-L in the vicinity of the SmA*-L –SmC* transition (Fig. 3.42d). Therefore, the SmA*-H phase is a chiral version of a conventional paraelectric SmA phase. With respect to the SmA*-L phase, the antiferroelectric-like switching suggests that it is the de Vires SmA* phase.

Figure 3.43 shows a binary phase diagram for mixtures of **MOHBC** with its enantiomer. The SmA*-L phase is replaced with the SmC* phase with decreasing optical purity, suggesting that the SmA*-L phase has an SmC-like layer structure. A

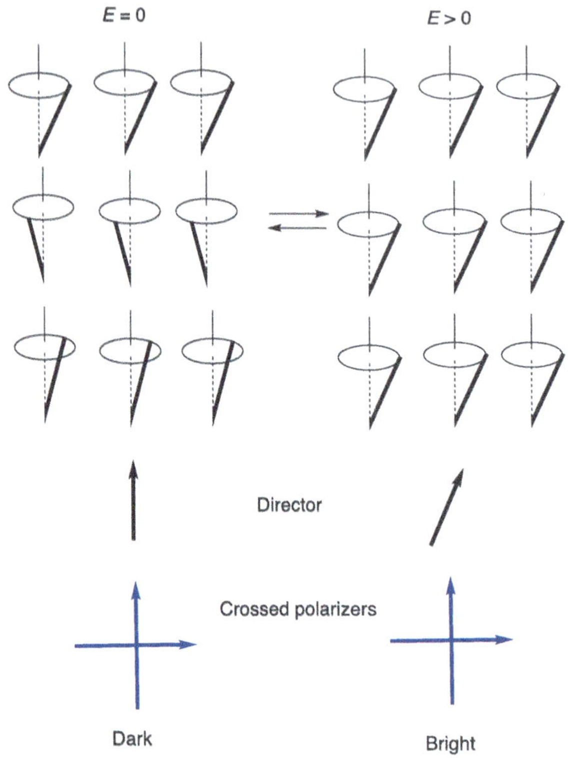

Fig. 3.39 Schematic illustration of electro-optical switching in the de Vries SmA* phase of **SFDPB** doped with **FLC-1**

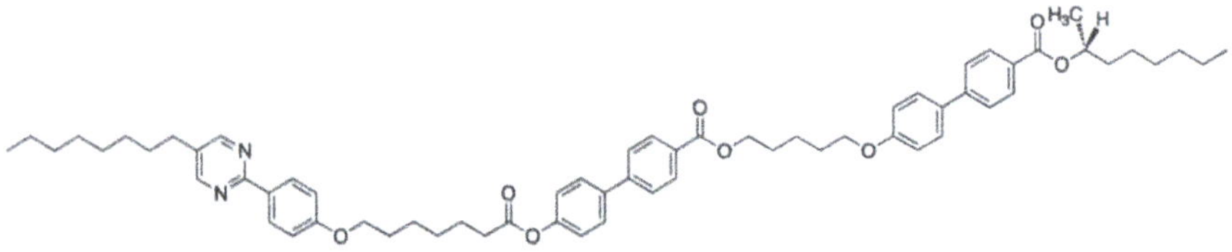

Fig. 3.40 Molecular structure of the chiral trimer **MOHBC**

double-peak polarization generally results from the electric field-induced antiferroelectric–ferroelectric transition. If molecules form the de Vries I-like layer structure, each layer possesses polarization, but the overall polarization is canceled. The double-peak polarization observed in the SmA*-L phase can be explained by the electric filed-induced de Vries SmA*–ferroelectric SmC* transition as show in Fig. 3.44.

Design of de Vries chiral smectic liquid crystals for device applications has been conducted [76, 77]. Displays utilizing the de Vries A phase offer the following advantages: (1) they have a fast-switching speed, (2) they show an in-plane switching

Fig. 3.41 Optical textures of **MOHBC** between homogeneously aligned glass plates in the SmA*-L phase with ± 8 V μm^{-1} or without an electric field. Reprinted with permission from [75]. Copyright (2012) American Chemical Society

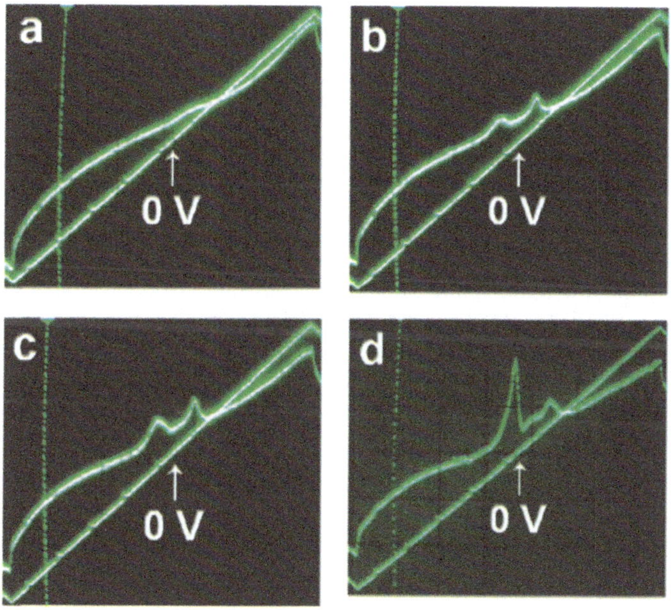

Fig. 3.42 Electrical response in the switching current of **MOHBC a** at $T - T_{IA^*} = -3K$ in the SmA*-H phase, **b** $T - T_{IA^*} = -7K$ in the SmA*-L phase, **c** $T - T_{IA^*} = -0K$ in the SmA*-L phase, and **d** $T - T_{IA^*} = -4.5K$ in the SmA*-L phase near the SmA*-L–SmC* phase transition upon application with a triangle-wave switching field of ± 8 V μm^{-1} at a frequency of 15 Hz. Reprinted with permission from [75]. Copyright (2012) American Chemical Society

with wide viewing angle, and (3) they easily produce gray levels. Furthermore, the randomization of the azimuthal distribution in the de Vries phase can decrease their threshold voltage.

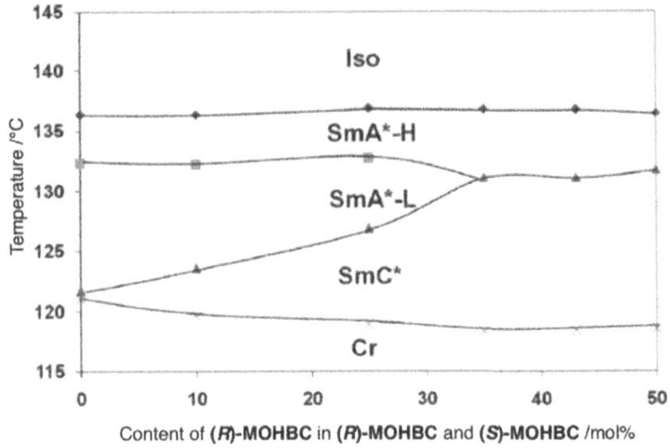

Fig. 3.43 Binary phase diagram for mixtures of **MOHBC** with its enantiomer. Reprinted with permission from [75]. Copyright (2012) American Chemical Society

3.6 Polar Order in an Optically Uniaxial Smectic A Phase

According to the phase symmetry rule [5], the macroscopic polarization in a smectic A phase is zero. However, the hypothesis stated above was overturned by Reddy et al. [24]. An achiral bent-core compound **W586** exhibited a ferroelectric property in the SmA phase. Figure 3.45a shows the molecular structure and phase-transition temperatures. The compound showed two uniaxial SmA phases with a partially intercalated layer structure (SmA_d) (Fig. 3.45b). Bistable ferroelectric switching was observed in the low-temperature SmA phase (SmA_dP_F) below the paraelectric SmA phase (SmA_d). Figure 3.46 shows the *P*s of **W586** in the SmA_dP_F phase as a function of temperature [79]. The hysteresis loop characteristic to ferroelectricity is shown in the figure. The ferroelectricity was also confirmed by optical second-harmonic generation activity. Guo et al. proposed a model for the molecular arrangement of **W586** in the SmA_dP_F when an electric field larger than the threshold electric field (*E*th) is applied (Fig. 3.45b) [79]. The direction of that polarization is perpendicular to the molecular long axis. The formation of sublayers through the aggregation of carbosilane groups weakens the interlayer interactions, allowing for the parallel order of dipole moments of bent-core molecules in the adjacent layers.

According to Zannoni's prediction [33], Kashima et al. prepared a taper-shaped molecule (**DFPYCB**) possessing three mesogenic units, i.e., phenylpyrimidine, cyanobiphenyl, and difluroterphenyl [80]. The trimer **DFPYCB** exhibited an enantiotropic SmA phase with a wide temperature range (Fig. 3.47). The layer spacing in the SmA phase and the molecular length are 27.1 Å and 61 Å, respectively, indicating that the SmA phase has an intercalated structure.

A ferroelectric-like electrooptical switching was observed in the SmA phase. When applying an AC field of 15 μm^{-1} perpendicular on the sample in a homogeneously aligned cell, the texture became completely dark (Fig. 3.48). This reveals that the director is oriented perpendicular to both glass substrates due to the interaction between the molecular dipole along the long axis and the electric field. The peak of the polarization reversal current was observed not only in the SmA phase but also the isotropic liquid. Figure 3.49 shows the switching polarization in the SmA phase

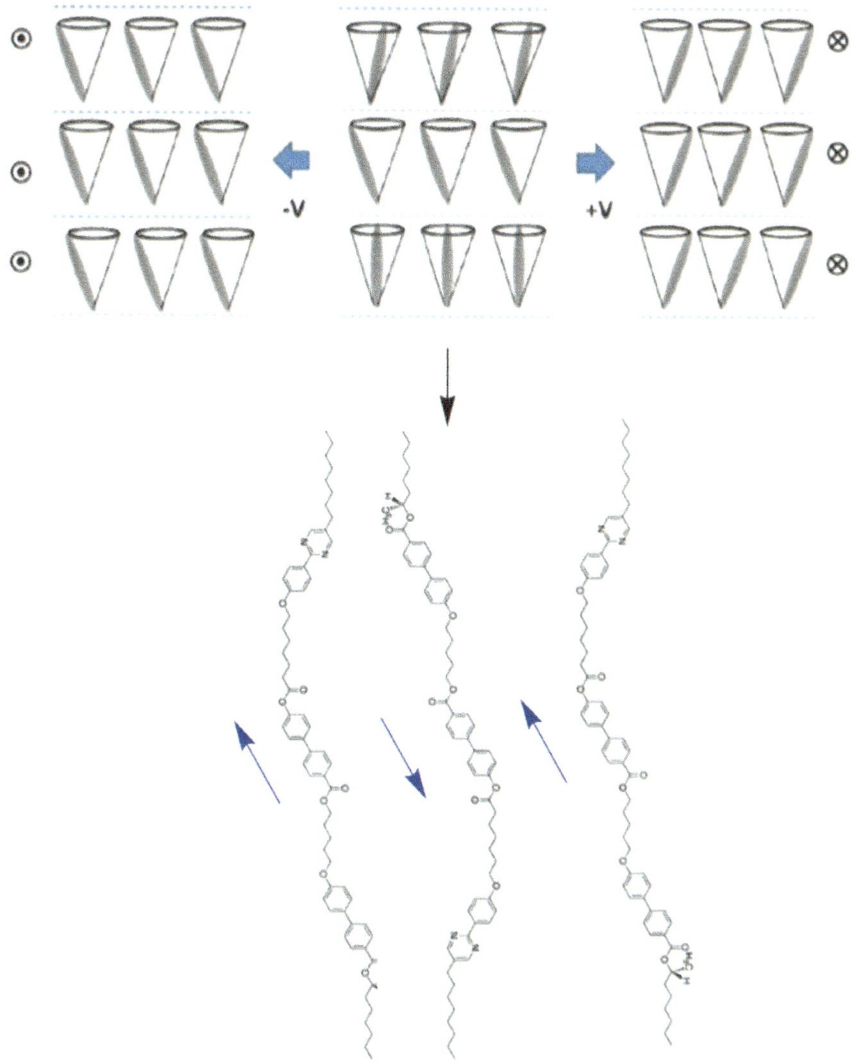

Fig. 3.44 Model for the double-peak polarization

and the isotropic liquid as a function of the electric field. It appears when the applied electric field is larger than a certain value and saturates at approximately at 160 nC cm^{-2}. According to Francescangeil et al. [81] and Weissflog et al. [82], the switching behavior is explained in terms of polar clusters. The molecules form a smectic cluster in which a polar order exists. The orientation of the polar clusters becomes random, resulting in the macroscopic polarization being canceled out to zero. When an electric field above the threshold is applied, the polar clusters are reoriented so that the

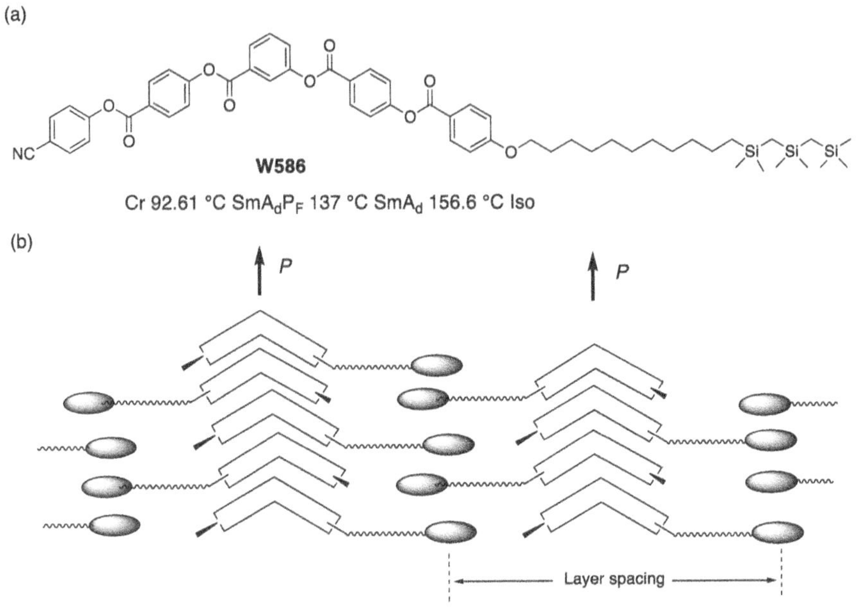

Fig. 3.45 **a** Molecular structure and phase-transition temperatures of **W586**. **b** Model for the molecular organization of **W586** in the SmA_dP_F phase with $E > E$th

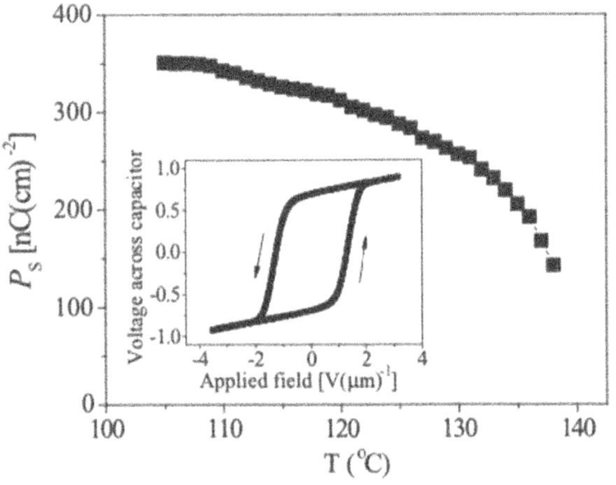

Fig. 3.46 Spontaneous polarization as a function of temperature. The hysteresis loop recorded at 125 °C is shown in the inset. Reprinted with permission from [79]. Copyright (2011) American Physical Society

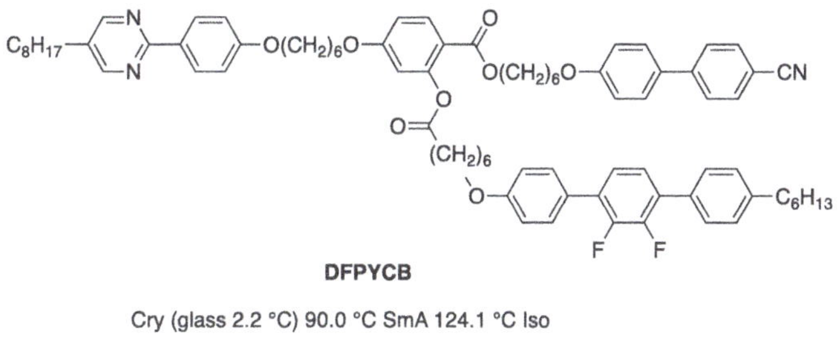

Fig. 3.47 Molecular structure and phase-transition temperatures of **DFPYCB**

polar vector of each cluster is parallel to the electric field. It should be noticed that the taper-shaped **DFPYCB** can exhibit the flexoelectric polarization (see Fig. 3.12a). Therefore, we cannot exclude the possibility that it is a ferroelectric SmA phase.

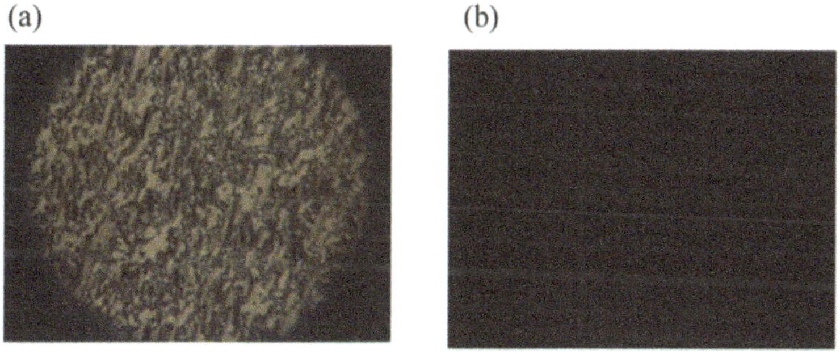

Fig. 3.48 **a** An optical texture of a sample in a 2 μm homogeneous aligned cell at 120 °C in the SmA phase, and **b** that on application with a triangle wave switching field of 15 V μm^{-1} at a frequency of 1 Hz. Reproduced from [80] with permission from the Royal Society of Chemistry

Fig. 3.49 Switched polarization in the SmA phase at 118 °C and that in the isotropic liquid at 150 °C as a function of electric field at a frequency of 1 Hz. Reproduced from [80] with permission from the Royal Society of Chemistry

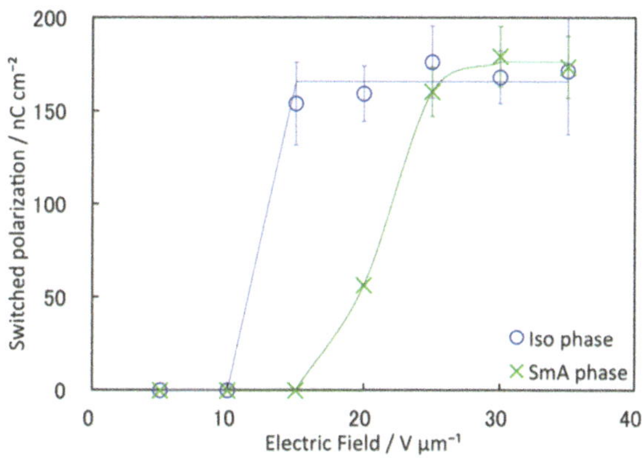

Figure 3.50 shows a molecular organization model for the polar cluster in the SmA phase of **DFPYCB**. The taper-shaped molecules are packed in the intercalated form in which head-to-tail recognition is induced by intermolecular phenylpyrimidine–cyanobiphenyl interactions. The molecular organization enables the dipole moments of taper-shaped molecules to align parallel between neighboring molecules. The formation of polar clusters is associated with the emergence of ferroelectricity in the nematic, which will be discussed in Sect. 3.7.

Khan and Mukherjee described the Iso–SmA phase transition in the trimer **DFPYCB** using Landau theory [83]. They demonstrated that experimental results such as the switched polarization show a good qualitative agreement with the theoretical model.

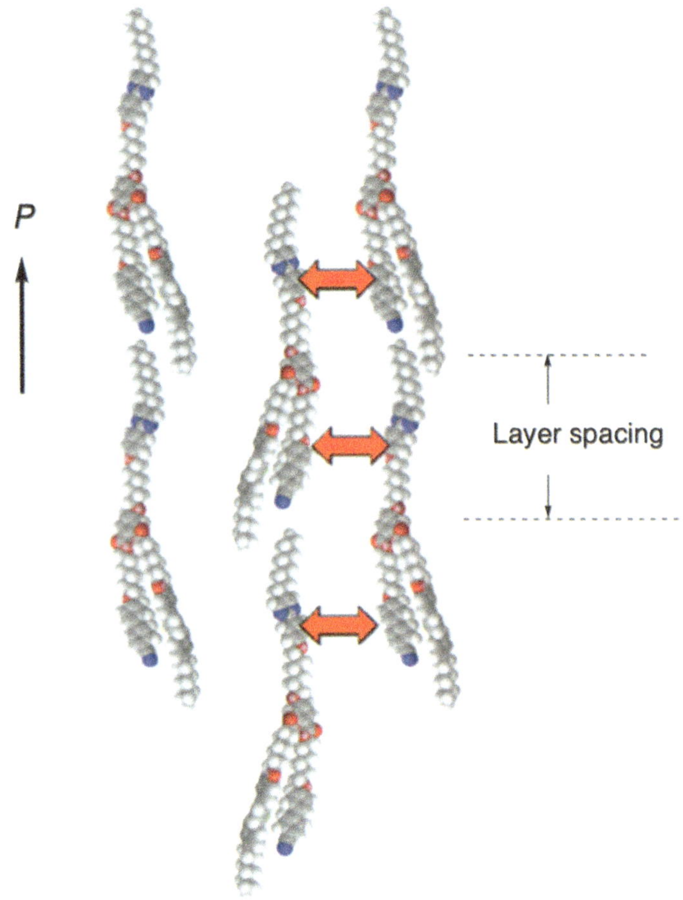

Fig. 3.50 Molecular organization model for the polar ordering of **DFPYCB** in the SmA phase. Reproduced from [80] with permission from the Royal Society of Chemistry

3.7 Ferroelectric Nematic Liquid Crystals

The discovery of ferroelectric nematic liquid crystals had a huge impact on liquid crystal science. Born envisioned the ferroelectric ordering of rod-shaped molecules in a nematic phase [84], however, no experimental proof was reported for the next 100 years [85]. In 2017, Mandle et al. [25, 26]. and Nishikawa et al. [27] reported independently and simultaneously new rod-shaped liquid crystals exhibiting two distinct nematic phases. Let me follow Mandle's papers to describe what they found. They found that the polar rod-shaped compound exhibits the N–N transition associated with a small enthalpy change (Fig. 3.51). Figure 3.52 shows photomicrographs of the optical textures of the nematic phase and the low-temperature nematic phase denoted as Nx. Cooling the compound from the isotropic liquid to the nematic phase, the schlieren texture appeared (Fig. 3.52a). With further cooling, a predominantly green colored texture was interrupted by aperiodic pink regions (Fig. 3.52b, c) and the pink domains increased (Fig. 3.52d). The Nx phase also exhibited a schlieren texture, with qualitatively high fluidity and visible Brownian motion (Fig. 3.52f). The complex texture at the phase transition (Fig. 3.52e) suggests the possibility the coexistence of a smectic phase, but defects associated smectic phases were not observed. Furthermore, calorimetry and X-ray scattering experiments support the N–Nx transition without appearance of a smectic phase. The X-ray scattering measurements also reveals that those phases are nematic. Both the nematic phases exhibited weak scattering in the small angle region, indicating that cybotactic smectic fluctuations exist in the N and Nx phases. The authors noted that the lateral methoxy unit induces lateral separation between molecules thereby suppressing smectic phase formation. The values of the dielectric anisotropy were ~8.5 in the N phase at 165 °C and ~6.2 in the Nx phase at 75 °C. The authors examined the molecular ordering in these nematic phases by comparing measured and simulated values of dielectric anisotropy. In the Maier–Meier equations for the simulation the effective dipole moment is the molecular dipole moment attenuated by the Kirkwood 'g' factor, which describes the degree of antiparallel/parallel correlations between molecules in a liquid crystal. The factors were ~0.262 in the N phase and ~0.117 in the Nx phase. The antiparallel ordering is greater in the Nx phase than in the N phase, which explains for the lower dielectric anisotropy in the Nx phase. The authors concluded that the onset of the lower temperature phase is driven by the formation of antiparallel molecular associations. Subsequently, Mandle et al. synthesized derivatives of the polar compound and investigated the structure–property relationship [26]. Among these, a compound with the terminal ethoxy group shortened to a methoxy group significantly increased the N–Nx transition temperature (Fig. 3.53). This compound was later named **RM743** and its detailed physical properties have been examined. Later Mandle et al. in collaboration with the Ljublijana group reported that the Nx phase of **RM734** is locally polar but exhibits antiferroelectric splay behavior on long scales [86–90].

Chen and Clark et al. demonstrated ferroelectricity in the N phase of **RM734** by observing electro-optical response to the application of tiny in-plane electric

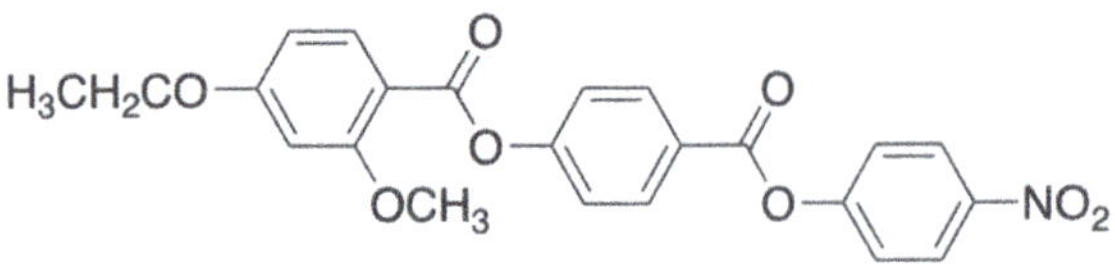

Cry 139.0 (34.8) [Nx 85.6 (0.2)] N 182.1 (0.6) Iso

Fig. 3.51 Transition temperatures (°C) and associated enthalpies of transition (kJ mol^{-1}, in parentheses) for the polar rod-shaped compound. Square brackets denote a monotropic transition

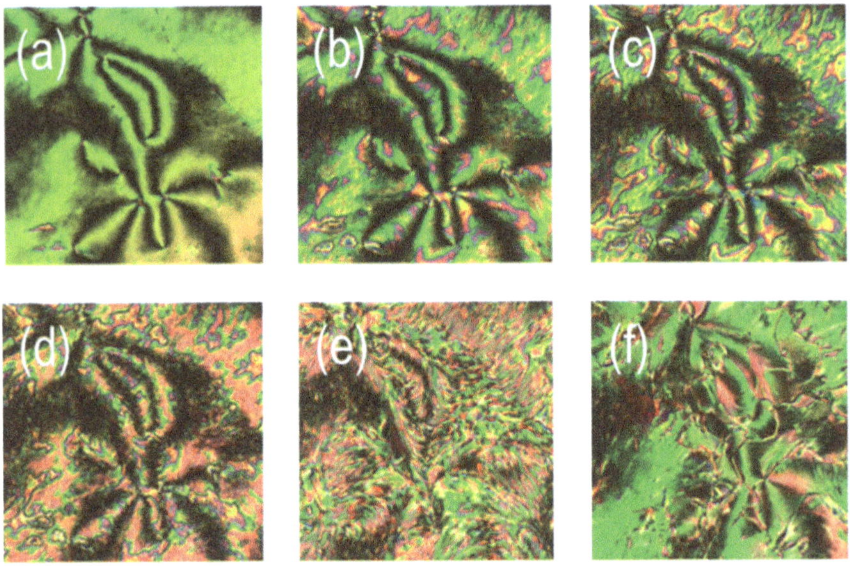

Fig. 3.52 Photomicrographs of the optical textures of the N phase (a: 130 °C, b: 110 °C, c: 100 °C, d: 90 °C) and the Nx phase (e: 83 °C, f: 80 °C) of the compound on glass slides with a cover glass between crossed polarizers. Reproduced from [25] with permission from the Royal Society of Chemistry

Fig. 3.53 Transition temperatures (°C) and associated enthalpies of transition (kJ mol^{-1}, in parentheses) for **RM734**. Square brackets denote a monotropic transition

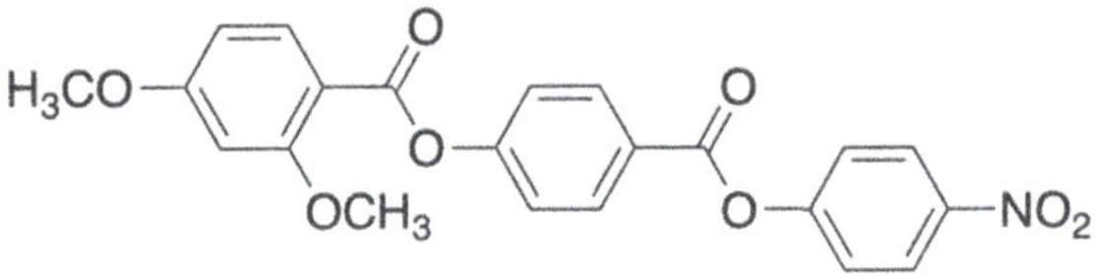

RM734

Cry 139.8 (29.9) [Nx 132.7(0.2)] N 187.9 (0.5) Iso

fields [91]. Let us follow their observations. Figure 3.54A presents the field-free initial state showing three domains separated by domain walls. The director ***n*** in each domain is aligned along the buffing direction z. Application of an ultra-small, positive test field $Ez = 1$ V cm^{-1} induces a birefringence color change resulting from in-plane reorientation of ***n*** in the center domain, leaving the upper and lower domains unchanged (Fig. 3.54B). Application of $Ez < 0$ induces an in-plane reorientation of ***n*** in the upper and lower domains (Fig. 3.54C). These observations demonstrate that: (1) the domains with opposite polarity are formed in the absence of an electric field and (2) their polar direction is reoriented with the field. The authors measured the polarization density ***P*** of **RM734** as a function of temperature (Fig. 3.55). It saturates at ~6 μC cm^{-2}, which is the largest ever measured as for fluid or glassy material. This polarization is close to the average value obtained by assuming perfect polar aliment of molecular dipoles of **RM734** in the nematic phase. Furthermore, atomic molecular dynamics simulation reveals short-range polar molecular interactions that favor ferroelectric ordering, including a tendency for head-to-tail association into polar, chain-like assembles having polar lateral correlations.

Nishikawa and Kikuchi et al. designed the polar liquid crystal **DOI**, and found a ferroelectric-like molecular ordering in the unusual nematic phase [27]. The phase transition is complicated and depends on the thermal history. They observed three high-fluidity mesophases denoted as M1, M2, and MP during the first cooling run below the clearing temperature (173.6 °C). The phase transition temperatures (°C) and associated enthalpies of transition (kJ mol^{-1}, in parentheses) are as follows: M1 84.5 (0.003) M2 68.8 (0.2) MP. The small-angle XRD in the mesophases showed a small broad peak at $2\theta = 4.0^{\circ}$. The periodicity length is 2.2 nm, which corresponds to the molecular length of **DIO**. In the wide-angle region, nematic-like diffuse scattering was observed for the three phases. The M1 phase is a conventional nematic phase. An anomalously large dielectric anisotropy of about 10,000 were observed in

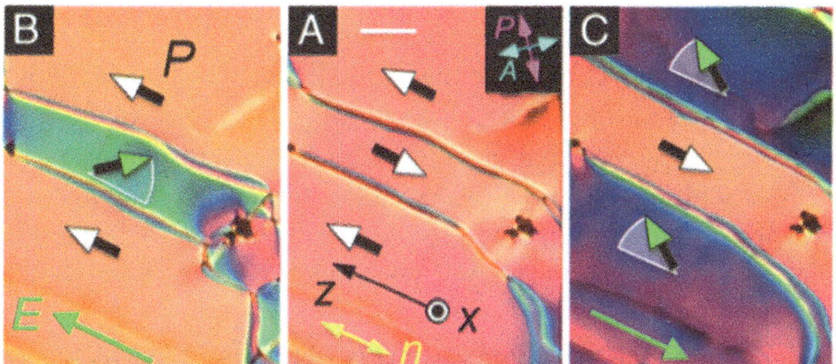

Fig. 3.54 Photomicrographs of the electro-optical switching in the N_F phase of **RM734** at 120 °C. A: Field-free initial state. B: Field-induced state with application of $Ez > 0$. C: Field-induced state with application of $Ez < 0$. The sample was confined in a 11 μm homogeneously aligned cell between crossed polarizers. Scale bar = 30 μm. Reproduced from [91]

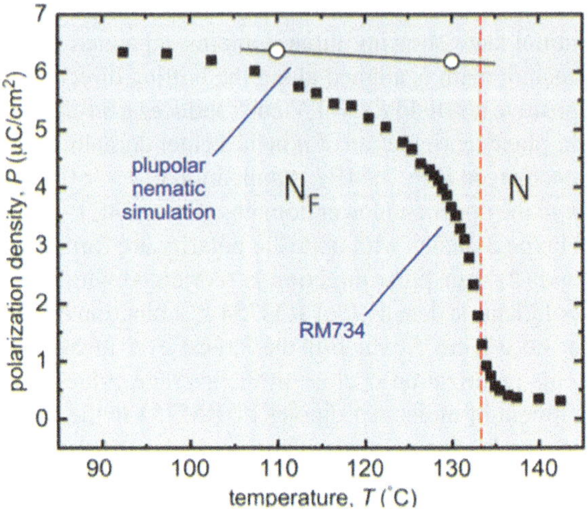

Fig. 3.55 The polarization density ***P*** of **RM734** measured on cooling (black squares). The open circles are calculated values of ***P*** of the plupolar nematic. Reproduced from [91]

the MP phase. The permanent dipole moment values for **DIO** and **5CB** are estimated to be 9.43 D and 5.94 D, respectively. The huge dielectric anisotropy of **DIO** indicates that the molecular ordering of **DIO** in the DP phase quite different from that of **5CB** in the N phase. The authors performed polarization-switching measurements under an applied triangular wave voltage at a frequency of 200 Hz. A large polarization ($P \approx 500$ nC cm^{-2}) was detected at 75 °C in the M2 phase. In the MP phase, a fairly large polarization (up to 4.4 μC cm^{-2} at 50 °C) was observed. **DIO** exhibited a parallelogram-shaped *P–E* hysteresis loop in the MP phase. Second-harmonic generation (SHG) measurements were performed for **DIO** to detect the macroscopic polar order. At $E = 0$ Vμm^{-1}, the M2 phase is SHG-inactive. In the MP phase at 65 °C, SHG intensity starts to increase at Eth ≈ 0.06 Vμm^{-1} and then saturates at Esa ≈ 0.6 Vμm^{-1}. Eth and Esa are threshold and saturated electric fields, respectively. The change in POM texture also corresponds to this voltage dependence of SHG intensity. The authors conclude that the MP phase with high fluidity exhibits a ferroelectric-like orientational order. They propose a possible model for the unusual polar behavior using a two-stage ordering model: i) the formation of ferroelectric-like polar order with centrosymmetry breaking inside the domains at Eth, and ii) the subsequent reorientation of the director and formation of large conglomerates of polar domains at Esa [27]. The MP phase later came to be called ferroelectric nematic (N_F) phase.

Chen et al. reported that the NF phase of **RM734** is miscible with that of **DIO** [92]. Both the N_F phases exhibit a small XRD scattering with a periodicity corresponding to their molecular length. The cybotactic smectic fluctuations might play an important role in the formation of the polar ordering in each domain for the polar rod-like molecules, **DOI** and **RM734**. A lot of researches are now progress to clarify the origin of the ferroelectric properties [91, 93–98] and to understand the structure–property relationships [99–103].

The structure of the unidentified polar M2 phase of **DOI** has been investigated. Chen and Clark et al. reported that this phase has antiferroelectric smectic order as a prelude to the ferroelectric nematic, and they termed the M2 phase as smectic Z_A phase [104]. In the SmZ_A phase, the nematic director and polarization are oriented parallel to smectic layer planes, and the polarization alternates in sign from layer to layer with 180 Å period. Figure 3.56 shows structural models for the N, SmZ_A, and N_F phases [105].

Kumari and Lavrentovich et al. investigated the grand states of ferroelectric nematic LCs by observing POM textures of the N_F phase of **DOI** and **RM734** [106]. Each sample was confined in a cell in which the bottom layer on the glass substrate was rubbed unidirectionally. Both cells showed qualitatively similar behavior. When the sample of **DOI** was cooled to the N_F phase, the homogeneous texture of the N phase transformed into a pattern of elongated domains. When the analyzer rotated

Fig. 3.56 Schematics of **a** the nematic (N) phase, **b** the antiferroelectric smectic Z (SmZ_A), and **c** the ferroelectric nematic (N_F) phase. Reproduced from [105]

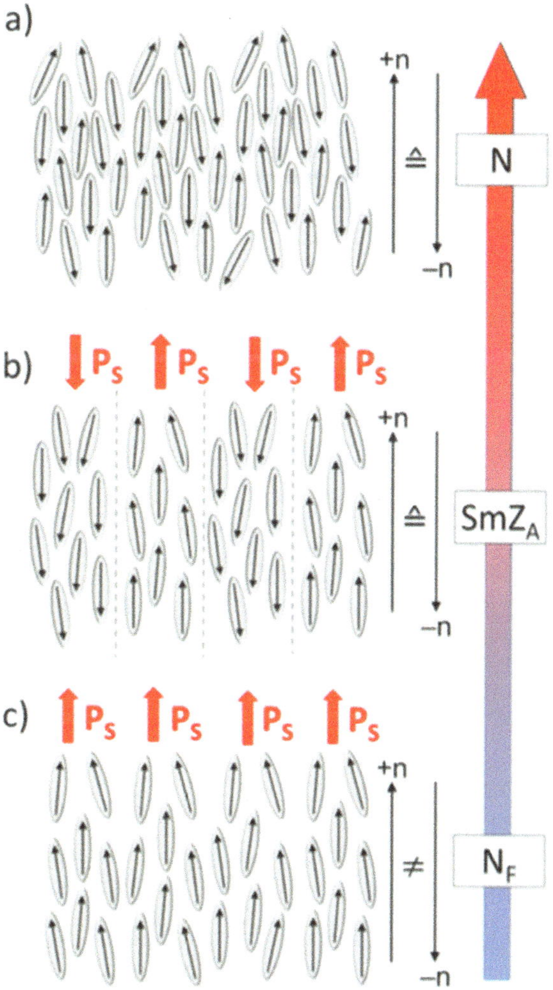

clockwise or counterclockwise, the contrast of the neighboring domains alternated, indicating that the optical axis and the polar direction in the N_F domains twist along the axis perpendicular to the cell as shown in Fig. 3.57. The samples exhibited approximately equal amounts of left- and right-handed domains. The domains were removed by applying a dc electric field. After the field was switched off, the domains reappeared. They demonstrated that the ground state of a ferroelectric nematic liquid crystal that is unconstrained by the surface-imposed torques is chiral. The sandwich-like cells split into an array of elongated domains with alternating left- and right-handed helical twists. Their findings show that the polar orientational order of molecules could trigger chirality in soft matter without intrinsic chirality [106].

The materials that form the N_F phase at room temperature are also being developed [107, 108]. Manabe et al. reported a polar rod-like liquid crystal showing a ferroelectric nematic phase at room temperature [107]. The compound exhibited a direct phase transition from the isotropic liquid to the N_F phase. The molecular structure and phase transition temperatures are shown in Fig. 3.58.

Gibbs, Hobbs, and Mandle demonstrated that systematic fluorination is a powerful design strategy toward ferroelectric nematic liquid crystals (Fig. 3.59) [109]. The authors note the highest stability of the N_F phase to be in materials with specific fluorination patterns rather than maximum fluorination. Electronic structure calculations indicate that molecules possessing a more oscillatory distribution of electrons

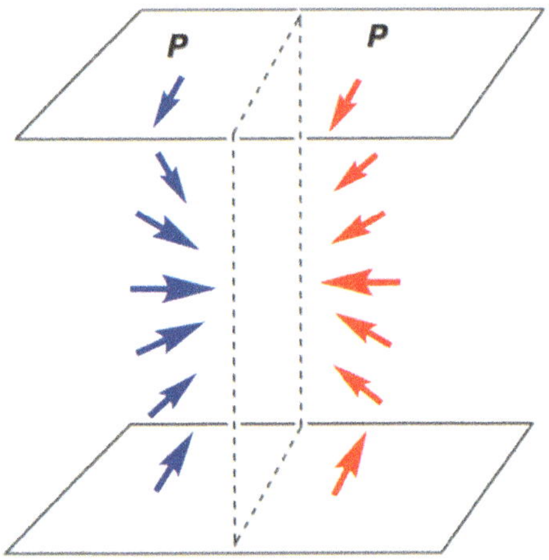

Fig. 3.57 Schematic illustration for left- and right-handed twists of polarization director *P* in neighboring domains of the N_F phase [106]

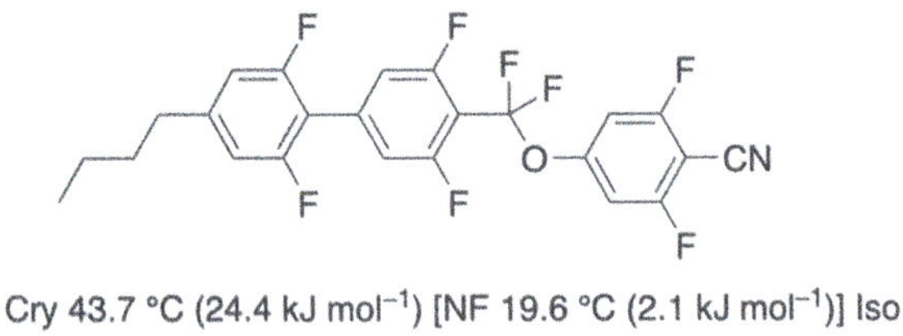

Fig. 3.58 The molecular structure, transition temperatures, and associated enthalpies of transition in parentheses for **RM734**. Square brackets denote a monotropic transition

across their surfaces have a higher propensity to form ferroelectric nematic phases. Furthermore, they state the dominant mode of interaction between molecules to be offset π–π stacking rather than the dipole–dipole interactions.

In the N_F phase, maintaining large polar domains in the same direction in the equilibrium state is difficult without an external field due to the fluidity [110, 111]. The orientation in the N_F phase of **DIO** relaxed under the zero electric field. On the other hand, the polar SmA phases possesses high symmetry but is expected to maintain a large-size polar domain due to the layered structure. The ferroelectric SmA have been discovered for ferroelectric nematic materials [112, 113]. Kikuchi et al. reported ferroelectric smectic A liquid crystals exhibiting the longitudinal ferroelectricity [112]. Figure 3.60 shows one of the reported compounds. It exhibited enantiotropic SmA_F and $SmA_{F'}$ phases.

Figure 3.61 presents POM textures, 2D small angle X-ray diffractions, and electric displacement-electric field (*D*–*E*) hysteresis curves in the N and SmA_F phases of the compound. An angular mosaic-like texture (Fig. 3.61a) and clear diffraction spots along the equatorial line parallel to the rubbing direction (Fig. 3.61b) support that the compound exhibits a smectic A phase. The layer spacing in the SmA phase and the molecular length are 25 Å and 22.5 Å, respectively, indicating that the SmA phase has a monolayer structure. The compound showed the *D*–*E* hysteresis along the director direction (Fig. 3.61c) and the SHG activity. Therefore, the SmA phase was identified

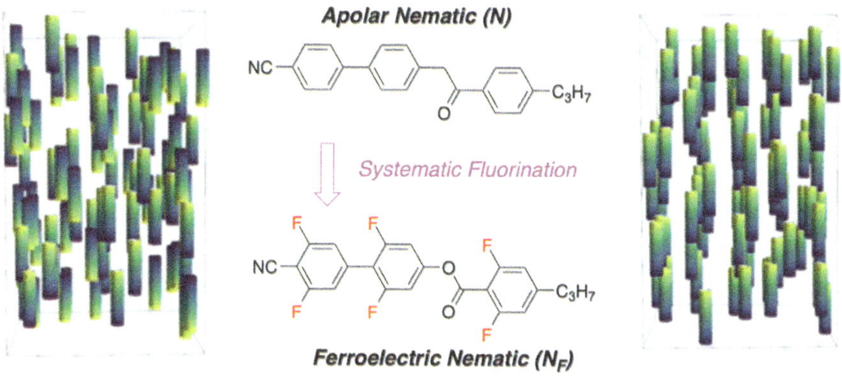

Fig. 3.59 The systematic fluorination transforms from an apolar nematic phase to a ferroelectric nematic phase. Reproduced from [109]

Fig. 3.60 Molecular structure of the ferroelectric SmA compound

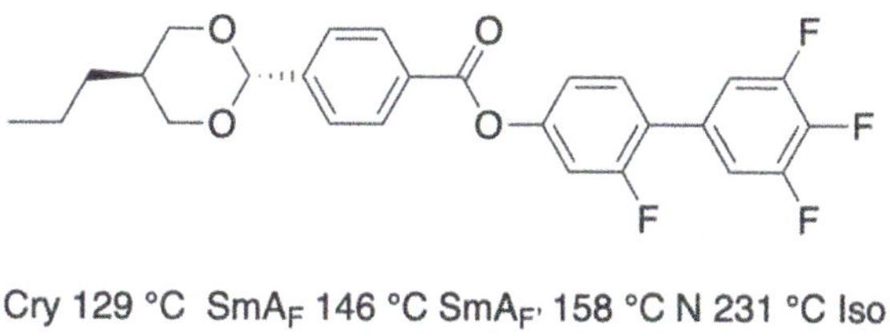

as a ferroelectric SmA phase exhibiting longitudinal ferroelectricity. Comparison of the observed magnitude of *Ps* (3–4 μC cm^{-2}) and the calculated molecular dipole moment μ (8.69 D) indicates that the dipole moment in the SmA_F phase is oriented almost perfectly along the long axis of the molecule. The macroscopically poled orientation was maintained in the SmA_F phase. This is because *Ps*-induced distortions are suppressed by the layer structure. The authors concluded that the ferroelectric smectic A phases uniquely possess high symmetry, fluidity, and memory ability.

Nishikawa and Araoka et al. investigated the ferroelectric smectic structure of tolan-based polar molecules, exhibiting the N_F–SmA_F phase transition and found that the chiral ground state (see Fig. 3.57) is inherited in the SmA_F phase of **nBOE-NO$_2$** [114]. Figure 3.62 shows their molecular structures. The chiral ground state consists of left/right-twits if smectic blocks with polarization in neighboring SmA_F domains as shown in Fig. 3.63. The formation of the spontaneous twist is driven by depolarization. On the other hand, **nDIOLT** preferred a non-twisted structure. **nBOE-NO$_2$** molecules possess a polar nitro group and a bulky 2,6,7-trioxabicycloocytane unit. The authors discussed the differences in the ground state between **nBOE-NO$_2$** and **nDIOLT** in terms of factors such as depolarization, molecular steric hindrance, and elasticity.

In addition to the orthogonal SmA_F phase, Strachan et al. [115], Kikuchi et al. [116], and Hobbs et al. [117] reported independently the observation of a ferroelectric smectic C (SmC_F) phase. Figure 3.64 shows schematic illustrations of the N_F, SmA_F, and SmC_F phases. The direction of polarization in the SmC_F phase is parallel to the smectic director. Figure 3.65 presents the molecular structure of a liquid crystal **PPZGU-3-F** exhibiting the SmC_F phase. The SmC_F phase occurs without helix formation. Figure 3.66 shows POM textures of **PPZGU-3-F** in the SmA, SmA_F, and SmC_F phases.

Figure 3.67 shows current response traces in the (a) SmA_F and (b) SmC_F phases of **PPZGU-3-F**. The LC samples were held in 4 μm thick cells with no alignment layer. The measurements were undertaken applying a triangular wave voltage at a frequency of 8 Hz. A clear single peak was observed in the SmA_F phase, while a second peak of a smaller magnitude to the main signal emerged in the SmC_F phase. The maximum values of polarization in the SmA_F and SmC_F phases are ~3.0 μC cm^{-2} and 3.3 μC cm^{-2}, respectively. The similar response behavior was also reported by Strachan et al. [115]. They discussed the switching process as follows. The process is a two-step switching associated with tilt removal as shown in Fig. 3.68. First, the polarization within the layers aligns with the electric field when a high voltage is applied. The tilt angle is removed and the phase becomes orthogonal. Second, when the electric field is reduced, molecular tilt is regained, which causes the small current peak. Finally, after reversal and increase of the applied voltage, the polarization switches and realigns with the applied electric field, giving rise to the main recorded current peak.

The high spontaneous polarization can break mirror symmetry to produce frustrated helical ferroelectric phases such as the twist bend ferroelectric nematic phase (N_{TBF}) [118, 119] and the heliocoidal polar smectic C ($SmC^H{}_P$) phase [97]. Gorecka et al. reported a polar liquid crystal exhibiting N_F, N_{TBF}, and $SmC^H{}_P$ phases, and

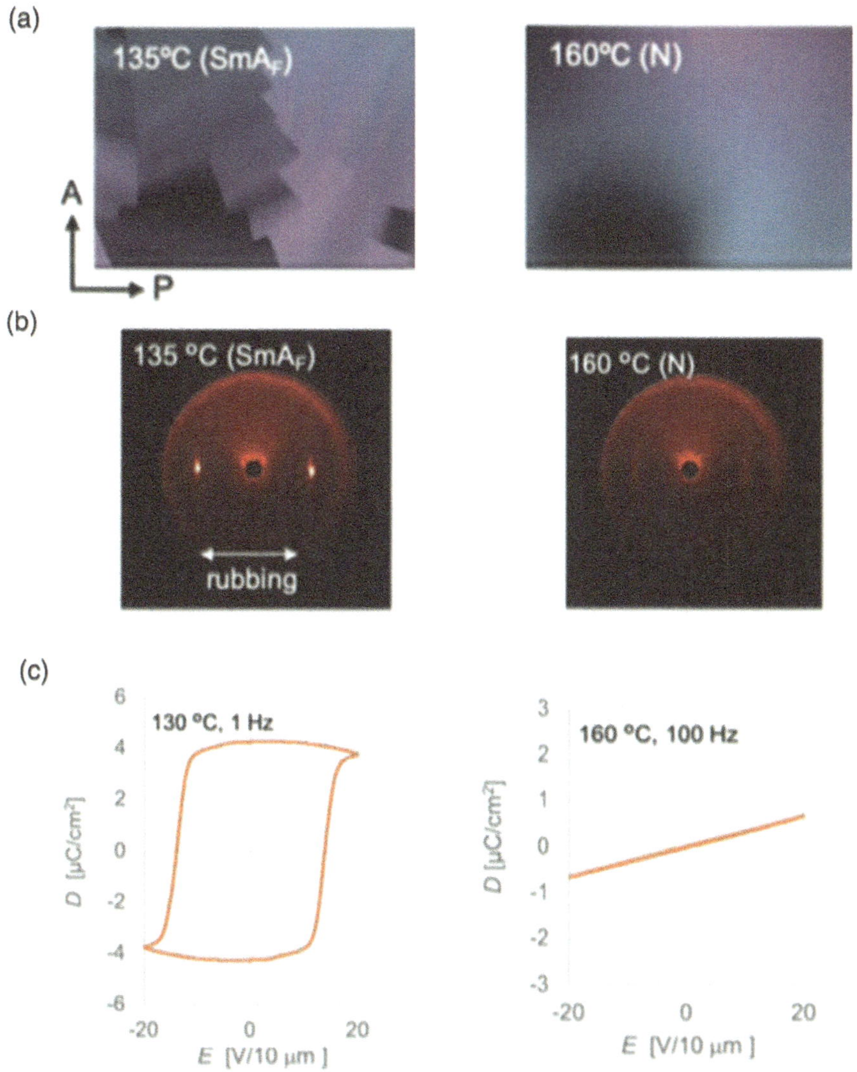

Fig. 3.61 POM textures, 2D small angle X-ray diffraction patterns, electric displacement-electric field (*D–E*) hysteresis curves in the N and SmA$_F$ phases of the compound. **a** Polarized optical textures observed at 135 °C (SmA$_F$) and 160 °C (N). **b** 2D SAXD patterns at 135 °C (SmA$_F$) and 160 °C (N). **c** D–E curves at 130 °C (SmA$_F$) and 160 °C (N). Reproduced from [112]

demonstrated evidence for the sequence of ferroelectric nematic and ferroelectric smectic phase, both of which exhibit spontaneous helical formation [120]. The molecular structure is shown in Fig. 3.69. The compound exhibited a phase sequence of Iso–N–SmZ$_A$–N$_F$–N$_{TBF}$–SmC$^H{}_P$ upon cooling from the isotropic liquid. The saturated spontaneous polarization in the SmC$^H{}_P$ phase is ~4.5 μC cm^{-2}. Based on the

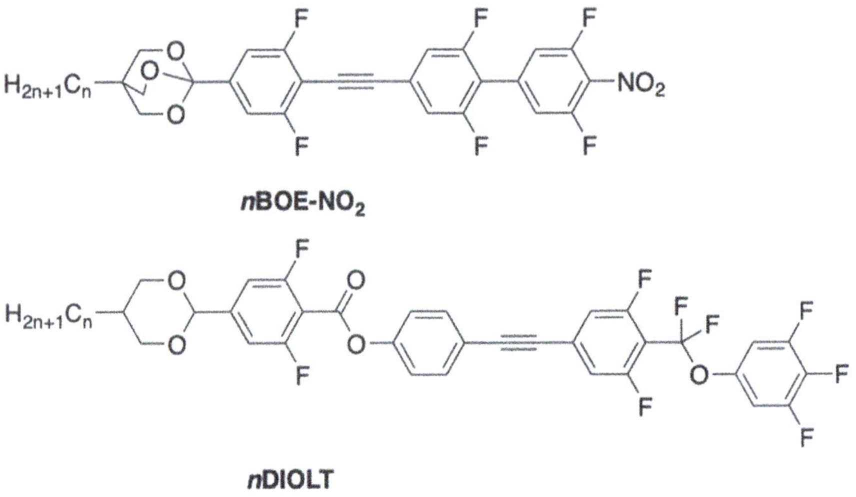

Fig. 3.62 The molecular structures of n**BOE-NO**$_2$ and n**DIOLT**

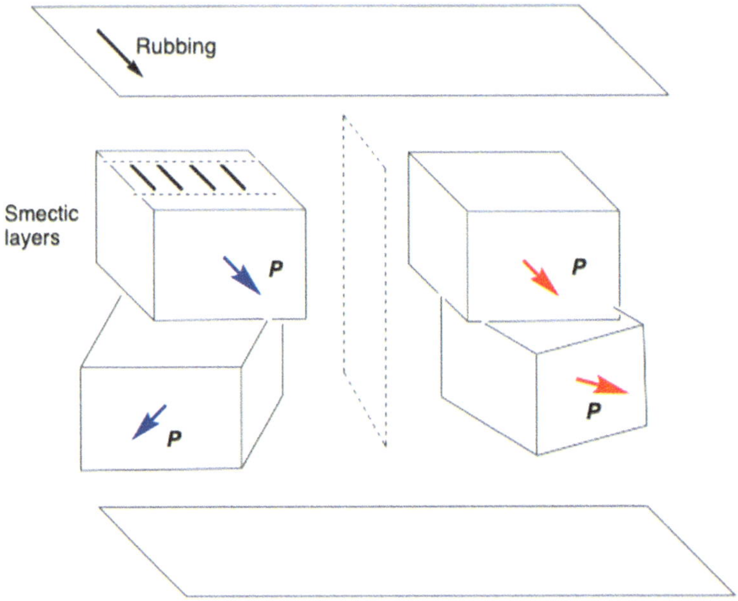

Fig. 3.63 Schematic illustration for left- and right-handed twisted smectic blocks with polarization director P in neighboring domains of the SmA$_F$ phase [114]

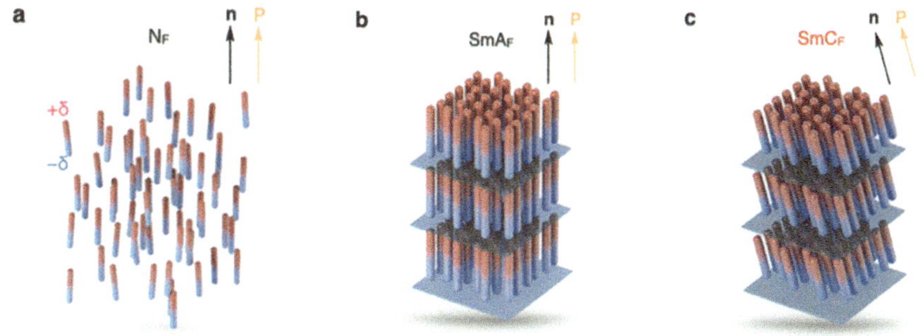

Fig. 3.64 Schematic illustrations of a) ferroelectric nematic (N_F), b) ferroelectric smectic A (SmA_F), and c) ferroelectric smectic C (SmC_F) phases. Reproduced from [116]

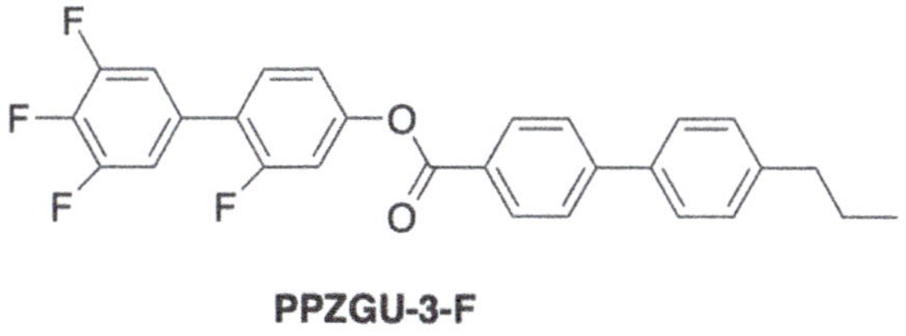

Fig. 3.65 Molecular structure, transition temperatures (°C), and associated enthalpies of transition (kJ mol^{-1}, in parentheses) for **PPZGU-3**. Square brackets denote a monotropic transition

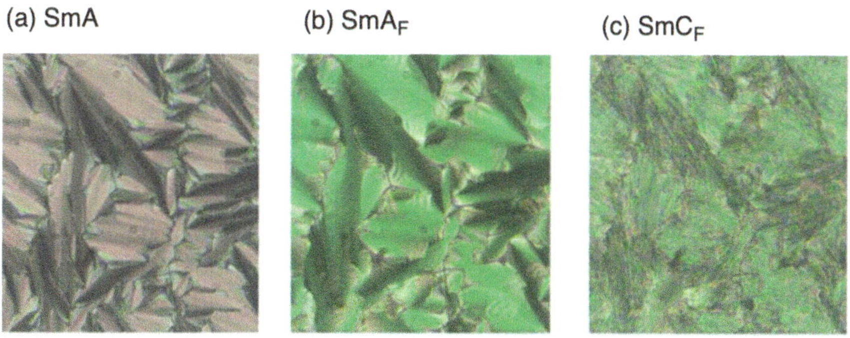

Fig. 3.66 POM micrographs of the **a** SmA (175 °C), **b** SmA_F (130 °C) and **c** SmC_F (110 °C) phases observed for **PPZGU-3-F** within a thin cell treated for homeotropic anchoring. Reproduced from [117]

assumption that all dipoles aligned perfectly, the maximum value is calculated to be 4.6 μC cm^{-2}. The stripes related to the helical pitch can be observed by POM only in the vicinity of the N_{TBF}–N_F phase transition, where the helix unwinds (Fig. 3.70). Figure 3.71a shows temperature dependencies of N_{TBF} and N_F helical pitch values

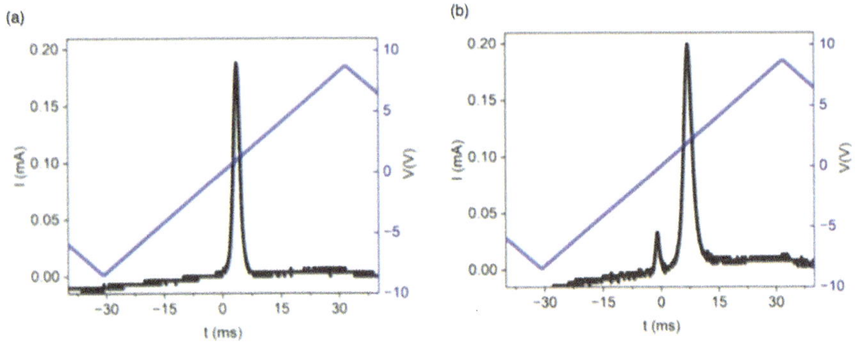

Fig. 3.67 **a** Current response trace measured at 8 Hz for **PPZGU-3-F** in the SmA_F phase at 136 °C and **b** that in the SmC_F phase at 104 °C. Reproduced from [117]

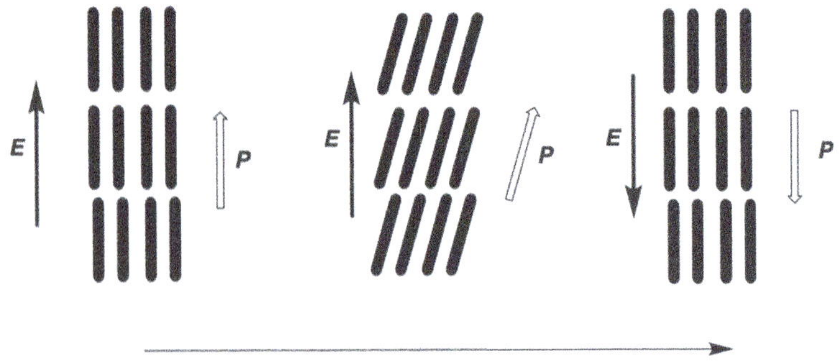

Fig. 3.68 The proposed switching process in the SmC_F phase [115]

detected by the light diffraction experiment. The N_{TBF} and $SmC^H{}_P$ phases have submicron periodicity due to the helical pitch. The helix in the N_{TBF} phase unwinds on heating as the N_F phase is approached. Figure 3.71b shows models of the helical arrangement of molecules in the N_{TBF} and $SmC^H{}_P$ phases. Furthermore, a clear pattern with a periodicity of 600–700 nm was observed by atomic force microscopy (AFM) as shown in Fig. 3.71c. The authors demonstrated that a strong internal electric field is reduced by a spontaneously formed helical structure.

3.8 Concluding Remarks

We have briefly reviewed ferroelectric SmC* materials for display applications. Particularly, compounds possessing de Vries-like characteristic form a defect-free alignment, reducing the manufacturing cost of FLCDs to expand their applications. We have shown the dynamic behavior of molecules in smectic phases based on the

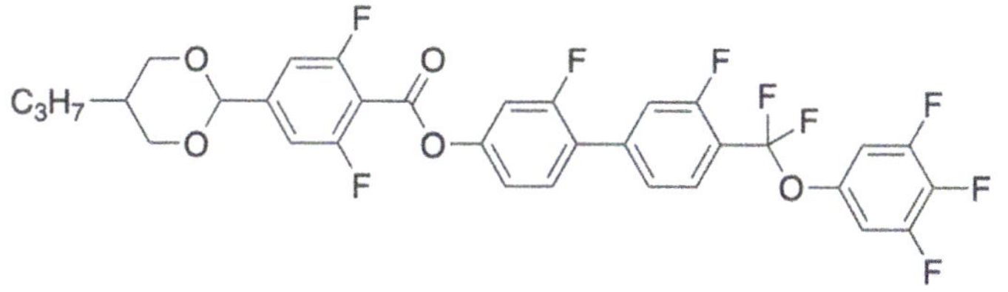

Fig. 3.69 Molecular structure, transition temperatures (°C), and associated enthalpies of transition (kJ mol^{-1}, in parentheses) for the LC compound. Square brackets denote a monotropic transition

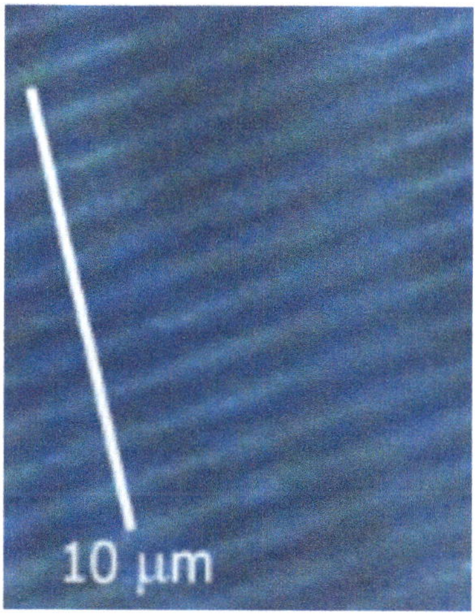

Fig. 3.70 The texture of the N_{TBF} phase in the vicinity N_{TBF}–N_F phase transition at 58 °C. The sample was hold in a planar cell with parallel rubbing. The cell thickness was 3 μm. Reproduced from [120]

^{13}C NMR studies, giving the better understanding on structure–property relationship in smectic liquid crystals. The NMR studies help us to design supermolecular liquid crystal trimers, such as, **MOHBC** and **DFPYCB**. The chiral trimer **MOHBC** exhibited an antiferroelectric-like switching in the chiral smectic A phase. This may be due to the electric-field-induced transition from the de Vries-like disordered state to the ferroelectric state. The achiral trimer **DFPYCB** exhibited an electric-filed induced polarization in the intercalated SmA phase. This electro-optical response is explained in terms of the random orientation of polar smectic clusters.

The ferroelectricity in a helical SmC* phase is improper ferroelectric. The surface stabilization realizes a ferroelectric liquid crystal. The discovery of ferroelectric nematic liquid crystals had a huge impact on liquid crystal science. Lots of N_F liquid crystals have been prepared. The structure–property relationship in N_F liquid crystals has been actively investigated, leading to the discovery of ferroelectric SmA and SmC phases. In those ferroelectric smectic phases, the polarization direction is parallel to their smectic director. A strong internal electric field in the polar nematic and smectic phases might be reduced by the random orientation of polar domains, or by the formation of a helical structure. At present, the large polar domain in a single direction in the equilibrium state is difficult to be maintained. Such challenges exist,

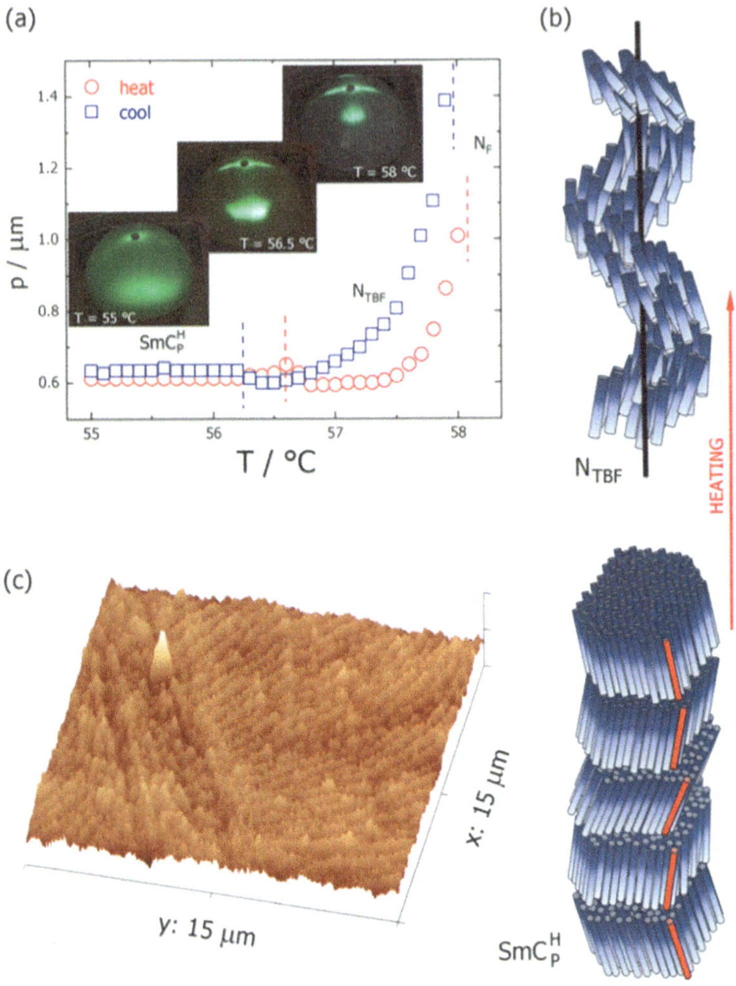

Fig. 3.71 **a** Temperature dependencies of N_{TBF} and N_F helical pitch values. The light diffraction patterns are shown in the inset. **b** Models of the heliconical arrangement of molecules in the N_{TBF} and $SmC^H{}_P$ phases. **c** AFM image of the $SmC^H{}_P$ phase supercooled to room temperature. Reproduced from [120]

but the remarkable ferroelectric properties of the N_F phase, coupled with nematic fluidity, provide not only fundamental significance but also huge potentiality for liquid crystal-based technologies, for example, capacitors, memories, sensors, optical devices, actuators, and energy conversion devices.

References

1. Meyer, R.B., Liebert, L., Strzelecki, L., Keller, P.: Ferroelectric liquid crystals. J. Phys. Lett. **36**, L69–L71 (1975)
2. Clark, N.A., Largerwall, S.T.: Submicrosecond bistable electro-optic switching in liquid crystals. Appl. Phys. Lett. **36**, 899–901 (1980)
3. Meyer, R.B.: Ferroelectric liquid crystals; a review. Mol. Cryst. Liq. Cryst. **40**, 33–48 (1977)
4. Goodby, J.W., Blinc, R., Clark, N.A., Lagerwall, S.T., Osipov, A., Pikin, S.A., Sakurai, T., Yoshino, K., Zeks, B.: Ferroelectric Liquid Crystals, Principles, Properties and Applications, 1st edn. Gordon and Breach, Philadelphia, PA, USA (1991)
5. Lagerwall, S.T.: Ferroelectric liquid crystals. In: Demus, D., Goodby, J.W., Gray, G.W., Spiess, H.W., Vill, V. (eds.) Handbook of Liquid Crystals, 1st edn. vol. 2B, pp. 515–664. Wiley-VCH Verlag GmbH, Weinheim, Germany (1998)
6. Guo, Q., Yan, K., Chigrinov, V., Zhao, H., Tribelsky, M.: Ferroelectric liquid crystals: physics and applications. Crystals **9**, 470 (2019)
7. Yoshizawa, A.: Ferroelectric smectic liquid crystals. Crystals **14**, 350 (2024)
8. Bahr, C.: Smectic liquid crystals: ferroelectric properties and electroclinic effect. In: Kitzerow, H.-S., Bahr, C. (eds.) Chirality in Liquid Crystals, pp. 223–250. Springer, New York, NY, USA (2001)
9. Walba, D.M.: Fast ferroelectric liquid-crystal electrooptics. Science **270**, 250 (1995)
10. Lagerwall, S.T.: Ferroelectric liquid crystal displays and devices. In: Goodby, J.W., Collings, P.J., Kato, T., Tschierske, C., Gleeson, H.F., Raynes, P. (eds.) Handbook of Liquid Crystals, 2nd edn, vol. 8, pp. 213–236. Wiley-VCH Verlag GmbH, Weinheim, Germany (2014)
11. Chandani, A.D.L., Gorecka, E., Ouchi, Y., Takezoe, H., Fukuda, A.: Antiferroelectric chiral smectic phases responsible for the tristable switching in MHPOBC. Jpn. J. Appl. Phys. **28**, L1265–L1268 (1989)
12. Nishiyama, I.: Antiferroelectric liquid crystals. Adv. Mater. **6**, 966–970 (1993)
13. Fukuda, A., Takanishi, Y., Isozaki, T., Ishikawa, K., Takezoe, H.: Antiferroelectric chiral smectic liquid crystals. J. Mater. Chem. **4**, 997–1016 (1994)
14. Miyachi, K., Fukuda, A.: Antiferroelectric liquid crystals. In: Demus, D., Goodby, J.W., Gray, G.W., Spiess, H.W., Vill, V. (eds.) Handbook of Liquid Crystals, 1st edn, vol. 2B, pp. 665–691. Wiley-VCH Verlag GmbH, Weinheim, Germany (1998)
15. Yamada, Y., Fukuda, A.: Antiferroelectric liquid crystal displays and their materials. In: Goodby, J.W., Collings, P.J., Kato, T., Tschierske, C., Gleeson, H.F., Raynes, P. (eds.) Handbook of Liquid Crystals, 2nd edn, vol. 8, pp. 263–302. Wiley-VCH Verlag GmbH, Weinheim, Germany (2014)
16. Inui, S., Iimura, N., Suzuki, T., Iwane, H., Miyachi, K., Takanishi, Y., Fukuda, A.: Thresholdless antiferroelectricity in liquid crystals and its application to displays. J. Mater. Chem. **6**, 671–673 (1996)
17. Niori, T., Sekine, J., Watanabe, J., Furukawa, T., Takezoe, H.: Distinct ferroelectric smectic liquid crystals consisting of banana shaped achiral molecules. J. Mater. Chem. **6**, 1231–1233 (1996)
18. Sekine, T., Niori, T., Sone, M., Watanabe, J., Choi, S.-W., Takanishi, Y., Takezoe, H.: Origin of helix in achiral banana-shaped molecular systems. Jpn. J. Appl. Phys. **36**, 6455–6463 (1997)
19. Reddy, R.A., Tschierske, C.: Bent-core liquid crystals: polar order, superstructural chirality and spontaneous desymmetrisation in soft matter systems. J. Mater. Chem. **16**, 907–961 (2006)
20. Takezoe, H., Takanishi, Y.: Bent-core liquid crystals: their mysterious and attractive world. Jpn. J. Appl. Phys. **45**, 597–625 (2006)
21. Cestari, M., Diez-Berart, S., Dunmur, D.A., Ferrarini, M.R., de la Fuente, M.R., Jackson, D.J.B., Lopez, D.O., Luckhurst, G.R., Perez-Jubindo, M.A., Richardson, R.M., et al.: Phase behavior and properties of the liquid-crystal dimer 1,7 -bis(4-cyanobiphenyl-4 -yl)heptane: a twist-bend nematic liquid crystal. Phys. Rev. E **84**, 031704 (2011)
22. Tschierske, C., Ungar, G.: Mirror symmetry breaking by chirality synchronization in liquids and liquid crystals of achiral molecules. ChemPhysChem **17**, 9–26 (2016)

23. Yoshizawa, A.: The formation of supramolecular chiral materials from achiral molecules using a liquid-crystalline system: symmetry breaking, amplification, and transfer. Crystals **14**, 97 (2024)
24. Reddy, R.A., Zhu, C., Shao, R., Korblova, E., Gong, T., Shen, Y., Garcia, E., Glaser, M.A., Maclennan, J.E., Walba, D.M., Clark, N.A.: Spontaneous ferroelectric order in a bent-core smectic liquid crystal of fluid orthorhombic layers. Science **332**, 72–77 (2011)
25. Mandle, R.J., Cowling, S.J., Goodby, J.W.: A nematic to nematic transformation exhibited by a rod-like liquid crystal. Phys. Chem. Chem. Phys. **19**, 11429–11435 (2017)
26. Mandle, R.J., Cowling, S.J., Goodby, J.W.: Rational design of rod-like liquid crystals exhibiting two nematic phases. Chem. A Eur. J. **23**, 14554–14562 (2017)
27. Nishikawa, H., Shiroshita, K., Higuchi, H., Okumura, Y., Haseba, Y., Yamamoto, S., Sago, K., Kikuchi, H.: A fluid liquid-crystal material with highly polar order. Adv. Mater. **29**, 1702354 (2017)
28. Chen, X., Korblova, E., Dong, D.P., Wei, X.Y., Shao, R.F., Radzihovsky, L., Glaser, M.A., Maclenan, J.E., Bedrov, D., Walba, D.M., et al.: First-principles experimental demonstration of ferroelectricity in a thermotropic nematic liquid crystal: Polar domains and striking electro-optics. Proc. Natl. Acad. Sci. U.S.A. **117**, 14021–14031 (2020)
29. Lavrentovich, O.D.: Ferroelectric nematic liquid crystal, a century in waiting. Proc. Natl. Acad. Sci. U.S.A. **117**, 14629–14631 (2020)
30. Blinov, L.M., Beresnev, L.A.: Ferroelectric liquid crystals. Sov. Phys. Usp. **27**, 492–514 (1984)
31. Lalanne, J.R., Buchert, J., Destrade, C., Nguyen, H.T., Marcerou, J.P.: Slowing down of molecular rotation at the smectic-A–smectic-C* transition of liquid crystals. Phys. Rev. Lett. **62**, 3046–3049 (1989)
32. Kremer, F., Vallerien, S.U., Kapitza, H., Zentel, R., Fischer, E.W.: Constant molecular rotation at the smectic-A to smectic-C* transition in ferroelectric liquid crystals. Phys. Rev. A **42**, 3667–3669 (1990)
33. Berardi, R., Ricci, M., Zannoni, C.: Ferroelectric nematic and smectic liquid crystals from Tapered molecules. ChemPhysChem **7**, 443–447 (2001)
34. Yoshizawa, A., Kikuzaki, H., Fukumasa, M.: Microscopic organization of the molecules in smectic A and chiral (racemic) smectic C phases: dynamic molecular deformation effect on the SmA to SmC* (SmC) transition. Liq. Cryst. **18**, 351–366 (1995)
35. Yoshizawa, A., Nishiyama, I., Kukuzaki, H., Noriko Ise, N.I.: C-13 NMR and X-ray investigations of phase transitions in an antiferroelectric liquid crystal. Jpn. J. Appl. Phys. **31**, L860–L863 (1992)
36. Nishiyama, I., Goodby, J.W.: A non-chiral swallow-tailed liquid crystal exhibiting a smectic C phase that has an antiferroelectric structure. J. Mater. Chem. **2**, 1015–1023 (1992)
37. Ouchi, Y., Yoshioka, Y., Ishii, H., Seki, K., Kitamura, M., Noyori, R., Takanishi, Y., Nishiyama, I.: J. Mater. Chem. **5**, 2297–2304 (1995)
38. Walba, D.M., Clark, N.A.: Molecular design of ferroelectric liquid crystals. Ferroelectrics **84**, 65–72 (1988)
39. Walba, D.M.: Ferroelectric liquid crystals–a unique state of matter. In: Mallouck, T.E. (eds.) Advances in the Synthesis and Reactivity of Solids. JAI Press, vol. 1, pp. 173–235. Greemwich, CT, USA (1991)
40. Lemieux, R.P.: Chirality transfer in ferroelectric liquid crystals. Acc. Chem. Res. **34**, 845–853 (2001)
41. Kelly, S.M.: Synthesis of chiral smectic liquid crystals. In: Demus, D., Goodby, J.W., Gray, G.W., Spiess, H.W., Vill, V. (eds.) Handbook of Liquid Crystals, 1st edn, vol 2B, pp. 493–514. Wiley-VCH Verlag GmbH: Weinheim, Germany (1998)
42. Sakurai, T., Mikami, N., Higuchi, R., Honma, M., Yoshino, K.: Synthesis ad properties of ferroelectric liquid crystals with large spontaneous polarization. Ferroelectrics **85**, 469–478 (1988)
43. Kelly, S.M., Buchecker, R., Fünfschhilling, J.: A-Fluoro esters incorporating a cyclohexane ring: some new chiral dopants for ferroelectric mixtures. J. Mater. Chem. **4**, 1689–1697 (1994)

44. Sakashita, K., Shindo, M., Nakauchi, J., Uematsu, M., Kageyama, Y., Hayashi, S., Ikemoto, T., Mori, K.: Novel ferroelectric liquid crystalline d-valerolactone derivatives with very large spontaneous polarization. Mol. Cryst. Liq. Cryst. **199**, 119–127 (1991)
45. Yoshizawa, A., Nishiyama, I., Fukumasa, M., Hirai, T., Yamane, M.: New ferroelectric liquid crystal with large spontaneous polarization. Jpn. J. Appl. Phys. **28**, L1269–L1270 (1989)
46. Shiratori, N., Yoshizawa, A., Nishiyama, I., Fukumasa, M., Yokoyama, A., Hirai, T., Yamane, M.: New ferroelectric liquid crystals having 2-fluoro-2-methyl alkanoyloxy group. Mol. Cryst. Liq. Cryst. **199**, 129–140 (1991)
47. Lemieux, R.P.: Molecular recognition in chiral smectic liquid crystals: the effect of core–core interactions and chirality transfer on polar order. Chem. Soc. Rev. **36**, 2033–2045 (2007)
48. Goodby, J.W.: Synthesis of non-chiral smectic liquid crystals. In: Demus, D., Goodby, J.W., Gray, G.W., Spiess, H.W., Vill, V. (eds.) Handbook of Liquid Crystals, 1st edn, vol. 2A, pp. 411–440. Wiley-VCH Verlag GmbH, Weinheim, Germany (1998)
49. Hird, M.: Designing principles, synthesis, and properties of smectic C host materials for ferroelectric liquid crystals. In: Goodby, J.W., Collings, P.J., Kato, T., Tschierske, C., Gleeson, H.F., Raynes, P. (eds.) Handbook of Liquid Crystals, 2nd ed, vol. 8, pp. 237–261. Wiley-VCH Verlag GmbH, Weinheim, Germany (2014)
50. Yokoyama, A., Nishiyama, I., Yoshizawa, A.: 6-Alkyl-2-(4-alkyloxyphenyl)quinoline: a new smectic C base material. Ferroelectrics **148**, 139–145 (1993)
51. Skraup, Z.H.: Eine syntheses der chinolins. Monatshefte Für Chem. **1**, 316–318 (1880)
52. Ishikawa, K., Hashimoto, K., Takezoe, H., Fukuda, A., Kuze, E.: A practical method of preparing thin homogeneous ferroelectric smectic cells for electro-optical microsecond switches (II): SmA liquid crystal growth under a temperature gradient. Jpn. J. Appl. Phys. **23**, L211–L213 (1984)
53. Hiji, N., Ouchi, Y., Takezoe, H., Fukuda, A.: Determination of chevron direction and sign of the boat-shaped disclination in surface-stabilized ferroelectric liquid crystals. Jpn. J. Appl. Phys. **27**, L1–L4 (1988)
54. Clark, N.A., Rieker, T.P.: Smectic-C "chevron", a planar liquid-crystal defect: Implications for the surface-stabilized ferroelectric liquid-crystal geometry. Phys. Rev. A **37**, 1053–1056 (1988)
55. Mochizuki, A., Yoshihara, T., Iwasaki, M., Nakatsuka, M., Takanishi, Y., Ouchi, Y., Takezoe, H., Fukuda, A.: Electro-optical switching of bookshelf structure Sc*cells. Inst. Image Inf. Telev. Eng. **14**, 21–27 (1990)
56. Takanishi, Y., Ouchi, Y., Takezoe, H., Fukuda, A., Mochizuki, A., Nakatsuka, M.: Spontaneous formation of quasi-book-shelf layer structure in new ferroelectric liquid crystals derived from a naphthalene ring. Mol. Cryst. Liq. Cryst. **199**, 111–118 (1991)
57. Janulis, E.P., Osten, D.W., Radcliff, M.D., Novack, J.C., Tristani-Kendra, M., Epstein, K.A., Keyes, M., Johnson, G.C., Savu, P.M., Spawn, T.D.: Fluorinated liquid crystals: an update. In: SPIE 1665, liquid crystal materials, devices, and applications. SPIE, Bellingham, DC, USA (1992)
58. Rieker, T.P., Janulis, E.P.: Dimerlike smectic-A and -C phases in highly fluorinated thermotropic liquid crystals. Phys. Rev. E **52**, 2688–2691 (1995)
59. Lagerwall, J.P.F., Giesselmann, F.: Current topics in smectic liquid crystals. ChemPhysChem **7**, 20–45 (2006)
60. De Vries, A.: X-ray photographic studies of liquid crystals: II. Apparent molecular length and thickness in three phases of ethyl-p-ethoxybenzal-p-aminobenzoate. Mol. Cryst. Liq. Cryst. **11**, 361–383 (1970)
61. De Vries, A.: Experimental evidence concerning two different kinds of smectic C to smectic A transitions. Mol. Cryst. Liq. Cryst. **41**, 27–31 (1977)
62. De Vries, A.: The description of the smectic A and C phases and the smectic A-C phase transition of TCOOB with a diffuse-cone model. J. Chem. Phys. **71**, 25–31 (1979)
63. Spector, M.S., Heiney, P.A., Naciri, J., Weslowski, B.T., Holt, D.B., Shashidhar, R.: Electroclinic liquid crystals with large induced tilt angle and small layer contraction. Phys. Rev. E **61**, 1576–1584 (2000)

64. Song, Q., Nonnenmacher, D., Giesselmann, F., Lemieux, R.P.: Tuning 'de Vries-like' properties in siloxane- and carbosilane-terminated smectic liquid crystals. J. Mater. Chem. C **1**, 343–350 (2013)
65. Kaur, S., Barthakur, A., Mohiuddin, G., Guta, S.P., Dhara, S., Pal, S.K.: Observation of "de Vries-like" properties in bent-core molecules. Chem. Sci. **13**, 2249–2257 (2022)
66. Ishida, N., Takanishi, Y., Yamamoto, J., Yoshizawa, A.: Amphiphilic liquid crystal exhibiting the smectic A to smectic C phase transition without layer contraction. Appl. Phys. Express **4**, 021701 (2011)
67. Giesselmann, F., Zugenmaier, P., Dierking, I., Lagerwall, S.T., Stebler, B., Kaspar, M., Hamplova, V., Glogarova, M.: Smectic-A*–smectic-C* transition in a ferroelectric liquid crystal without smectic layer shrinkage. Phys. Rev. E **60**, 598 (1999)
68. Clark, N.A., Bellini, T., Shao, R.-F., Coleman, D., Bardon, S., Link, D.R., Maclennan, J.E., Chen, X.-H., Wand, M.D., Walba, D.M., Rudquist, P., Lagerwall, S.T.: Electro-optic characteristics of de Vries tilted smectic liquid crystals: Analog behavior in the smectic A* and smectic C* phases. Appl. Phys. Lett. **80**, 4097–4099 (2002)
69. Naciri, J., Carboni, C., George, A.K.: Low transition temperature organosiloxane liquid crystals displaying a de Vries smectic A phase. Liq. Cryst. **30**, 219–225 (2003)
70. Kapernaum, N., Walba, D.M., Korblova, E., Zhu, C., Jones, C., Shen, Y., Clark, N.A., Gisselmann, F.: On the origin of the "giant" electroclinic effect in a "de Vries"-type ferroelectric liquid crystal material for chirality sensing applications. ChemPhysChem **10**, 890–892 (2009)
71. Prasad, S.K., Rao, D.S.S., Sridevi, S., Lobo, C.V., Rarana, B.R., Naciri, J., Shashidhar, R.: Unusual dielectric and electrical switching behavior in the de Vries smectic A phase of two organosiloxane derivatives. Phys. Rev. Lett. **102**, 147802 (2009)
72. Ghosh, S., Nayek, P., Roy, S.K., Majumder, T.P., Zurowska, M., Dabrowski, R.: Double-peak polarization current response in the unusual SmA* phase of a fluorinated high-tilt antiferroelectric liquid crystal. Eur. Phys. Lett. **89**, 16001 (2010)
73. Garoff, S., Meyer, R.B.: Electroclinic effect at the A-C phase change in a chiral smectic liquid crystal. Phys. Rev. Lett. **38**, 848–851 (1977)
74. Abdulhalim, I., Moddel, G.: Switching behviour and electro-optic response due to the soft mode ferroelectric effect in chiral smectic A liquid crystals. Liq. Cryst. **9**, 493–518 (1991)
75. Tsuji, D., Takanishi, Y., Yamamoto, J., Yoshizawa, A.: Chiral liquid crystal trimer exhibiting an optically uniaxial smectic phase with a double-peak polarization. J. Phys. Chem. C **116**, 8678–8687 (2012)
76. Schubert, C.P.J., Müller, C., Wand, M.D., Giesselmann, F., Lemieux, R.P.: Electroclinic effect in a chiral carbosilane-terminated 5-phenylpyrimidine liquid crystal with 'de Vries-like' properties. Chem. Commun. **51**, 12601–12604 (2015)
77. Swaminathan, V., Panov, V.P., Panov, A., Rodriguez-Lojo, D., Stevenson, P.J., Gorecka, E., Vij, J.K.: Design and electro-optic investigations of de Vries chiral smectic liquid crystals for exhibiting broad temperature ranges of SmA* and SmC* phases and fast electro-optic switching. J. Mater. Chem. C **8**, 4859–4868 (2020)
78. Link, D.R., Natale, G., Shao, R., Maclennan, J.E., Clark, N.A., Körblova, E., Walba, D.M.: Spontaneous formation of macroscopic chiral domains in a fluid smectic phase of achiral molecules. Science **278**, 1924–1927 (1997)
79. Guo, L., Gorecka, E., Pociecha, D., Vaupotic, N., Cepic, M., Reddy, R.A., Gornik, K., Araoka, F., Clark, N.A., Walba, D.M., Ishikawa, K., Takezoe, H.: Ferroelectric behavior of orthogonal smectic phase made of bent-core molecules. Phys. Rev. E **84**, 031706 (2011)
80. Kashima, S., Chiba, M., Takanishi, Y., Yamamoto, J., Yoshizawa, A.: Polar order of an achiral taper-shaped liquid crystal in the uniaxial smectic A phase. J. Mater. Chem. C **6**, 5521–5527 (2018)
81. Francescangeli, O., Stanic, V., Torgova, S.I., Strigazzi, A., Scaramuzza, N., Ferrero, C., Dolbnya, I.P., Weiss, T.M., Berardi, R., Muccioло, L., Orlandi, S., Zannoni, C.: Ferroelectric response and induced biaxiality in the nematic phase of a bent-core mesogen. Adv. Funct. Mater. **19**, 2592–2600 (2009)

82. Weissflog, W., Baumeister, U., Tamba, M.-G., Pelzl, G., Kresse, H., Friedemann, R., Hempel, G., Kurz, R., Roos, M., Merzweiler, K., et al.: Unexpected liquid crystalline behaviour of three-ring bent-core mesogens: Bis(4-subst.-phenyl) 2-methyl-iso-phthalates. Soft Matter **8**, 2671–2685 (2012)
83. Khan, B.C., Mukherjee, P.K.: Isotropic to smectic-A phase transition in taper-shaped liquid crystal. J. Mol. Liq. **329**, 115539 (2021)
84. Born, M.: Über anisotrope Flüssigkeiten. Versuch einer Theorie der flüssigen Kristalle und des elektrischen Kerr-Effekts in Flüssigkeiten. Sitzungsber. Preuss. Akad Wiss. **30**, 614–650 (1916)
85. Lavrentovich, O.D.: Ferroelectric mematic liquid crystal. A century in waiting. Proc. Natl. Acad. Sci. USA **117**, 14629–14631 (2020)
86. Sebastian, N., Cmok, L., Mandle, R.J., et al.: Ferroelectric-ferroelastic phase transition in a nematic liquid crystal. Phys. Rev. Lett. **124**, 037801 (2020)
87. Mertelj, A., Cmok, L., Sebastian, N., et al.: Splay nematic phase. Phy. Rev. X **8**, 041025 (2018)
88. Mandle, R.J., Mertelj, A.: Orientational order in the splay nematic ground state. Phys. Chem. Chem. Phys. **21**, 18769–18772 (2019)
89. Goodby, J.W.: Nano-objects – sculpting and shape in molecular material design. Liq. Cryst. **46**, 1901–1924 (2019)
90. Connor, P.L.M., Mandle, R.J.: Chemically induced splay nematic phase with micron scale periodicity. Soft Matter **16**, 324–329 (2020)
91. Chen, X., Korblova, E., Dong, D., et al.: First-principles experimental demonstration of ferroelectricity in a thermotropic nematic liquid crystal: Polar domains and striking electro-optics. Proc. Natl. Acad. Sci. U.S.A. **117**, 14021–14031 (2020)
92. Chen, X., Zhu, Z., Magrini, M.J., et al.: Ideal mixing of paraelectric and ferroelectric nematic phases in liquid crystals of distinct molecular species. Liq. Cryst. **49**, 1531–1544 (2022)
93. Mandle, R.J., Sebastian, N., Martinez-Perdiguero, J., Mertelj, A.: On the molecular origins of the ferroelectric splay nematic phase. Nat. Commun. **12**, 4962 (2021)
94. Madhusudana, N.V.: Simple molecular model for ferroelectric nematic liquid crystals exhibited by small rodlike mesogens. Phys. Rev. E **104**, 014704 (2021)
95. Erkoreka, A., Mertelj, A., Huang, M., Aya, S., Sebastian, N., Martinez-Perdiguero, J.: Collective and non-collective molecular dynamics in a ferroelectric nematic liquid crystal studied by broadband dielectric spectroscopy. J. Chem. Phys. **159**, 184502 (2023)
96. Osipov, M.A.: Dipole-dipole interactions and the origin of ferroelectric ordering in polar nematics. Liq. Cryst. **51**, 2349–2355 (2024)
97. Gibb, C.J., et al.: Spontaneous symmetry breaking in polar fluids. Nat. Commun. **15**, 5845 (2024)
98. Marchnko, A.A., et al.: Polar self-organization of ferroelectric nematic-liquid-crystal molecules on atomically flat Au(111) surface. Phys. Rev. Lett. **132**, 098101 (2024)
99. Li, J., et al.: Development of ferroelectric nematic fluids wit giant-ε dielectricity and nonlinear optical properties. Sci. Adv. **7**, eabf5047 (2021)
100. Mandle, R.J.: A new order of liquids: polar order in nematic liquid crystals. Soft Matter **18**, 5014–5020 (2022)
101. Cruickshank, E.: The emergence of a polar nematic phase: a chemist's insight into the ferroelectric nematic phase. ChemPlusChem **89**, e202300726 (2024)
102. Song, Y., et al.: Ferroelectric nematic liquid crystals showing high birefringence. Adv. Sci. **12**, 2414317 (2025)
103. Cruickshank, E., Imrie, C.T., et al.: The role of fluorine substituents in the formation of the ferroelectric nematic phase. J. Mater. Chem. C **13**, 3902–3916 (2025)
104. Chen, X., Clark, N.A., et al.: The smectic Z_A phae: Antiferroelectric smectic order as a prelude to the ferroelectric nematic. Proc. Natl. Acad. Sci. U.S.A. **120**, e2217150120 (2023)
105. Nacke, P., Manabe, A., Clark, N.A., Giesselmann, F., et al.: New examples of ferroelectric nematic materials showing evidence for the antiferroelectric smectic-Z phase. Sci. Rep. **14**, 4473 (2024)

106. Kumari, P., Basnet, B., Lavrentovich, M.O., Lavrentovich, O.D.: Chiral ground states of ferroelectric liquid crystals. Science **383**, 1364–1368 (2024)
107. Manabe, A., Bremer, M., Kraska, M.: Ferroelectric nematic phase at and below room temperature. Liq. Cryst. **48**, 1079–1086 (2021)
108. Parton-Barr, C., Gleeson, H.F., Mandle, R.J.: Room-temperature ferroelectric nematic liquid crystal showing a large and diverging density. Soft Matter **20**, 672–680 (2024)
109. Gibb, C.J., Hobbs, J., Mandle, R.J.: Systematic fluorination is a powerful design strategy toward fluid molecular ferroelectrics. J. Am. Chem. Soc. **147**, 4571–4577 (2025)
110. Chen, X., Korblova, E., Glaser, M.A., Maclennan, J.E., Walba, D.M., Clark, N.A.: Polar in-plane surface orientation of a ferroelectric nematic liquid crystal: polar monodomains and twisted state electro-optics. Proc. Natl. Acad. Sci. U.S.A. **118**, e2104092118 (2021)
111. Caim, F., Nava, G., Barboza, R., Clark, N.A., Korblova, E., Walba, D.M., Bellini, T., Lucchetti, L.: Surface alignment of ferroelectric nematic liquid crystals. Soft Matter **17**, 8130–8139 (2021)
112. Kikuchi, H., Matsukizono, H., Iwamatsu, K., Endo, S., Anan, S., Okunura, Y.: Fluid layered ferroelectrics with global $C_{\infty V}$ symmetry. Adv. Sci. **9**, 2202048 (2022)
113. Chien, X., Martinez, V., Nacke, P., Korblova, E., Manabe, A., Klasen-Memmer, M., Freychet, G., Zhernenkov, M., Glaser, M.A., Radzihovsky, L., et al.: Observation of a uniaxial ferroelectric smectic A phase. Proc. Natl. Acad. Sci. U.S.A. **119**, e2210062119 (2022)
114. Nishikawa, H., Okumura, Y., Kwaria, D., Nihonyanagi, A., Araoka, F.: Spontaneous twist of ferroelectric smectic blocks in polar fluids. Adv. Mater. **37**, 2501946 (2025)
115. Strachan, G., Gorecka, E., Szydlowska, J., Makal, A., Pociecha, D.: Nematic and smectic phases with proper ferroelectric order. Adv. Sci. **12**, 2409754 (2025)
116. Kikuchi, H., Nishikawa, H., Matsukizono, H., Iino, S., Sugiyama, T., Ishioka, T., Okumura, Y.: Ferroelectric smectic C liquid crystal phase with spontaneous polarization in the direction of the director. Adv. Sci. **11**, 2409827 (2024)
117. Hobbs, J., Gibb, C.J., Pociecha, D., Szydlowska, J., Gorecka, E., Mandle, R.J.: Polar order in a fluid like ferroelectric with a tilted lamellar structure—observation of a polar smectic C (SmC_P) phase. Angew. Chem. Int. Ed. **64**, e202416545 (2025)
118. Karcz, J., et al.: Spontaneous chiral symmetry breaking in polar fluid-heliconical ferroelectric nematic phase. Science **384**, 1096–1099 (2024)
119. Nishikawa, H., Okada, D., Kwaria, D., Nihonyanagi, A., Kuwayama, M., Hoshino, M., Araoka, F.: Emergent ferroelectric nematic and heliconical ferroelectric nematic states in an achiral "straight" polar rod mesogen. Adv. Sci. **11**, 2405718 (2024)
120. Gorecka, E., Pociecha, D., et al.: Spontaneous helix formation in a polar smectic phase. Mater. Horiz. **12**, 5352–5356 (2025)

Chapter 4
Blue Phases: Structures, Materials, and Applications

Abstract Liquid-crystalline blue phases have garnered significant attention due to their hierarchical phase structures exhibiting optical isotropy and their potential applications in electro-optical devices. These phases are classified into three categories: blue phase I (BPI), blue phase II (BPII), and blue phase III (BPIII). The frustrated structures of these systems are stabilized by defects that are dependent on chirality. Consequently, the disparity in driving force of phase formation between BPs and the conventional LCs necessitates a novel molecular design concept. The BPIII, in particular, is said to be in an amorphous state, and its structure has only recently been elucidated. We will discuss the recent findings that BPIII is a "spaghetti"-like tangled arrangement of double-twist cylinders with special dynamics. The present study explores the structure–property relationships that facilitate the stabilization of blue phases. Particularly, molecular biaxiality and flexibility have been identified as significant contributors to the stabilization of BPIII. Following a detailed exposition of the electro-optical effects in blue phases, we introduce the characteristics of displays using BPIII and tunable photonic crystals employing polymer-stabilized BPI. The blue phase liquid crystals are no longer regarded as a unique phase of liquid crystals that displays an extremely broad temperature range. Instead, it is now recognized as a foundation for the development of functional materials.

Keywords Liquid crystals · Blue phase · Chirality · Electro-optical effect · Display · Photonic crystals

4.1 Introduction

In general, liquid crystals exhibit birefringence due to optical anisotropy. However, there are exceptions: the cubic phases (see Chap. 2) and the blue phases are optically isotropic. There are two types of blue phases: cholesteric blue phases and smectic

---

Supplementary Information The online version contains supplementary material available at https://doi.org/10.1007/978-981-92-2524-8_4.

A. Yoshizawa, *Liquid-Crystalline Supermolecules*, The Materials Research Society Series, https://doi.org/10.1007/978-981-92-2524-8_4

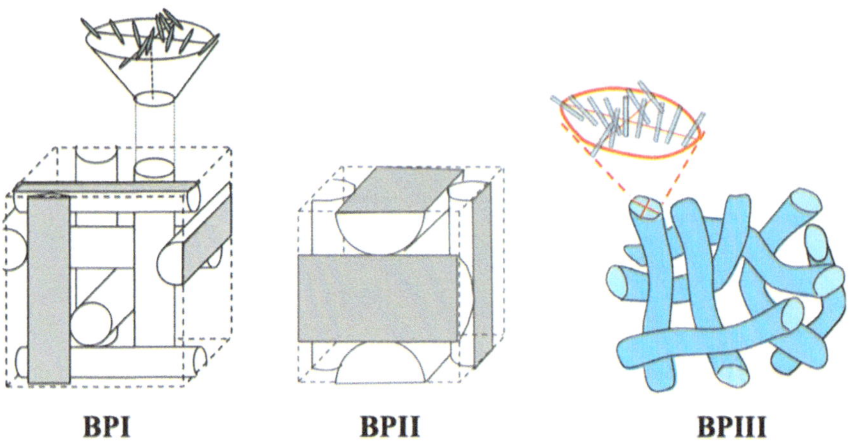

Fig. 4.1 Schematic representations of the blue phase structures. Reproduced from [9, 14] with permission from the Royal Society of Chemistry

blue phases. The cholesteric blue phases designated simply as blue phases (BPs) in this chapter are of particular interest because they have a fluid lattice for which the structure is stabilized by lattice defects [1–12]. Although the existence of the blue phases has been known since the discovery of liquid crystals, its narrow temperature range meant that for a long time it remained confined to the realm of basic research. Generally, blue phases are found in a very narrow temperature range (less than 2 K) between the isotropic liquid (Iso) and the chiral nematic (N*) phase of sufficiently short pitch. Blue phases consist of double-twist cylinders (DTCs). They are classified into three categories depending on the cylinders' packing structure: blue phase I (BPI), blue phase II (BPII), and blue phase III (BPIII) (Fig. 4.1) [6]. BPI and BPII have a body-centered cubic structure and a simple cubic one, respectively [2, 13]. On the other hand, BPIII is thought to have an amorphous structure.

The formation of DTCs inevitably accompanies line defects because the constituent molecules cannot fill the space uniformly. The presence of the line defects destabilizes the system. Research has been conducted to expand temperature range of the blue phase [9, 10]. Kikuchi et al. reported polymer-stabilized BPI in which the temperature range was extended to 60 K [15]. This resulted in the formation of a defect-free system by filling the defects with polymer, and the significant decrease in the BP–N* transition temperature broadened the temperature range of the BP. With respect to the molecular design of BPs, Coles and Pivnenko reported that a nematic mixture of bimesogenic compounds doped with a small quantity of a chiral additive showed BPI of an extremely wide temperature range (40 K) [16]. In addition, hydrogen-bonded assemblies for stabilizing cubic BPs were reported [17, 18]. However, rational molecular design for the blue phase has not yet been established. When a chiral substance is added to a nematic phase, that nematic phase transforms into the corresponding chiral nematic phase. The same applies to the formation of the ferroelectric SmC* phase (see Chap. 3). On the other hand, there are no achiral host liquid crystals that can be converted to the blue phase by addition of chiral substances.

Blue phases are recognized as offering the potential for application as fast light modulators, tunable photonic crystals, or next-generation displays. BPI and BPII consisting of three-dimensional periodic structures are potentially useful for application as tunable photonic crystals [19–21]. Fast tunable reflection was also observed in amorphous BPIII, which lacks a three-dimensional structure [22]. The cubic blue phases show stimulus responsiveness, i.e., electrical, photonic, mechanical, and chemical [12]. They can be used not only for electro-optical applications but also for chemical sensors. For example, cubic BPs that change their reflection wavelength in response to chemical stimulus can serve as the basis for a sensor used in calorimetry. Calorimetric sensors based on BP for detecting humidity [23] or toluene [24] have been reported. In display applications, blue phases are optically isotropic, so they appear black even without surface treatments such as an alignment layer. The response time is in the submillisecond range [25, 26]. In previous studies, a blue-phase liquid crystal display (BPLCD) mode using BPI was developed [27]; however, some serious issues such as hysteresis in the voltage–transmittance curves [28] and residual transmittance were identified. This is partially due to the three-dimensional cubic structure. On the other hand, BIII is believed to be amorphous and lack a three-dimensional structure. We reported high-contrast and hysteresis-free electro-optical response in BPIII without surface treatment [29, 30].

Blue phase liquid crystals are expected to lead to the development of functional soft materials [12, 31]. Although many BP compounds have been synthesized and their physical properties investigated to date, no clear structure–property relationship has been established. Furthermore, the structure of BPIII has remained unsolved. We will discuss structures of BPs, molecular design for BP LCs, and their applications.

4.2 Phase Structures

Figure 4.2a, b is optical textures of BPI and BPII under crossed polarizers, respectively. The platelet textures are characteristic of cubic BPs. Those vibrant colors are structural colors and are not caused by birefringence. Both BPI and BPII have cubic structures with lattice parameters of approximately visible wavelengths. Consequently, the diverse colors of the platelets originate from selective reflection at wavelengths corresponding to various crystal planes. On the other hand, BPIII lacks a periodic structure. Figure 4.2c is a foggy texture of BPIII. The cubic BPs were actively investigated by several experimental methods, such as the Kossel diagram technique [32, 33], and are well understood. By contrast, the structure of BPIII remains to be solved.

By measuring the temperature dependence of the Bragg peak wavelength of cholestryl nonanoate, Stegemeyer and Bergmann demonstrated the existence of two

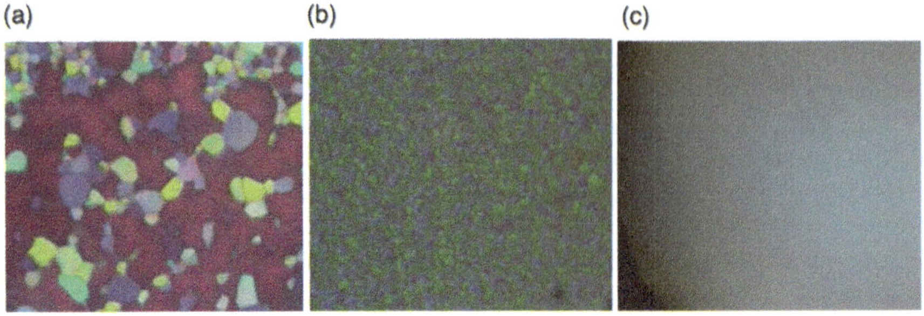

Fig. 4.2 Photomicrographs of **a** BPI, **b** BPII, and **c** BPIII. These photographs were taken in our laboratory

blue phases with cubic symmetry [1]. Study on Bragg reflections from monocrystals in mixtures of the eutectic nematic **E9** and the chiral dopant **CB15** yielded the following information [7, 33]. In the BPI, circularly polarized Bragg reflections occur at the (110), (200) and (211) planes. Those in the BPII occur at the (100) and (110) planes. By indexing the Bragg reflections in the cubic BPs, the following conclusions were drawn: (1) the (110), (200), and (211) series characteristic of BPI are features of the body-centered cubic Bravais lattice, and (2) the (100) and (110) series characteristic of BPII are consistent only with the simple cubic Bravais lattice. Within each cylinder, the molecules are twisted by 90° in the radial direction. In other words, the molecules are twisted from −45° at one end to +45° at the other. Hence, while the lattice constant of BPI is one pitch, that of BPII is half a pitch. In defect regions, the molecules are believed to adopt an isotropic orientation. Figure 4.3 shows illustration of defects for BPI and BPII. Recently, periodic pattern images of four types were found, attributed to (110), (100), (111) and (211) planes of a bcc lattice in polymer-stabilized BP I using confocal laser scanning microscopy [34].

Marcus used a reflection microscope to observe the blue phases of a polycrystalline sample of cholesteryl ester [35]. When the sample was cooled from the isotropic liquid, a distinct "fogy" texture appeared above the "mosaic" texture of the cubic blue phase. He named this amorphous blue phase as BPIII [7, 35]. BPIII is the highest-temperature blue phase. Previous studies have encountered some difficulties in the experimental investigation of BPIII, due to its narrow temperature range and the limited number of compounds exhibiting BPIII behavior. Nevertheless, the following important results have been obtained [4, 6].

1. The visual appearance of BPIII is foggy, and it is often bluish or grayish in color. BPIII appears to be closer in structure to the isotropic liquid.
2. BPIII reflects circularly polarized light and exhibits only one broad peak [36, 37].
3. A heat capacity peak between BPIII and the isotropic liquid is much larger than that between BPIII and BPII. Therefore, BPIII appears to be closer in structure to BPII than to the isotropic phase, in contrast to the visual appearance [38].
4. BPIII shows higher shear elasticity and viscosity than the BPII and isotropic phases [39].

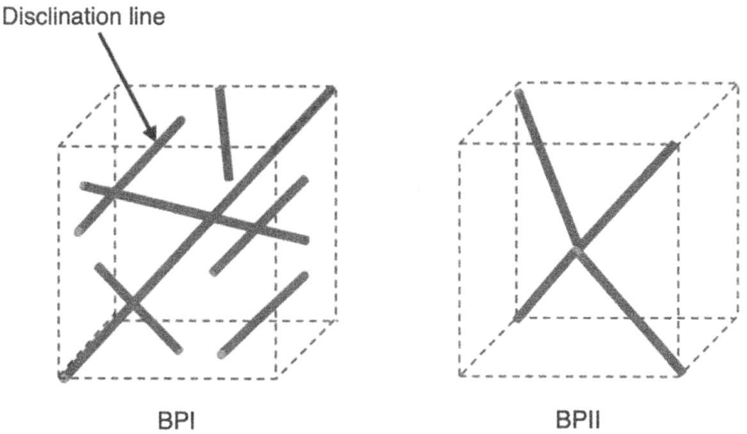

Fig. 4.3 Schematic illustration of defects for BPI and BPII

5. Transmission electron microscopy (TEM) observations of the BPIII of compound **CE4** reveal that BPIII consists of DTCs–each with a diameter only a fraction of that of a cholesteric pitch–entangles in a spaghetti-like structure. Although the DTCs appear to be aligned over distances of several micrometers, no long-range structure or periodicity is present [40].

Figure 4.4 shows TEM images of a fracture surface of **CE4** in the cholesteric, BPI and BPIII phases [40]. Molecular structure and transition temperatures of **CE4** are also shown. In the Ch phase, a one-dimensional periodic structure is observed (Fig. 4.4a). Each pair of bright and dark lines corresponds to half a helical pitch. The full pitch length is approximately 240 nm. The small arrows in Fig. 4.4b point to square patterns corresponding to fracture along a {100} plane of the cubic lattice in the BPI. A regular zigzag pattern is also visible near the large arrow. These regions exhibit regular periodicity in two directions in the BPI. The surface of the BPIII is densely covered with a random arrangement of filamentary-shaped objects ranging in diameter from 10 to 50 nm (Fig. 4.4c). Small groups of filaments extend for 0.1–1.0 μm in a given direction before ending or changing direction. These cylindrically shaped objects are arranged in a random and liquid-like fashion.

A number of models for BPIII, i.e., emulsion model [41], double-twist model [42], cubic domain model [43], and icosahedral model [44], were presented. Among these, we think that the double-twist model and the cubic-domain model are the most realistic. The double-twist model suggests that BPIII may consist of spaghetti-like tangles of DTCs. On the other hand, the cubic-domain model implies that BPIII may retain a localized cubic structure, though this is limited to short, correlated regions.

Recently, Henrich et al. reported, based on large-scale simulations of the blue phase of cholesteric liquid crystals, that BPIII is an amorphous network of disclination lines and is thermodynamically and kinetically stable over cubic BPs at intermediate chiralities [45]. Aperiodic structures are generated by placing a localized

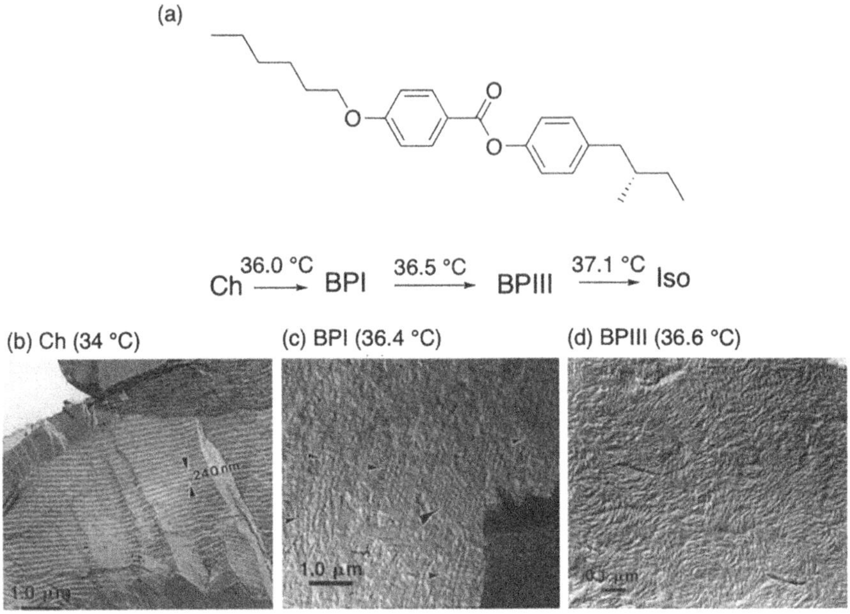

Fig. 4.4 **a** Molecular structure and transition temperatures of **CE4**. The authors referred that **CE4** is a British Drug House designation for chiral 4-(2-methylpentylphenyl)-4 -hexyloxybenzoate, though **CE4** is generally used as the abbreviation for chiral 4-(2-methylbutyl)phenyl 4-hexyloxybenzoate. We show here the molecular structure of (*S*)- 4-(2-methylbutyl)phenyl 4-hexyloxybenzoate. **b** TEM image of a freeze-fracture replica of **CE4** quenched from the Ch phase at 34 to −190 °C. **c** That quenched from the BPI at 36.4 °C. **d** That quenched from the BPIII at 36.6 °C. Reprinted with permission from [40]. Copyright (1996) American Physical Society

nucleus of BPI or BPII in a cholesteric matrix. For low chirality, the initial defects washed out to leave a cholesteric phase. Across a broad region of intermediate chiralities, the dilute double-twisted regions grow and rearrange dramatically to form a whole network of disclinations, which very slowly creeps to a final amorphous state (Fig. 4.5). The authors propose that an amorphous disclination network is locally close to BPII.

More recently, two interesting simulation studies on BPIII have been reported. The study reported by Paul and Saha supports the theoretical proposal that double-twist cylinders arrange themselves in a spaghetti-like configuration as a model for BPIII [46]. Pišljar and Muševič et al. reproduced the bulk BPIII of mutually enmeshed structures of singular defect lines (yellow) and skyrmion filament-DTCs (Fig. 4.6) [47]. They concluded as follows: BPIII is a topologically protected fluid composed of skyrmion filaments with unique dynamics that fluctuate violently. The fundamental building blocks of BPIII are skyrmion filaments. They form highly dynamic, topologically charged entities, which cannot be annihilated due to the elastic energy barrier.

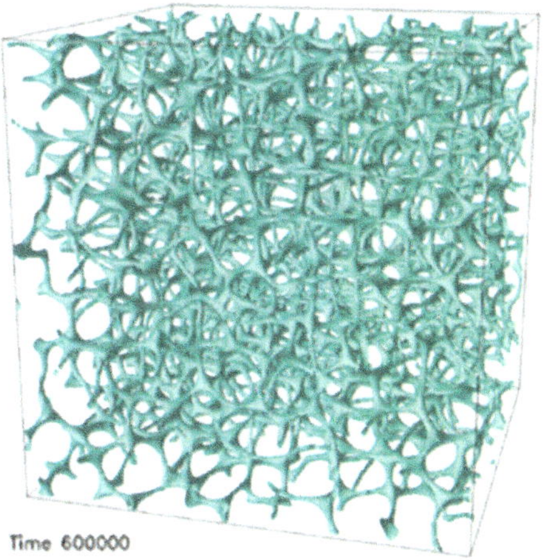

Fig. 4.5 End-state disclination network in BPIII. Reprinted with permission from [45]. Copyright (2011) American Physical Society

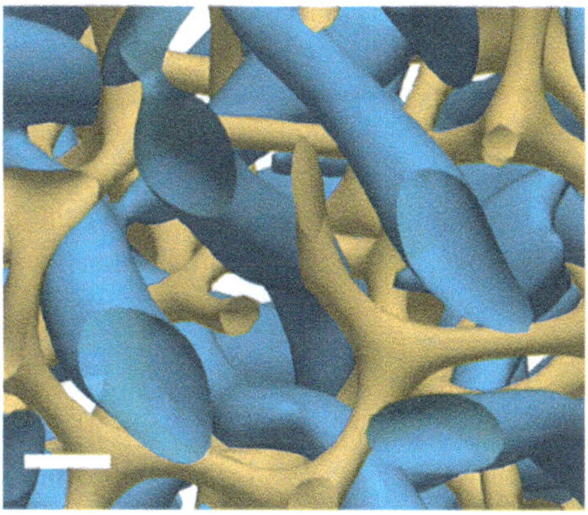

Fig. 4.6 Numerical simulations of bulk BPIII. Bulk BPIII is a topological fluid of skyrmionic filaments (blue), enmeshed with a 3D network of −1/2 chiral singular lines (yellow). The scale bar represents 50 nm. This figure is taken from [47]

Experimental investigations on the structure of BPIII have been conducted. One such study involved the observation of the polymer network arising from defects of BPIII. This was based on the method reported by Coles et al. [48]. As shown in Fig. 4.7, they fabricated a polymer template that replicates the defect structures of BPI. A mixture of chiral liquid crystals and reactive mesogens forms a BPI phase (Fig. 4.7a). The reactive mesogens were photopolymerized within the BPI (Fig. 4.7b). The template was obtained by removing the low-molecular-mass substances (Fig. 4.7c). When the template was refilled with an achiral nematic liquid crystal, a texture characteristic of cubic BP appeared over a wide temperature range (Fig. 4.7d).

Based on the assumption that the polymer network obtained from this manner faithfully reproduce the defect structures of the original blue phase, the morphologies of the polymer networks derived from the cubic BP and BPIII systems were investigated using field-emission scanning electron microscopy (FE-SEM) [49, 50]. Gandhi

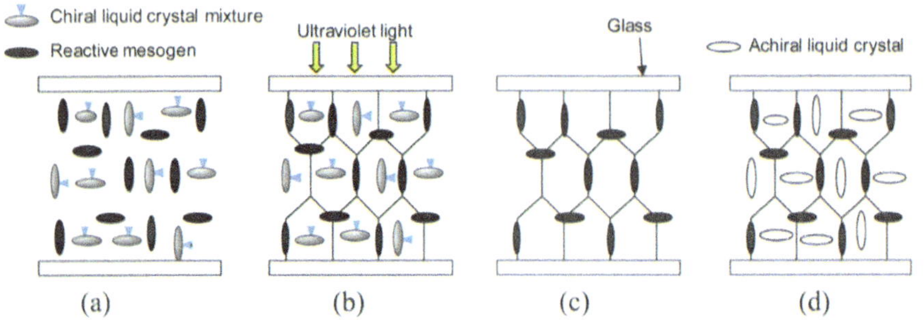

Fig. 4.7 Schematic diagram of the formation of the 3D nanostructured polymer network [48]. **a** Chiral LC mixture forms a blue phase. **b** Cell is exposed to UV light to photopolymerize the reactive mesogens. **c** Removing the liquid crystal, chiral dopant, and remaining reactive mesogen–photoinitiator mixture creates a porous cast. **d** An unopened cell is refilled with the achiral nematic liquid crystal. Reproduced from [9] with permission from the Royal Society of Chemistry

and Chin et al. reported the fabrication of porous polymer networks that mimic the 3D structure of BPIII at the nanoscale by imprinting reactive mesogen polymer networks along the topological defects of BPIII [50]. The BP mixture used to prepare the polymer network contained 4.41 wt% chiral dopant and 19.88 wt% reactive mesogen dissolved in a commercially available nematic mixture. Figure 4.8a, b show confocal laser scanning microscopy (CLSM) and SEM images of the BPIII-derived network, respectively. The authors noted that the SEM image of the BPIII-derived network matches the simulated structure of the amorphous disclination network in BPIII reported by Henrich et al. [45].

Recently, we reported SEM images of polymer networks derived from the cubic BP and BPIII systems [51]. We used a BP mixture consisting of the T-shaped liquid crystal [52] (49.3 wt%), the chiral dopant **S-811** (40.7 wt%), the reactive bifunctional monomer **RM257** (7.3 wt%), the monofunctional monomer **C12A** (2.4 wt%), and the photo-initiator **DMAP** (0.3 wt%). Their molecular structures are shown in Fig. 4.9. The phase transition temperatures of the BP mixture upon cooling were Iso 65.7 C BPIII 60.7 C cubic BPII 51.6 C N*. The reactive mesogen was photopolymerized within each BP. After removing low-molar-mass materials, the corresponding polymer networks were obtained.

Figure 4.10 shows the FE-SEM images of the BPII and BPIII networks. A marked difference is observed in the three-dimensional structure between them. The pore diameter is 170 ± 70 nm for the BPII network and 180 ± 70 nm for the BPIII network. These values correspond to the diameters of the double-twisted cylinders in each BP. Meanwhile, the width of the polymer chain widths in the BPII and BPIII polymer networks is approximately 80 nm and 130 ± 50 nm, respectively. The BPIII polymer network was thicker than the cubic one. A large number of reactive monomers has accumulated in the defect regions of BPIII, indicating that the defect structure of BPIII is in a disordered state characterized by fluidity. Consequently, the DTCs in BPIII easily rearrange, much like a liquid. These FE-SEM observations revealed that, whereas the DTCs in BPII are arranged in a regular pattern to form a

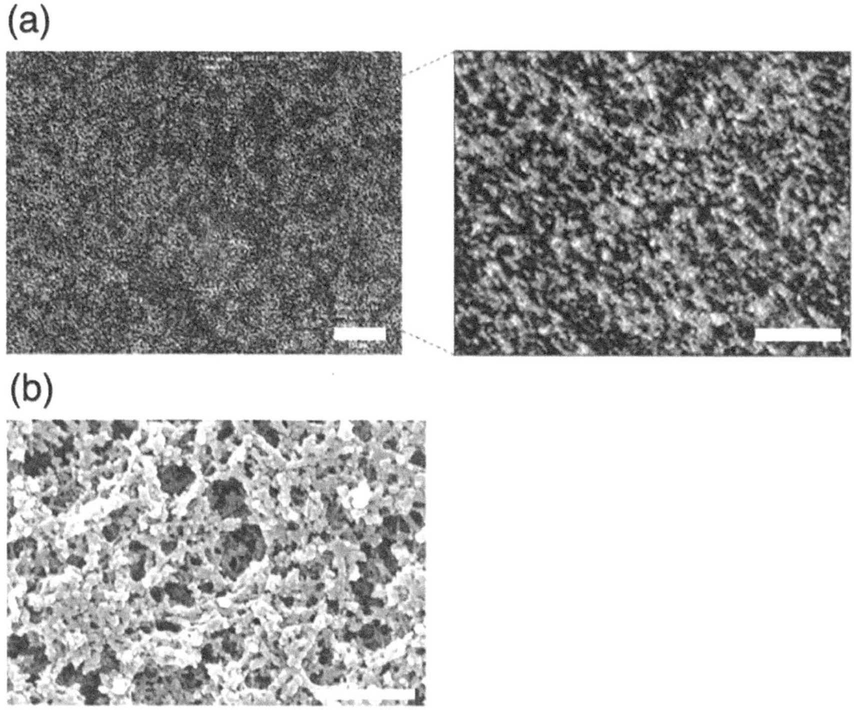

Fig. 4.8 **a** CLSM image of a scaffold nanoengineered from BPIII. The scale bars are 15 μm for the complete image and 3 μm for the enlarged region. **b** SEM image of a scaffold nanoengineered from BPIII. The scale bar is 500 nm. Reprinted with permission from [50]. Copyright (2016) John Wiley and Sons

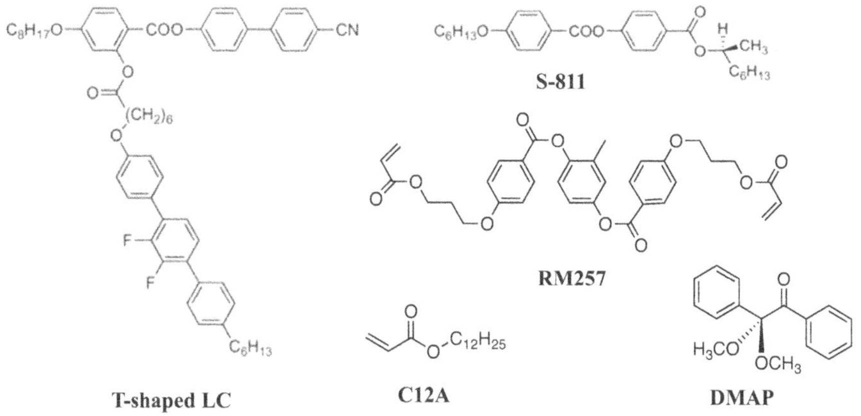

Fig. 4.9 **a** FE-SEM image of the polymer network derived from BPII and **b** that derived from BPIII

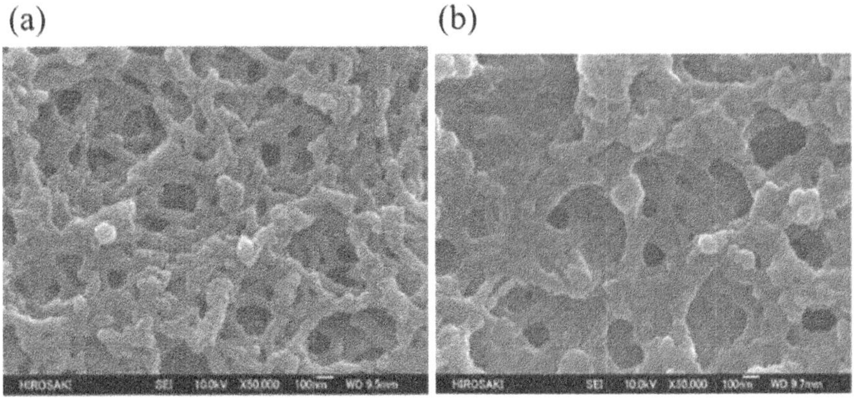

Fig. 4.10 **a** FE-SEM image of the polymer network derived from BPII and **b** that derived from BPIII. Reprinted with permission from [51]. Copyright (2023) Taylor & Francis

three-dimensional periodical structure, those in the BPIII are arranged in a random and entangled state. It is possible that the entropy-driven entangled arrangement of the DTCs contributes to the higher viscosity observed in BPIII. BPIII exhibits characteristic dynamic behavior, and the SEM image presented here shows the time-averaged structure of BPIII. This FE-SEM study supports the spaghetti-like double-twisted model rather than the cubic domain model. Recent simulations [46, 47] and our FE-SEM measurements [51] have revealed that BPIII consists of a spaghetti-like, entangled DTC arrangement that exhibits characteristic dynamic behavior.

4.3 Stabilizing Blue Phases

4.3.1 Numerical Study

The blue phase inevitably involves defects caused by double twists. Therefore, the following two important factors should be considered [53].

1. Stabilization through the formation of a double-twisted structure in which molecules twist in all directions instead of that of a single-twisted one [3].
2. Instability caused by the formation of defects necessary for a double-twisted cylinder to fill the space.

Based on the above, there are two approaches for stabilizing the blue phase: designing liquid crystal molecules that stabilize the double-twisted structure, and methods to reduce the instability caused by the presence of defects [53].

The cubic blue phases can be explained by both of the two principal continuum theoretical models for liquid crystals [54]; the Oseen-Frank theory for the director [55] and the Landau-de Gennes theory based on the Q-tensor order parameter [56].

From the director field, the locally preferred structure is a double-twist structure, however, this contradicts the overall topological requirements. The structure of the blue phase described by the Landau-de Gennes theory is a linear combination of biaxial helices selected to optimize the trade-off between bulk and gradient free energies. These theoretical models have successfully explained the origin, symmetry and general properties of the cubic blue phase. Recently, numerous compounds exhibiting a blue phase have been reported. The theoretical models described above are insufficient to explain the thermal stability of the cubic blue phase in these various compounds. Alexander and Yeomans published a study on the phase diagram of cholesteric liquid crystals within the framework of Landau-de Gennes theory [54]. They demonstrated that when $K_{33} = 2K_{11} = 2K_{22}$, the stable region of the blue phase narrows compared to cases where the elastic constants are equal. Furthermore, they pointed out that the stability of BPI is significantly improved in systems where the cholesteric phase undergoes a helical sense inversion [54, 57]. More recently, Fukuda [58] reported on the effect of changes in elastic constants on the stability of the cubic blue phases, based on the defect theory proposed by Meiboom et al. Meiboom et al. put forward a structure where the cylinders are placed along the three directions [59]. They wrote the free energy per unit length F_{disc} of a single disclination line, which consists of four terms [6, 59]:

$$F_{disc} = F_{el} + F_{surf} + F_{core} + F_{int}. \tag{4.1}$$

The first term is the elastic energy associated with the defect, which can be calculated using the Frank free energy outside a disclination core with elastic constants. The second term is an additional elastic term. The third term describes the difference in free energy between the isotropic and blue phases. The last term is an interfacial energy, characterized by surface tension between the core and the chiral material. According to Meiboom's defect mode [59], the free energy phase can be written as

$$F = a(T_1 - T)\pi R^2 + 2\sigma\pi R - \pi K + \left(\frac{1}{4}\right)\pi K \ln(R_{max}/R) \tag{4.2}$$

where coefficient a can be estimated from the latent heat of the cholesteric–isotropic phase transition. T_1 stands for the transition temperature of the BP–isotropic phase transition, T denotes the transition temperature of the BP–cholesteric phase transition, σ is the interfacial tension, R is the radius of the disclination core, R_{max} is the radius of the double twist cylinders, and K represents the typical elastic constant of bend in the BP mixture ($K = K_{33}$). The $\sigma\pi R$ term is much smaller than the other terms. When Eq. 4.2 yields a negative value, the blue phase becomes stable. In that context, only cases where the elastic constants are equal are considered is considered. Fukuda extended this theory to cases where the elastic constants are unequal and presented the following estimates for the BP temperature range [58]:

$$\Delta T = \left(\frac{\tilde{K}}{8aR_{max}^2}\right)\exp\{(2K_{11} + 0.57K_{22} - 0.50K_{33})/\tilde{K} - 1\}. \tag{4.3}$$

where

$$\tilde{K} = \left(\frac{1}{2}\right)(K_{11} + K_{33})\left[1 - \left(\frac{25}{72}\right)\left(\frac{K_{11} - K_{33}}{K_{11} + K_{33}}\right)^2\right]$$

He demonstrated that (1) the stability of the BP improves significantly as the bend elastic constant K_{33} decreases, and (2) the BP becomes more stable as the splay (K_{11}) and twist (K_{22}) elastic constants increase. On the other hand, $R_{\max}$ is proportional to the helical pitch. Therefore, Eq. 4.3 indicates that a large helical twisting power (HTP) is a major factore in broadening the temperature range of the BP.

Some theoretical reports have described the effects of stabilizing the defects on stabilization of BPs. Particle trapping sites and trapping potentials were calculated by Ravnik et al. [60]. Three basic regimes of trapping were identified for particles with weak anchoring. (i) Small particles are trapped inside regions, which exactly correspond to the disclination lines. Particles are well immersed inside the melted cores of the disclinations. (ii) Medium-sized particles are trapped within the cores of disclinations. However, the relative arrangements of the double-twist cylinders and the local curvature of the disclinations become relevant. Particles bulge out from the disclination cores. Therefore, the trapping sites become points of a regular periodic array. (iii) Large particles are not necessarily trapped within the disclinations. They maximize the coverage of disclination line regions and minimize the coverage of the double-twist regions. They induce the rewiring of disclinations, which may function as nucleation sites. It has been shown that maximizing the filling rate of trapping sites by particles reduces the total free energy of the system, thereby serving as a means to thermodynamically stabilize the confined blue phase.

Fukuda [61] discussed the stabilization of the blue phase in chiral liquid crystals through the introduction of guest components using numerical calculations based on the Landau-de Gennes theory. The discussion is based on the general conjecture that cores of disclination lines with higher free-energy density are replaced by the guest component. Based on the assumption that the guest component forms a sharp interface with the liquid crystal (strong segregation), it was shown that by appropriately selecting a material with low interfacial tension, a guest component of a volume fraction of less than 10% can drastically increase the thermodynamic stability temperature of the blue phase, which is consistent with experimental results [15]. Furthermore, he showed that although the stability of BPII was only slightly improved, the temperature range of BPI was significantly expanded by the guest component [62].

4.3.2 Material Design

The most successful approach for expanding the BP temperature range is polymer stabilization, as reported by Kikuchi et al. [15]. This is achieved by the polymer network filling defects in the blue phase to create a defect-free state. Blue phases are also stabilized by mixing surface-functionalized nanoparticles with chiral liquid crystals. The aggregation od the nanoparticles within a defect creates a defect-free state. Apart from filling the lattice defects, the molecular design of blue phase materials has been investigated.

Filling the Lattice Defects

Kitzerow et al. reported the first observation of a blue phase in chiral anisotropic polymer networks [63]. In a binary mixture composed of a chiral and a non-chiral diacrylate, they observed the typical structures of the BPI, BPII, and BPIII phases. By performing in-situ photopolymerization of the mixture, they obtained a polymer network exhibiting a BP structure. Kikuchi et al. reported on a polymer-stabilized blue phase for which the temperature range is extended to more than 60 K [15]. Furthermore, they demonstrated fast electro-optical switching for the stabilized BPI. Switching with the speed of 10^{-4} s is attributed to an electro-optic Kerr effect [64]. The polymer/liquid crystal composite was prepared by in situ photopolymerization of a small amount of acrylate monomers in a mixture of monomer, photoinitiator, and chiral liquid crystal in the blue phase state. The proposed mechanism is the following. That monomers are more miscible with an isotropic region than with a double-twist region. Consequently, the monomers migrate toward the defect regions. The monomers are concentrated selectively in the disclination core, being in an isotropic state, rather than in the double-twist cylinder. The polymerization of the monomers forms a specific polymer network, which can stabilize the lattice defects of the cubic BPI. The physical mechanisms of this polymer-stabilized blue phase is due to the preferential distribution of polymer chains within the blue phase, and is fundamentally different from the polymer-stabilized blue phase reported by Kitzerow et al. [63], in which the orientation of liquid crystal directors is stabilized by a polymer network formed in a desired orientation in the liquid crystal.

This polymer stabilization method has been applied not only to cubic BPI but also to amorphous BPIII [14, 50, 65, 66]. Let us introduce our results here [14]. We used a BPIII mixture containing a small amount of dimeric LC with an even-numbered methylene spacer (**8PY8OCB**) for the polymer stabilization. The sample consisted of the following components: 64.2 wt% nematic mixture, 3.4 wt% **8PY8OCB**, 11.9 wt% **ISO-(6OBA)$_2$**, 20 wt% binary monomer mixture (50 wt% **RM257** and 50 wt% **C12A**), and 0.5 wt% **DMAP**. The nematic mixture consists of 4-pentyl-4-cyanobiphenyl (50 wt%), 4-octyloxy-4-cyanobiphenyl (35 wt%) and 4-pentyl-4'-cyanoterphenyl (15 wt%). Molecular structures of the dimeric LC and **ISO-(6OBA)$_2$** are shown in Fig. 4.11. The phase transition temperatures of the BP mixture were Iso 75.2 °C BPIII 71.5 °C N* and those of the sample after photopolymerization were Iso 80 °C BPIII. No further phase transition was observed until 0 °C. Adding a small amount of **8PY8OCB** to the low-molecular-weight blue phase LC increases the Iso–BPIII transition temperature. The PS-BPIII exhibited electro-optical properties characteristic of BPIII.

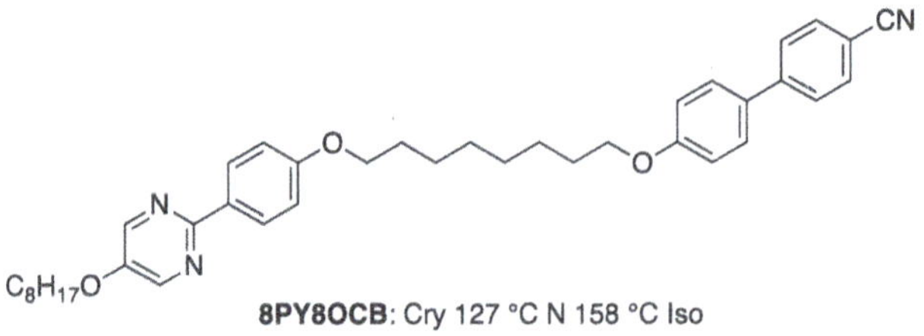

8PY8OCB: Cry 127 °C N 158 °C Iso

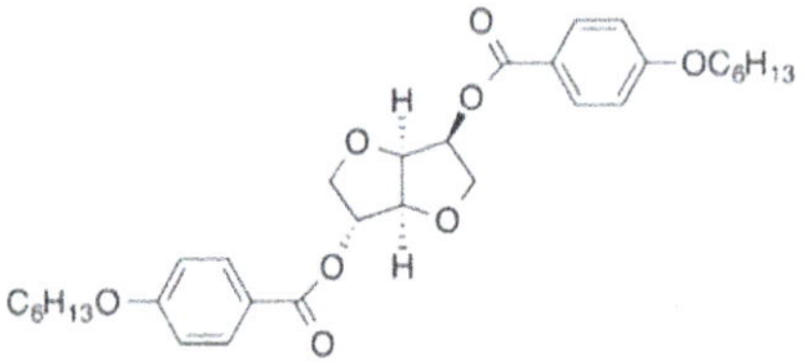

ISO-(6OBA)$_2$

Fig. 4.11 Figure 4.10 Molecular structures of **8PY8OCB** and **ISO-(6OBA)$_2$**

Yoshida and Ozaki et al. reported the expansion of the temperature range of blue phases by doping nanoparticles [67]. When spherical gold nanoparticles with an average diameter of 3.7 nm were added to a chiral liquid crystal mixture exhibiting BPII, the temperature range of the blue phase increased from 0.5 to 5 K. They proposed that nanoparticles accumulate along the lattice disclination lines, as presented in Fig. 4.12.

The technique of expanding the BP temperature range using nanoparticles was also applied to BPIII. CdSe nanoparticles, where the surface was treated with oleyl amine and trioctyl phosphine, were mixed with chiral liquid crystal (*S*)-4-(2-methylbutyl)phenyl 4-octylbiphenyl-4-carboxylate (**CE8**). The mixture exhibited a phase sequence of Iso–BPIII–BPII–BPI–N*–SmA* [68]. Figure 4.13 shows $T - T_{\mathrm{I\text{-}BPIII}}$ of **CE8** + CdSe mixtures as a function of the nanoparticle concentration

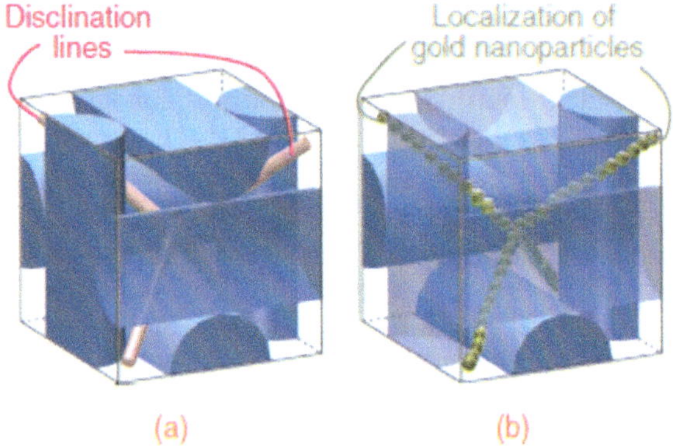

Fig. 4.12 **a** Schematic of the cubic BPII structure and **b** BPII stabilized by the nanoparticles trapped in the disclination lines. Reprinted figure with permission from [67]. Copyright (2009) IPO publishing

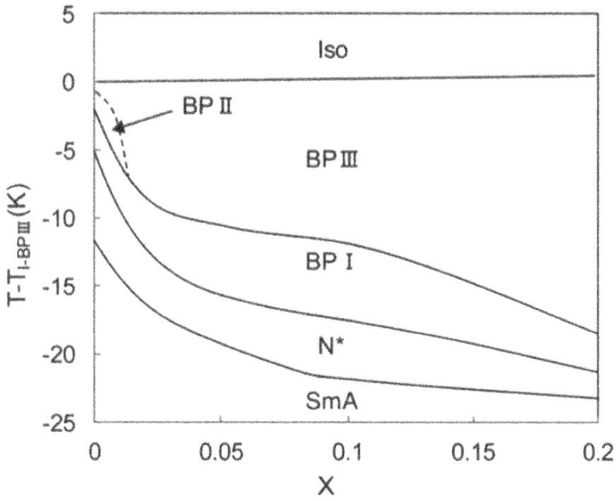

Fig. 4.13 *T*–*x* phase diagram for a mixture of **CE8** and CdSe obtained in cooling runs. Reprinted with permission from Ref. [68]. Copyright (2010) American Physical Society

defined as the mass ratio $x = m_{\mathrm{np}}/m_{\mathrm{lc}}$. The Iso–BPIII transition temperature remains almost the same as that of **CE8** for all the studied mixtures. The BPIII–BPI and BPI–N* transition temperatures decrease markedly with increasing x. The temperature range of BPIII is 20 K at $x = 0.20$.

Coles et al. demonstrated the fabrication of self-assembled three-dimensional nanostructures using polymer templating BPI [48], as described in Sect. 4.2. By refilling the cast with an achiral nematic liquid crystal, they created a template blue phase with unprecedented thermal stability in the range of −125 to 125 °C. This polymer templating method was also applied to stabilize BPIII [50].

Filling defects by polymers or nanoparticles makes it possible to freeze the blue phase structure without changing short-range order of the constituent molecules. However, this may affect the long-range correlations characteristic of the blue phase. To elucidate the relationship between structure and properties in the blue phase is necessary to systematically study BPLCs from the perspective of molecular design.

Molecular Design for Blue Phase Materials

Coles and Pivnenko reported that a nematic mixture of three homologues of a symmetric dimer doped with a small quantity of chiral additive **ISO-(6OBA)$_2$** shows BPI of an extremely wide temperature range (Fig. 4.14) [16]. The chiral mixture exhibited the following phase transition sequence upon cooling: Iso 57.72 °C BPIII 57.58 °C BPII 57.22 °C BPI 16.5 °C unidentified smectic phase; however, upon heating it exhibited the phase transition sequence: smectic–N*–Iso. The BPI was reportedly stable at room temperature for more than four months, irrespective of the sample thickness. Using symmetric and non-symmetric bimesogens, they prepared approximately 30 different mixtures that exhibit a blue phase in the temperature range of 40–50 K. These bimesogens were found to exhibit very high flexoelectro-optic ratios in the chiral nematic phase. The authors proposed that a large localized flexoelectric polarization, arising from orientations strain near the disclination line singularities in the blue phase, fixes lattice defects and stabilizes the blue phase. Macroscopically, because of symmetry, the net polarization will be zero when averaged around the singularity. Not only a dimeric structure but also fluorosubstitutents play an important role in stabilizing blue phases.

Overall molecular shape plays an important role in the BP stabilization. Non-symmetric chiral liquid–crystal dimers (**CBO*n*O.2MB**) show interesting odd–even effects of the number of methylene units in the spacer on their chirality-dependent phase transition behavior [69]. Figure 4.15 portrays the molecular structure and phase transition temperatures of **CBO*n*O.2MB**. Dimers with an odd-numbered spacer exhibited a phase sequence of Iso–BPI–N*–SmA, whereas those with an even-numbered spacer showed that of I–N*–SmA. The helical pitch for **CBO7O.2MB** is markedly smaller than that for **CBO6O.2MB** and **CBO8O.2MB**, which is consistent with the appearance of the BPI for **CBO7O.2MB**.

Since Nakata et al. discovered that doping with achiral bent-core molecules significantly broadens the temperature range of the blue phase [70], lots of bent-shaped compounds capable of stabilizing the blue phase have been designed. Choi et al. investigated the correlation between the elastic constants and thermal stability of BPI in chiral mixtures containing bent-core molecules, as shown in Fig. 4.16a [71]. Figure 4.16b shows the elastic constants K_{11} and K_{33} for conventional host nematic liquid crystal mixtures with varying concentrations of the bent-core molecules. When

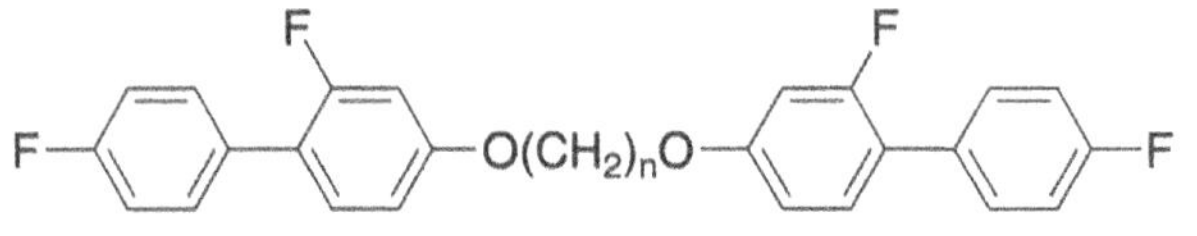

ISO-(6OBA)$_2$

n = 7: 30.4 wt%, *n* = 9: 35.1 wt%, *n* = 11: 30.6 wt%   3.9 wt%

Fig. 4.14 Chiral mixture consisting of symmetrical dimers doped with **ISO-(6OBA)$_2$**

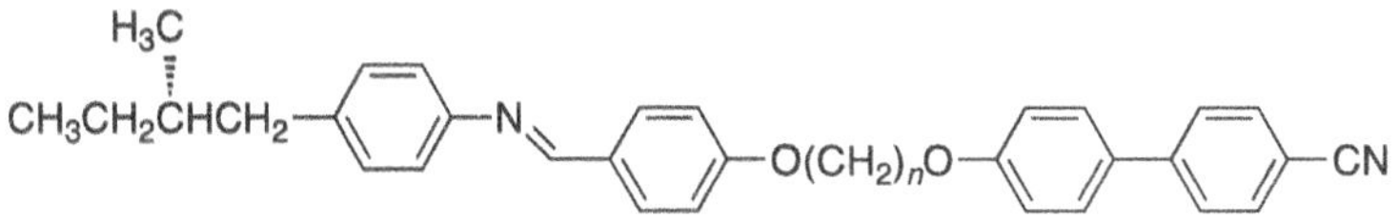

CBO*n*O.2MB

n = 6	Cry 137 °C (SmA 104 °C) N* 176 °C Iso
n = 7	Cry 97 °C (SmA 74 °C) N* 131.5 °C BPI 131.9 °C Iso
n = 8	Cry 128 °C SmA 143 °C)N* 165 °C Iso
n = 9	Cry 110 °C (SmA 102 °C) N* 132.2 °C BPI 132.8 °C Iso

Fig. 4.15 Phase transition temperatures (°C) on cooling for the nematic mixture doped with a high twisting chiral additive [69]

bent-core molecules are added to the host nematic LC, the elastic constant K_{11} (K_{33}) increases (decreases). Figure 4.16c portrays a phase diagram for the mixtures doped with 28 wt% of chiral compound **S-811** against the concentrations of the bent-core molecule. When K_{33} was larger than K_{11} as in **BC17** (17 wt% of the bent-core compound) and **BC20**, the temperature range of BPI was ~2 K. For **BC30**, K_{33} became much smaller than K_{11}, then the observed temperature range of BPI amounted to ~30 K. As K_{33} decreases relative to K_{11} the temperature range of BPI widens. These characteristics are well explained by the numerical study [58].

1,3,4-Oxazoline and thiophene have been used for the preparation of bent-core LCs [72–75]. Interestingly, some bent-core molecules stabilize the amorphous BPIII [76]. Yelamagad et al. designed LC dimers **USD-1-*n*** featuring an achiral bent core that is connected to a chiral, promesogenic cholesterol segment via a flexible spacer [77]. Their molecular structures and phase transition temperatures are presented in Fig. 4.17. The liquid-crystalline properties of the dimers depend significantly on the shape of the molecules' all-trans conformation as a result of the parity of the flexible spacer joining the two anisometric segments. **USD-1-3**, **USD-1-5**, and **USD-1-7**, which have even parity (the sum of n and C of CO) exhibit BPIII with a wide temperature range. Notably, **USD-1-3** exhibited enantiotropic BPIII with a temperature range of 22 K, whereas **USD-1-4**, which has odd parity, showed a phase sequence of Iso–Colr–Cr.

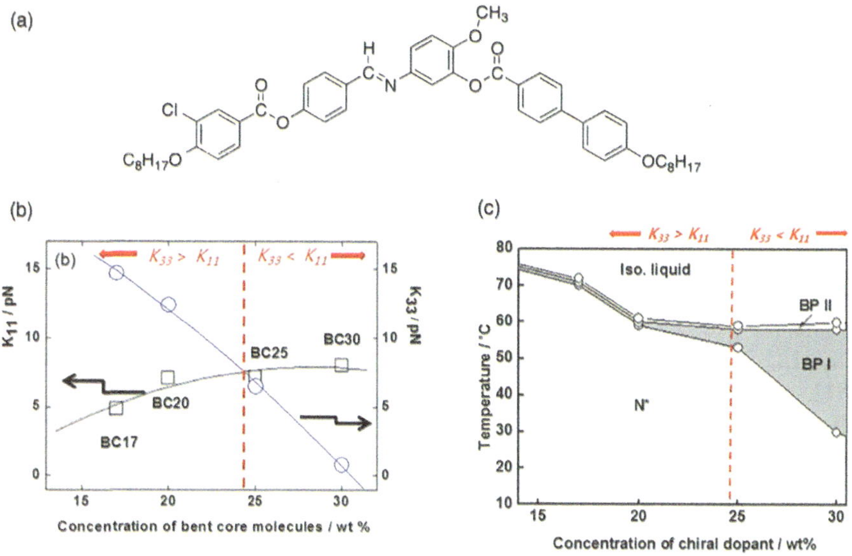

Fig. 4.16 **a** Molecular structure of the bent-core molecule. **b** Elastic constants K_{11} and K_{33} of the host NLC mixtures with different concentrations of the bent-core molecule, measured at the same reduced temperature (T_{NI-20} °C). **c** Phase diagram for the mixtures doped with 28 wt% of **S811** against the concentrations of the bent-core molecules. Reprinted with permission from [71]. Copyright (2011) Royal Society of Chemistry

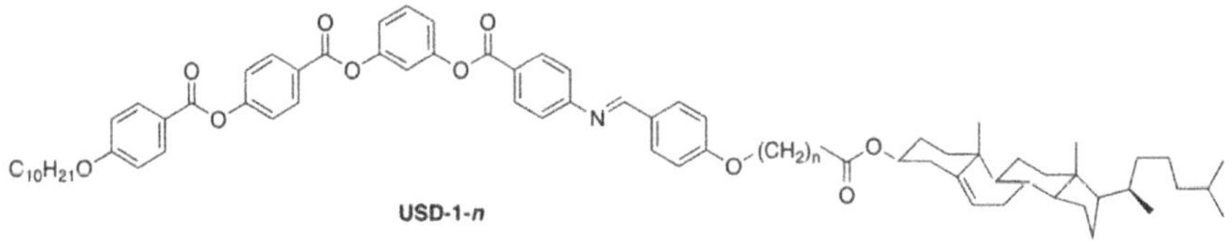

USD-1-*n*

USD-1-3: Iso 170.2 °C BPIII 148 °C N* SmX
USD-1-4: Iso 135.1 °C Col
USD-1-5: Iso 164 °C BPIII 148 °C N* 118.4°C Col
USD-1-7: Iso 151.3 °C BPIII 137 °C N* 114.7 Col

Fig. 4.17 Molecular structure, phase transition temperatures (°C), and enthalpies (J g^{-1}) of **USD-1-*n***. M_{X1} and M_{X2} are unknown tilted smectic mesophases. [a]Enthalpies of the BPIII to N* transition were too small to be detected. [b]A monotropic phase

Taushanoff et al. reported the stabilization of BPIII by achiral nematic bent-core liquid crystals (BCN LCs) (Fig. 4.18) [78]. BPIII was induced by doping the BCN LC with the chiral compound **ISO-(6OBA)$_2$**. When **ClPbis10BB** was doped with a small amount of the chiral compound, BPIII replaced the original nematic phase (see Fig. 4.19). Electro-optical switching was observed in the bent-core system in BPIII. The authors explained the stabilization of BPIII in BCN LCs as follows: The bent-core LC **ClPbis10BB** exhibits smectic clusters throughout the nematic region. These nanosized smectic clusters interrupt the long-range development of the BPI or BPII lattice structures, thereby randomizing their evolution and resulting in an amorphous structure (see Fig. 4.20).

Chiang and Lin et al. synthesized two chiral, bent-core 1,3,4-oxadiazole-based LCs (**OXD7*** and **OXD5B7F***) that display BPIII (34 K and 7 K), respectively as shown in Fig. 4.21 [79]. The authors noted that the HTP and the molecular biaxiality are important factors in stabilizing the BPIII.

Recently, Cordoyiannis et al. demonstrated that quantum dots can stabilize BPIII [80]. Khatun and Nari et al. reported the stabilization of BPIII using a twist–bend nematic LC and surface-functionalized quantum dots (QDs: CdSe (core)/ZnS (shell), which are tethered with octadecylamine) [81]. Adding a twist–bend nematic dimer **N_{TB} LC** (see Fig. 4.23) to a mixture of nematic liquid crystals and the chiral dopant **S-811** stabilizes cubic BPI (see Fig. 4.23a). Conversely, the mixture containing 20 wt% **Bent-core LC** (Fig. 4.22) exhibited cubic BPI with a thermal range of approximately 3.5 K, indicating that the **N_{TB} LC** is more effective in enhancing the thermal range of cubic BPI than the **Bent-core LC**. This is due to the NTB LC's ultralow bend elastic constant and inherent saddle-splay deformation. Doping with a small amount (0.005 wt%) of QDs reduces the free energy associated with the topological defects, leading to a complete transformation of the cubic BPI to an amorphous BPIII (Fig. 4.23b). The effect of QDs depends on the concentration of **S-811**. In the case of a low **S-811** concentration (e.g., 30–40 wt%), the addition of QDs has no effect on the BPIII thermal range.

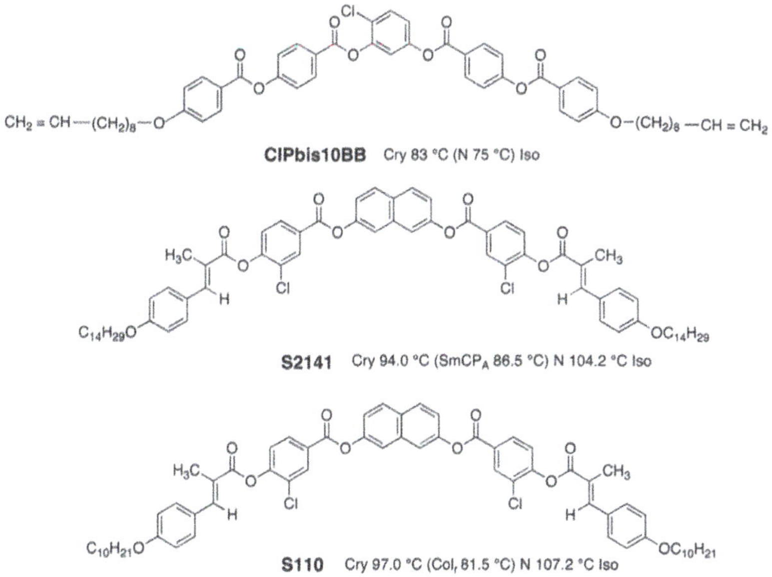

Fig. 4.18 Molecular structures and phase transition temperatures of the bent-core molecules

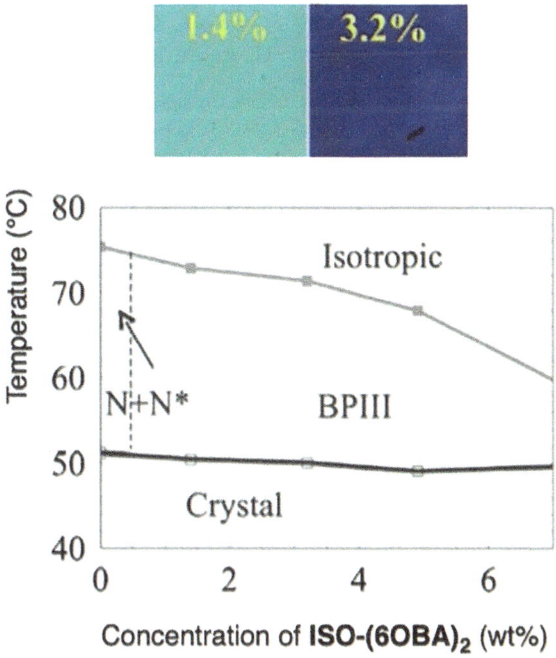

Fig. 4.19 Phase diagram of **ClPbis10BB** and **ISO-(6OBA)$_2$** on cooling. Optical textures of the BCN LC doped with the chiral compound (1.4 wt% and 3.2 wt%) are inserted. Reprinted with permission from [78]. Copyright (2010) Royal Society of Chemistry

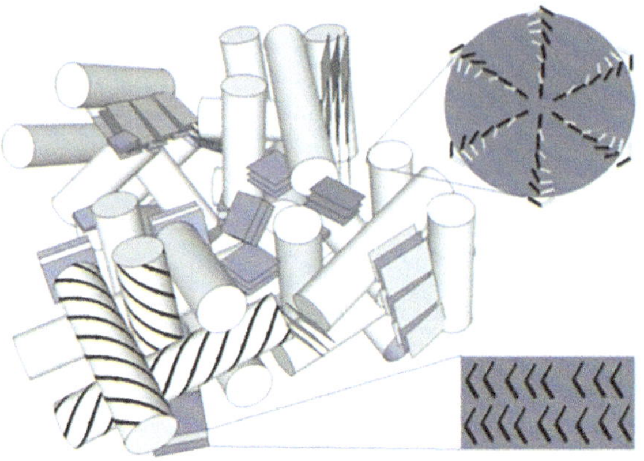

Fig. 4.20 Illustration of the smectic cluster-stabilized BPIII structure of BCN LCs. Reprinted with permission from [78]. Copyright (2010) Royal Society of Chemistry

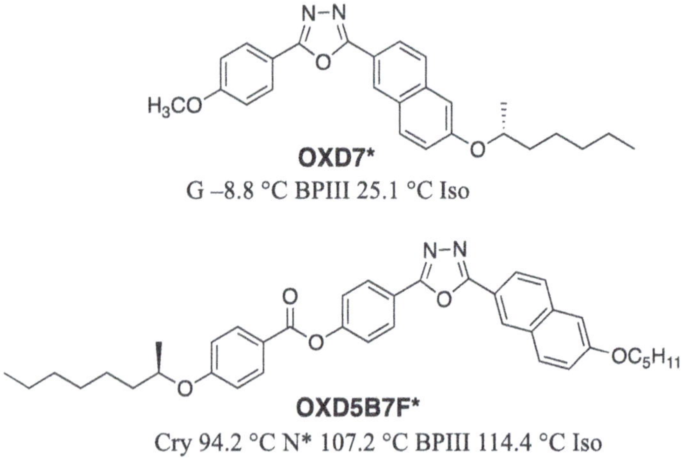

Fig. 4.21 Molecular structures and cooling phase transition temperatures **OXD7*** and **OXD5B7F***

It has been reported that cybotactic clusters stabilize the BP, even in polar bent-core LC systems (see Fig. 4.24) [82]. Bent-core compounds form a cybotactic structure in the N phase (Ncyb; see Chap. 2). When mixed with the room-temperature LC **5CB** and a chiral additive (**S-811**), they stabilize the blue phase at room temperature, achieving a temperature range of up to 22.9 K.

It is noteworthy that reports have been published on other systems, including those fluorine substituents [52, 83], rod-like systems [84], and hydrogen-bonding systems [17, 85]. As illustrate in Fig. 4.25, a self-assembled complex is formed by hydrogen bonding between a chiral fluorinated benzoic acid (**SFBA**) and a pyrimidine

Fig. 4.22 Molecular structures and phase transition temperatures of **N_{TB} LC** and **Bent-core LC**

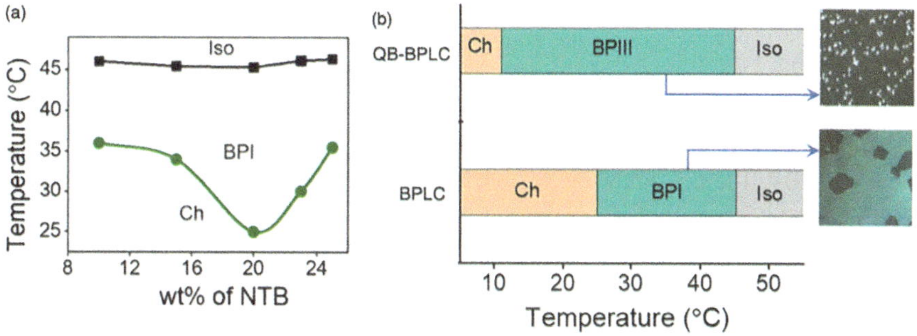

Fig. 4.23 **a** Dependence of **N_{TB} LC** concentration on the thermal range of cubic BPI. The concentration of **S-811** was kept constant at 45 wt%. **b** The phase sequence of the BPLC and QD-BPLC mixtures. The BPLC mixture consists of 45 wt% of **S-811**, 20 wt% of NTB LC, and 35 wt% of a commercial nematic mixture. The QB-BPLC mixture is a BPLC mixture doped with 0.005 wt% of QDs. Reprinted with permission from [81]. Copyright (2023) Royal Society of Chemistry

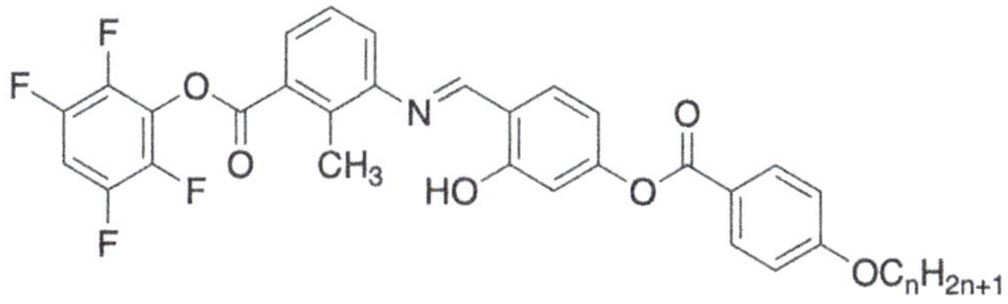

F4-8a Iso 108.0 N 91.0 Ncyb 80.9 SmC 69.2 Cry

F4-12a Iso 103.4 N 99.0 Ncyb 94.2 SmC 59.4 Cry

F4-14a Iso 102.3 N 98.3 Ncyb 92.0 SmC 62.0 Cry

Fig. 4.24 Molecular structures and phase transition temperatures on cooling of the polar bent-core LC **F4-*n*a**

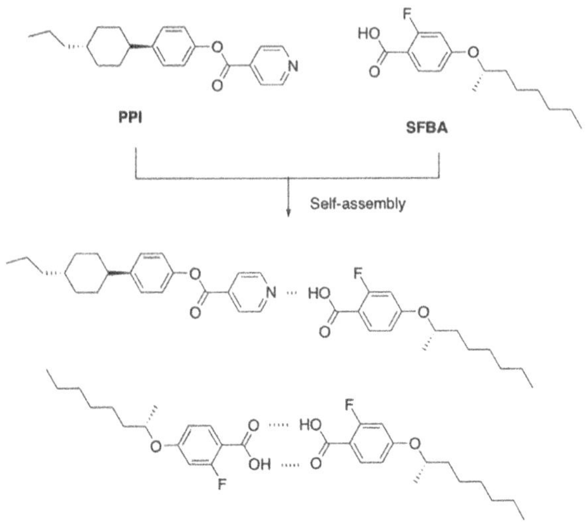

Fig. 4.25 Hydrogen-bonded self-assembled complex of chiral fluoro-substituted benzoic acid (**SFBA**) and pyrimidine derivative (**PPI**) [17]

derivative (**PPI**) [17]. A mixture of 66.7 mol% **SFBA** and 33.3 mol% **PPI** exhibited the widest BP II temperature range, reaching a maximum of approximately 23 K.

It is hypothesized that a substantial twisting power is imperative for the formation of blue phases [6]. According to Eq. 4.3, widening the BP temperature range necessitates a large helical twisting power. Conversely, our findings indicated an absence of a discernible correlation between HTP value and BP temperature range for specific binaphthyl derivatives [86, 87]. Takezoe et al. noted that a chiral dopant with a large HTP can easily induce BPs with a small dopant ratio, but that HTP values do not affect BP stabilization [88]. As posited by Alexander and Yeomans, the chiral nematic phase below the wide-temperature BPI undergoes helical twist inversion [54]. The question of how chirality stabilizes BPs is of particular interest. The following investigation will introduce the flexible supermolecular LC systems that produce a dynamic chiral origin in order to answer this question.

As demonstrated in Fig. 4.26, a chiral binaphthyl derivative, **BINAPTFT-9**, was found to exhibit a wide BP temperature range during a cooling process [89]. It is evident that this compound possesses an asymmetric axis within the binaphthyl unit. Moreover, the two mesogenic units have the capacity to form a twisted conformation. Upon undergoing a cooling process, compound **BINAPTFT-9** exhibited a platelet texture that is typically observed in BPs characterized by a cubic structure. Subsequently, a transition to a subsequent phase, designated SmX phase, occurred. The temperature range of the cubic BP is 32 K. The SmX phase exhibits a layer structure, with a layer spacing of 31 Å. The molecular length for the cisoid conformer is estimated through semi-empirical calculations with MOPAC-6/PM3 to be 35 Å. The SmX phase is hypothesized to possess a monolayered structure, in which the molecules form the cisoid conformer. Consequently, it is a rational expectation that the chiral compound **BINAPTFT-9** exists as a U-shaped cisoid conformer in the blue phase. It was observed that, upon subjecting the sample to heating during the X

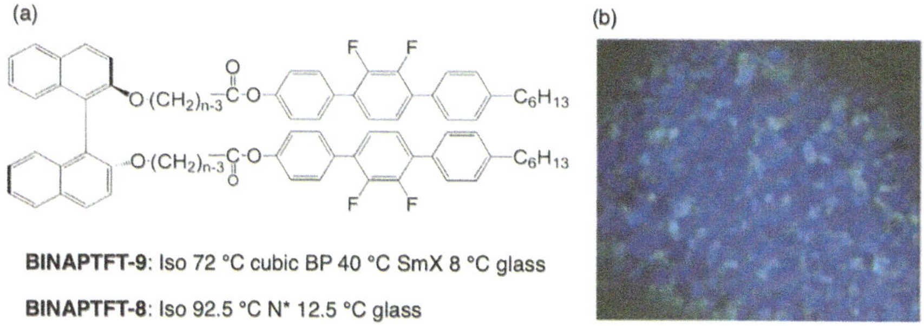

Fig. 4.26 **a** Molecular structure and transition temperatures on cooling of compounds **BINAPTFT-*n***. **b** Optical texture of compound of compounds **BINAPTFT-9** in the cubic BP at 60 °C. Reproduced from [89] with permission from the Royal Society of Chemistry

phase, a chiral nematic (N*) phase became apparent. As the temperature increased, a transition to the blue phase occurred in proximity to the clearing temperature. It is interesting to note that the N* phase was not present during the cooling process. Conversely, compound **BINAPTFT-8** exhibited a phase sequence of Iso–N*–glass [87].

The phase transition behavior of the U-shaped compounds is consistent upon the parity of a methylene number in the spacer. The temperature-dependent helical pitches of a mixture of each chiral compound and a host nematic liquid crystal, 4-(4-hexylphenyl)-1-(4-propyloxyphenyl)-2,3-difluorobenzene [87], were investigated. The host nematic compound contains the same mesogenic unit as the chiral binaphthyl derivatives. The selection of the host material is predicated on expectation of favorable core–core interaction between a terphenyl moiety of the host LC and each chiral compound. The helical pitches in the chiral nematic phase of a mixture of each chiral compound and the nematic LC (50:50 mol%) were observed as a function of temperature, and the results are shown in Fig. 4.27. Compound **BINAPTFT-8** with even-numbered spacers induced left-handed (LH) helical structure with short helical pitches in the N* phase of the mixture with the host LC (Fig. 4.27a). On the other hand, compound **BINAPTFT-9** with odd-numbered spacers exhibited temperature-dependent helical twist inversion (Fig. 4.27b), which is in good agreement with the theoretical prediction [54]. **BINAPTFT-9** induced much longer helical pitches than **BINAPTFT-8**. Therefore, high twisting power is not a prerequisite for the appearance of a blue phase. Furthermore, the helical twist inversion suggests that **BINAPTFT-9** exhibiting a BP has two helical origins. It can be explained in terms of competition between the two twisting powers.

The present study explores the impact of the dynamic behavior of compound **BINAPTFT-9** on the manifestation of the blue phase. In the blue phase, the compound **BINAPTFT-9** is believed to exist as a cisoid form possessing two origins of the twisting power, the binaphthyl axis and the twisted conformation of the mesogenic units (Fig. 4.28a). The former is defined as molecular chirality, while the latter is characterized as dynamic chiral origin. As the system cools to the SmX phase,

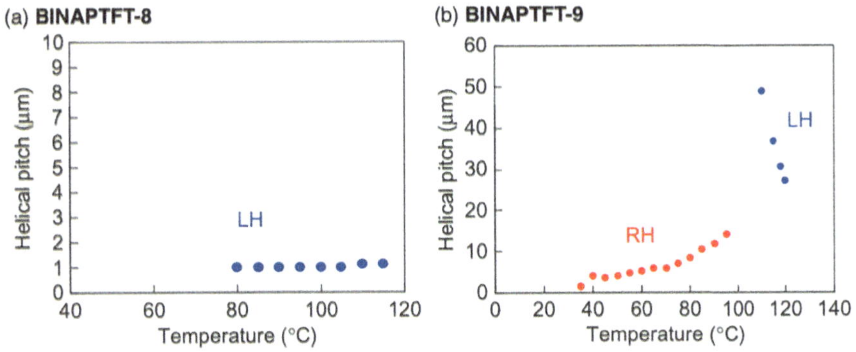

Fig. 4.27 Temperature dependency of N* helical pitch values induced by each chiral compound in mixtures with 4-(4-hexylphenyl)-1-(4-propyloxyphenyl)-2,3-difluorobenzene. Reprinted with permission from [87]. Copyright (2011) Taylor & Francis

the mesogenic units in the cisoid form exhibit a preference for parallel alignment due to the favorable packing in each layer, as illustrated in Fig. 4.28b. During the heating process from the SmX phase, the molecule may adopt a cisoid configuration, wherein the two mesogenic units remain aligned in parallel (Fig. 4.28c). The cisoid conformer is incompatible with two helical origins. Consequently, the N* phase emerges in place of the blue phase. On further heating, the two mesogenic cores can result in the formation of a twisted conformation in the cisoid form (Fig. 4.28d), thereby inducing the N*–BP transition. We posit that competition between the two helical twisting powers destabilizes the formation of a single twist structure, thereby producing a blue phase [90].

The subsequent inquiry pertains to the underlying mechanism by which the U-shaped molecule contributes to the stabilization of the BP. It is hypothesized that the formation of double-twist structures can be stabilized by a molecule that produces more than two helical helices instead of a single helix. Typically, a rod-like molecule will rotate freely around its long axis. Consequently, all short axes perpendicular to the long axis may be indistinguishable. A rod-like molecule is characterized by the presence of a single, clearly identifiable short axis. The incorporation of chirality into a rod-like nematic molecule results in the formation of a helical axis along the short axis within the chiral nematic phase, as illustrated in Fig. 4.29a. Conversely, a board molecule may possess two distinguishable axes. The interaction between chirality and the two short axes, which have been observed to stabilize double-twist structures (Fig. 4.29b).

Tanaka and Yoshizawa prepared a homologous series of catechol derivatives (**I-*n***) and one of biphenyl derivatives (**II-*n***) (Fig. 4.30) [91]. The phase transition behavior of each U-shaped compound doped with 10 wt% of a chiral dopant, **ISO-(6OBA)$_2$**, was investigated. The emergence of the blue phases was examined in the context of helical twisting power, elasticity, and molecular biaxiality.

As illustrated in Fig. 4.31a, the relationship between HTP and the BP temperature range for **I-*n***, **II-*n***, and **III** is evident. As the spacer length increases, the HTP for

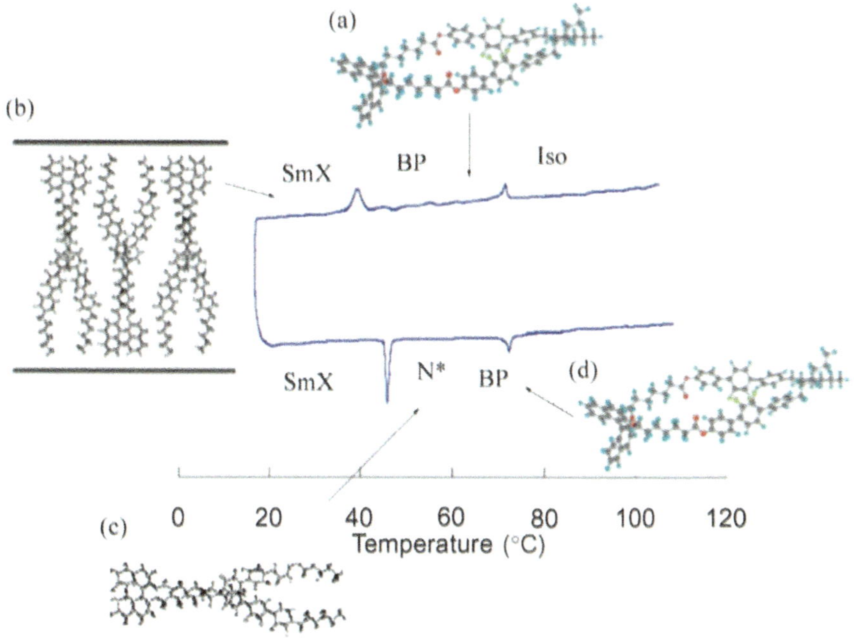

Fig. 4.28 Molecular organization models for the BP, SmX and N* phases of compound **BINAPTFT-9**. Reprinted with permission from [87]. Copyright (2011) Taylor & Francis

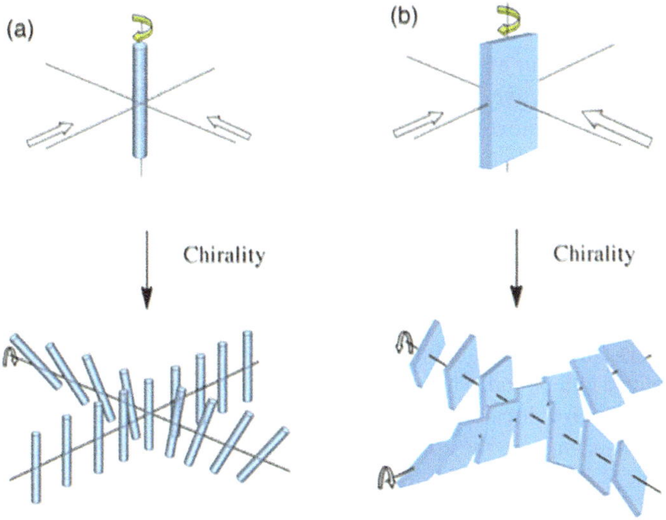

Fig. 4.29 Helical structures of **a** a rod-like molecule and **b** a board molecule. Reproduced from [91] with permission from the Royal Society of Chemistry

Fig. 4.30 Molecular structures of U-shaped compounds **I-*n*** and **II-*n***, and the corresponding monomeric compound **III** [91, 92]

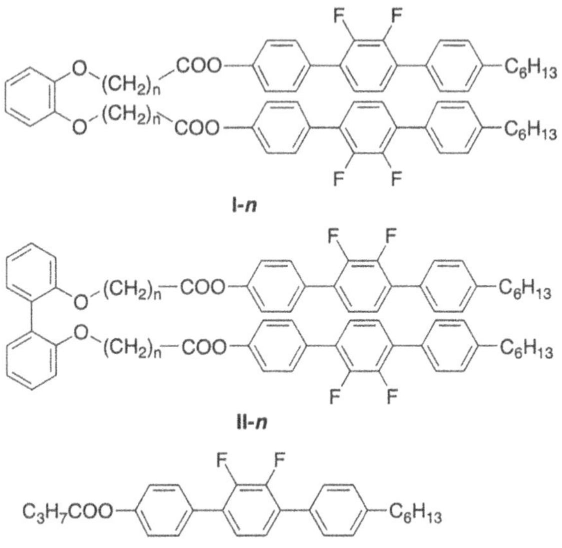

the **I-*n*** series increases exhibits a continuous increase, while the HTP for **II-*n*** series demonstrates a continuous decrease. These results suggest a discrepancy in chirality transfer between **I-*n*** and **II-*n*** series. In the context of **I-*n*** series, the transfer of chirality transfer is believed to occur from the chiral dopant to the mesogenic units. The twisting power for **I-7** is nearly equivalent to that of **III**. Conformational freedom of each mesogenic unit of **I-*n*** has been observed to increase with increasing spacer length. The most effective chirality transfer may occur for **I-7** with the longest spacer. Conversely, in the context of **II-*n*** series, the transfer of chirality is believed to occur not only from the chiral dopant to the mesogenic units but also from the chiral dopant to the biphenyl unit. Increasing spacer length has been shown to enhance conformational freedom of the biphenyl unit, thereby diminishing the efficacy of chirality transfer from the chiral dopant to the biphenyl unit. It is particularly intriguing that the stability of the blue phases in the U-shaped system cannot be explained in terms of the HTP. The researchers observed splay (K_{11}) and bend (K_{33}) elastic constants in an N phase of a host nematic liquid crystal (JNC-0001) doped with 10 wt% of each U-shaped compound. The induction of a marked decrease in K_{33} was observed in both **I-6** and **II-6**. It can be posited that a decrease in K_{33} stabilizes the BP. This finding is consistent with the effects of elastic constants on stabilizing BP for a bent-core system [71]. The mixture containing **II-8** also exhibited the same K33 value as the mixture containing **I-6**. However, **II-8** did not stabilize any blue phase. It is noteworthy that rigid board molecules may possess two short axes, as illustrated in Fig. 4.29b. It is hypothesized that such a molecule possesses two molecular planes that are mutually perpendicular. Each plane possesses a specific length to breadth ratio, denoted by *L*/*D*. Taking *I-6* as an example, it has a plane with a short axis parallel to the phenyl ring and a plane with a short axis perpendicular to it. The L/D ratio for the former is 2.14, and that for the latter is 4.01. We define $(L/D)_{large}/(L/D)_{small}$ as the biaxial parameter. As illustrated in Fig. 4.31b, a direct correlation exists between the biaxial parameter and the BP temperature range for the **I-*n*** and **II-*n*** series. An intriguing discovery emerged from the study: a U-shaped compound, characterized by the largest biaxial parameter, elicited the most extensive BP range within the chiral mixture among each series. This observation lends further credence to the hypothesis that the interplay between chirality and molecular biaxiality serves to stabilize the double-twist structures of BPs.

The U-shaped compounds have been found to stabilize cubic BP more effectively than BPIII. The stabilization of amorphous BPIII was initially achieved through the use of a chiral T-shaped compound (**T-1**) [93] (Fig. 4.32). Compound **T-1** demonstrated BPIII with a temperature range of 13 K during the cooling process. Sato and Yoshizawa's seminal contribution to the field involved the introduction of a polar group into a chiral T-shaped system, a maneuver that enables the coupling of the molecule with an electric field. This step was followed by the meticulous preparation of compound **T-2**, which possessed a terminal cyano group and exhibited BPIII and N* on cooling, as depicted in Fig. 4.32. The results demonstrated high-contrast

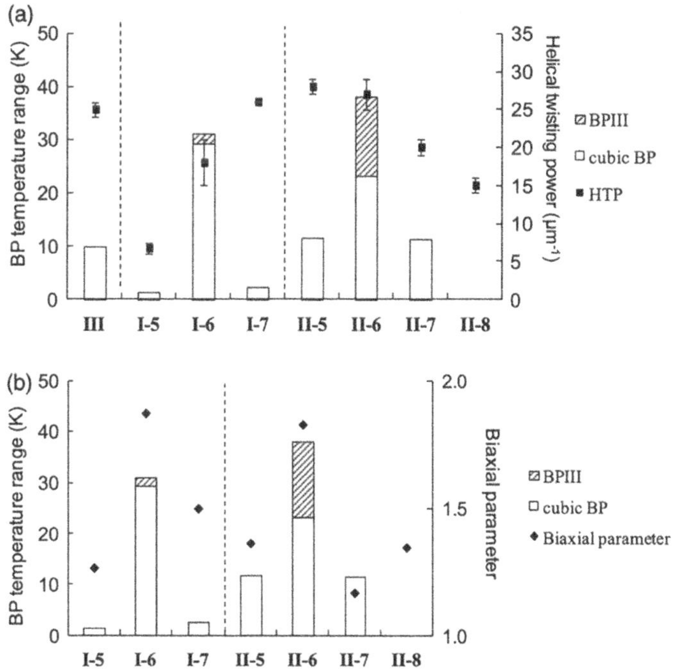

Fig. 4.31 **a** Relation of helical twisting power and the BP temperature range for **I-*n***, **II-*n***, and **III**. **b** Relation of the biaxial parameter and BP temperature range for **I-*n*** and **II-*n*** series. Reproduced from [91] with permission from the Royal Society of Chemistry

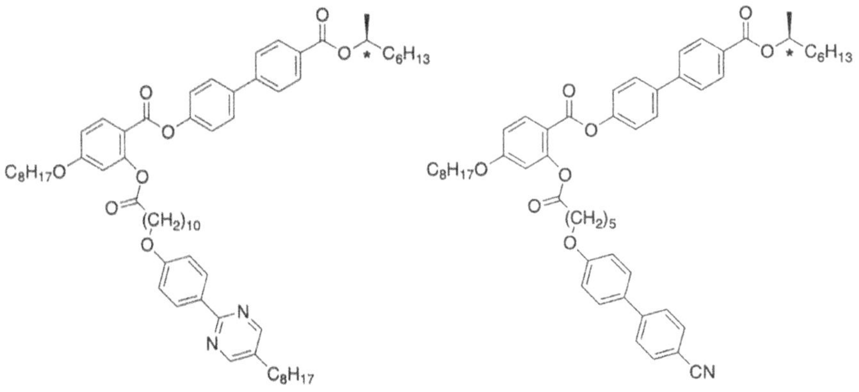

Fig. 4.32 Molecular structures and phase transition temperatures of **T-1** and **T-2**

electro-optical switching in BPIII [29]. T-shaped compounds shave been shown to stabilize BPIII, and the effect of BPIII stabilization is explained in terms of molecular biaxiality [9, 93]. A dielectric study of **T-2** in Iso and BPIII revealed that two different modes of rotation around the short axis can exist [94], thereby supporting the inference that the molecular biaxiality stabilizes DTCs in the BPIII.

In addition, certain H-shaped compounds induced BPIII in their mixtures doped with 15 wt% of **ISO-(6OBA)$_2$** [95]. Their molecular structures and transition temperatures are illustrated in Fig. 4.33. Mixture A, containing **H-1**, exhibited a cooling phase sequence of Iso 89.6 °C BPIII 75.5 °C N* 16.8 °C. Mixture B, which contains **H-2**, exhibited that of Iso 36.7 °C BPIII 5.5 °C glass. To facilitate a comparative analysis of the oligomers and their corresponding monomers, a mixture of 85 wt% **CPOB** (see Fig. 4.33) and 15 wt% **ISO-(6OBA)$_2$** was prepared. This chiral mixture (Mixture C) displays a phase sequence of Iso 202.4 °C cubic BP 195.5 °C N* 49.2 °C Cry on cooling. The optical textures of BPIII for Mixture B and cubic BP for Mixture C are shown in Fig. 4.34. Compounds **H-1** and **H-2**, due to their increased flexibility, result in a substantially broader range of BP temperatures in comparison to the monomeric compound **CPOB**. Specifically, they have been shown to stabilize amorphous BPIII. The flexibility plays an important role in stabilizing BPIII. It is hypothesized that the chiral dopant induces a transiently twisted conformation of each H-shaped compound in the BPIII, as illustrated in Fig. 4.35. The structure–property relationship in T-shaped and H-shaped compounds indicates that molecular biaxiality and flexibility are critical factors in the formation of BPIII with DTCs in a random and tangle configuration. The molecular biaxiality stabilizes the DTCs. The inherent disorder of a flexible molecule may, in itself, induce the entropy-driven tangled arrangement of the DTCs in amorphous BPIII.

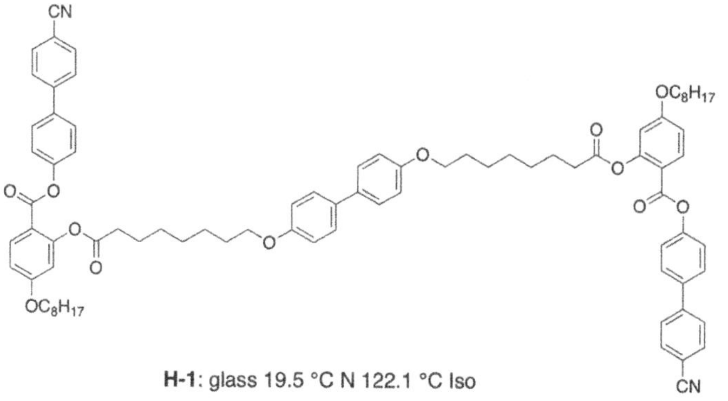

H-1: glass 19.5 °C N 122.1 °C Iso

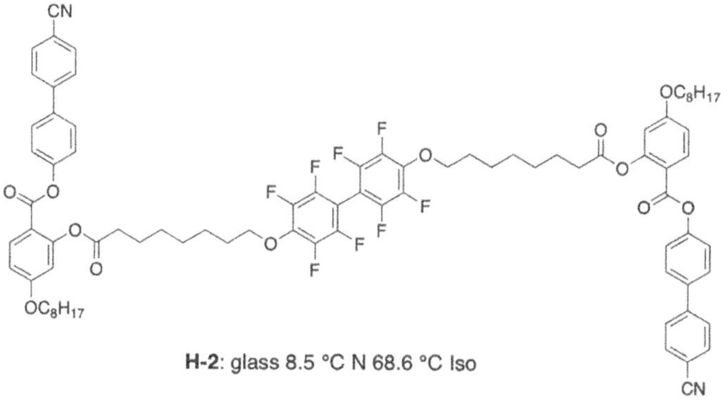

H-2: glass 8.5 °C N 68.6 °C Iso

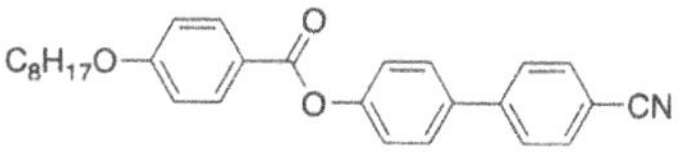

CPOB: Nre 118.6 °c SmA 195.8 °C N 233.9 °C Iso

Fig. 4.33 Molecular structures and phase transition temperatures on cooling of **H-1**, **H-2**, and **CPOB**

4.4 Electro-Optical Effects

In the context of electric field effects in cubic blue phases, the application of an external field can induce three distinct transformations. The first transformation is a rapid reorientation of the molecules due to the electro-optic Kerr effect. The second transformation is a distortion of the lattice, known as electrostriction. The third transformation is a phase transition to lower symmetry phases [6]. The application of an electric field has been demonstrated to induce a tetragonal blue phase, designated as BPX, as well as a two-dimensional hexagonal blue phase (BPH^{2D}) and a three-dimensional hexagonal blue phase (BPH^{3D}) [6, 7]. In the presence of larger electric fields, the blue phases may undergo a transformation to the chiral nematic phase

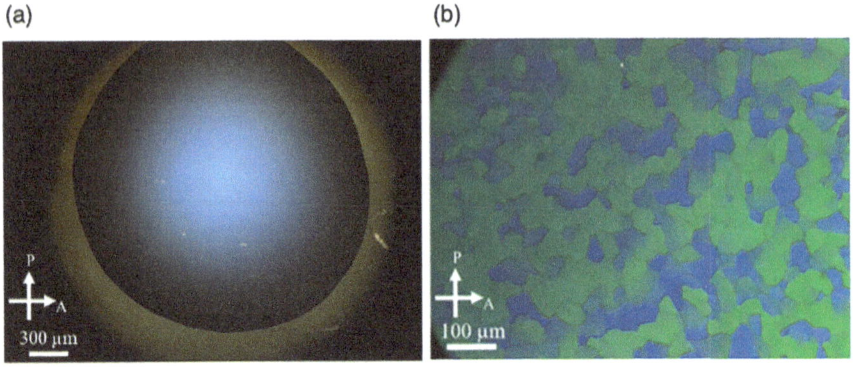

Fig. 4.34 Optical textures of **a** BPIII of mixture B at 23.4 °C and **b** cubic BP of Mixture C at 196.7 °C under crossed polarizers. Reproduced from [95] with permission from the Royal Society of Chemistry

Fig. 4.35 Model for the induced transiently twisted conformation of compound **H-1**. Reproduced from [95] with permission from the Royal Society of Chemistry

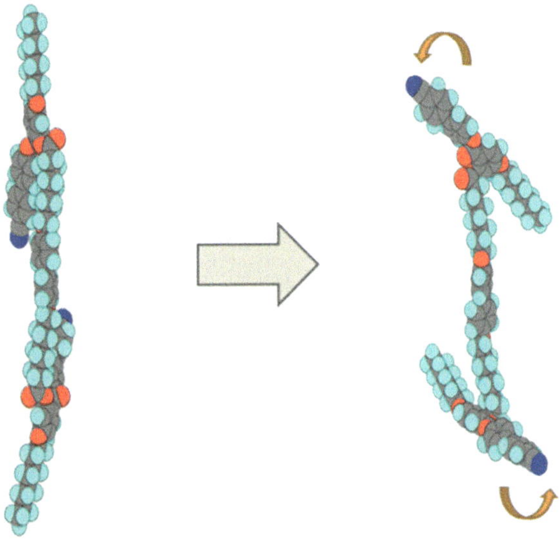

and ultimately to the nematic phase. The primary focus of this chapter is electro-optical switching for application to display devices. A polymer-stabilized BPLCD mode using the Kerr effect has undergone extensive development, however, a practical BPLCD has yet to be obtained due to unresolved critical problems such as high voltage and voltage-dependent hysteresis. The development of both devices and materials had been undertaken to address these challenges. The present study introduces the numerical analysis of electric-field-induced switching in cubic blue phases. Subsequently, recent advancements in BPLCDs will be presented.

4.4.1 *Numerical Study of Electric-Field-Induced Switching in Blue Phases*

Tiribocchi et al. conducted a numerical analysis of the steady states and switching hydrodynamics of BPI and BPII cells subjected to an electric field [96]. The observation of three distinct regimes was made when the field was operational, as illustrated in Fig. 4.36. In the presence of very weak fields, accompanied by strong anchoring at the boundaries, the blue phases demonstrated minimal alteration. For intermediate fields, the disclinations underwent a twisting (BPI) or unzipping (BPII) phenomenon. For very large voltages, the network underwent dissolution within the bulk of the cell. A particularly salient finding of this assessment is that a BPII cell can recover its original structure when the field is switched off, whereas a BPI cell was found to be more readily trapped more easily into metastable configurations. Consequently, BPI is anticipated to manifest a more pronounced hysteresis in voltage-dependent transmittance in comparison to BPII.

Fukuda and Žumer investigated the relaxational dynamics of the orientational order of BPI in a planar cell. They enforced normal alignment in response to the application and cessation of an external electric field [97]. It was demonstrated that the dynamics of the disclination lines in the liquid crystal are contingent on the field strength and the history of the application of the electric field. When the field is not strong, the disclination networks inherent in BPI undergo a reduction in size along the field direction without undergoing a change in its topology. However, the presence of a stronger field gives rise to complex reorganization processes of disclination lines, resulting in a final profile that may manifest as disclination lines of varying shapes localized around the midplane of the cell or a uniform alignment along the field direction devoid of disclination lines. Furthermore, the relaxation dynamics of the cell when the field is switched off were studied. The relaxation process and the resulting profile are contingent on the initial profile of the cell immediately preceding the cessation of the field. In the context of a compressed disclination network of BPI under weak field conditions ($E = 10\ \text{V}\ \mu\text{m}^{-1}$), the network exhibits a reversible switching behavior between two states, thereby recovering the original profile. However, the orientation profiles are found to be strongly influenced by a stronger field ($E = 20\ \text{V}\ \mu\text{m}^{-1}$), exhibiting nontrivial and complex reorganization processes. These processes are found to be dependent on the initial profile of the relaxation.

4.4.2 *Displays*

In comparison to conventional nematic liquid crystal displays, BPLCDs offer several distinct advantages [27, 98]: (1) No surface alignment is required; (2) the response time is in the submillisecond range; (3) the viewing angle is wide and symmetric; and (4) cell-gap insensitivity is exhibited. As stated in the review by Wu et al. [26],

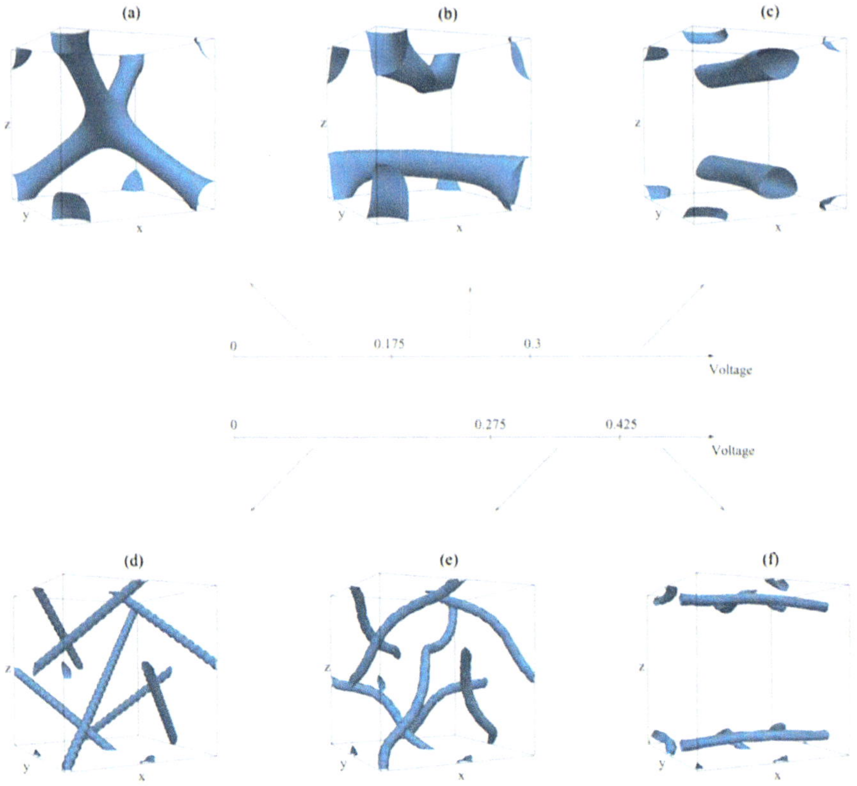

Fig. 4.36 Steady state disclination networks for BPI and BPII under a varying field, and strong anchoring of the zero-field structure on the surface. BPII structures are shown in the first row, BPI are in the bottom row. Reprinted with permission from [96]. Copyright (2011) Royal Society of Chemistry

the display performances, such as contrast ratio, response time and driving voltage, are determined by several parameters. Achieving a high contrast ratio necessitates the presence of both a high bright state, attributable to large birefringence, and a pronounced black state. When a small electric field (E) is applied to a BPLC composite, the LC molecules tend to be reoriented parallel to the electric field if $\Delta\varepsilon > 0$, and perpendicular to the electric field if $\Delta\varepsilon < 0$, where $\Delta\varepsilon$ is the dielectric anisotropy of the LC. The molecular reorientation that occurs as a result of this process leads to a phenomenon of birefringence in BPLC. That induced birefringence can be described by the Kerr effect as

$$\Delta n_{ind} = \lambda K E^2 \tag{4.4}$$

where λ is the wavelength and K is the Kerr constant. The Kerr constant is proportional to the square of the coherence length of the molecules (Fig. 4.37). As the electric field strength increases, two additional effects manifest: electrostriction, which involves the distortion of the cubic blue phase lattice; and a phase transition to a chiral nematic phase, ultimately resulting in the nematic phase [5, 99–104]. The Kerr effect is characterized by its rapid response time. In contrast, the subsequent effects, namely the lattice distortion and phase transition, exhibit a significantly slower response time. The lattice distortion takes tens of milliseconds, while the phase transition takes several seconds. In the absence of significant field strength, the phase transition can be disregarded. In the case of the polymer-stabilized BP(PS-BP), it is hypothesized that the polymers present within the defects contribute to the preservation of the cubic structure in the presence of an electric field. the polymers existing in the defects are thought to maintain the cubic structure against the electric field. Consequently, reorientation of molecules is attributed to the Kerr effect. A large Kerr constant can achieve large induced birefringence for a high contrast ratio.

The Kerr constant can be approximated by Eq. 4.5 [99]:

$$K \sim \frac{\Delta n_{ind}}{\lambda E^2} \approx \Delta n \cdot \Delta \varepsilon \left(\frac{\varepsilon_0 P^2}{k\lambda (2\pi)^2} \right) \tag{4.5}$$

where Δn, $\Delta \varepsilon$, and k respectively represent the birefringence, dielectric anisotropy and elastic constant, and where P is the pitch length for a host BPLC. While nearly all BPLCs used for display applications exhibit a positive Kerr constant, a polymer-stabilized BPLC with a negative Kerr constant was reported [105].

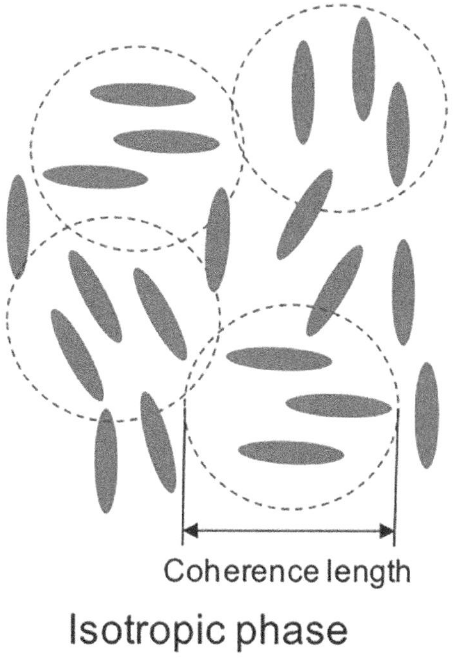

Fig. 4.37 Molecules with coherence length in the isotropic phase. Reproduced from [9] with permission from the Royal Society of Chemistry

According to Wu et al. [26], the temperature dependence of the Kerr constant is described as follows: Assuming the pitch length is independent of the temperature, the temperature-dependent Kerr constant can be described as shown in Eq. 4.6.

$$K \approx \Delta n \cdot \Delta\varepsilon\left(\frac{\varepsilon_0 P^2}{k\lambda(2\pi)^2}\right) \sim \Delta n_0 S \cdot \left(\frac{S}{T}\right)\cdot\left(\frac{1}{S^2}\right)\cdot\left(\frac{\varepsilon_0 P^2}{\lambda(2\pi)^2}\right) \sim \alpha \cdot \left(\frac{1}{T}\right) \tag{4.6}$$

The Kerr constant is directly proportional to the reciprocal temperature, and α is the constant that is material-dependent. To ensure the fulfillment of the boundary condition at the clearing point (T_c), the Eq. 4.6 is rewritten as Eq. 4.7. The Kerr constant exhibits a linear decrease in proportion to reciprocal temperature, reaching a state of nullification as the temperature attains the clearing temperature.

$$K = \alpha \cdot \left(\frac{1}{T} - \frac{1}{T_c}\right) \tag{4.7}$$

In the context of optically isotropicity, blue phases are characterized by this property. However, Bragg reflection in the visible range produces a platelet texture. In the case of BPI, the lattice parameter corresponds to one helical pitch. Therefore, the helical pitch should be sufficiently short to maintain Brag reflection in the UV region. To achieve a large Kerr constant, materials with large $\Delta n\Delta\varepsilon$ should be obtained.

One benefit of BPLC is its fast response time. According to Gleeson and Coles [104], rise and decay times can be expressed respectively as Eqs. 4.8 and 4.9.

$$T_{rise} = T_{decay}/\{(V/V_c)^2 - 1\} \tag{4.8}$$

$$T_{decay} = \gamma_1/(E_c^2\varepsilon_0\Delta\varepsilon) \tag{4.9}$$

Here, V_c is the critical voltage to unwind the chiral nematic pitches, $\gamma 1$ represents the rotational viscosity, and E_c is the critical electric field. Equation 4.9 is rewritten as follows.

$$T_{decay} = \gamma_1 P^2/\left(k\pi^2\right) \tag{4.10}$$

In that equation, k stands for the twist elastic constant of BPLC. Therefore, to decrease response times, BPLC should have low V_c, low γ_1, short P, and large k. However, shorter P may need higher V_c.

To achieve a low-power-consumption display, low operating voltage is an important factor. The operating voltage (V_{on}) is governed by both the device structure and material as follows [25].

$$V_{on} = A/(K)^{1/2} \tag{4.11}$$

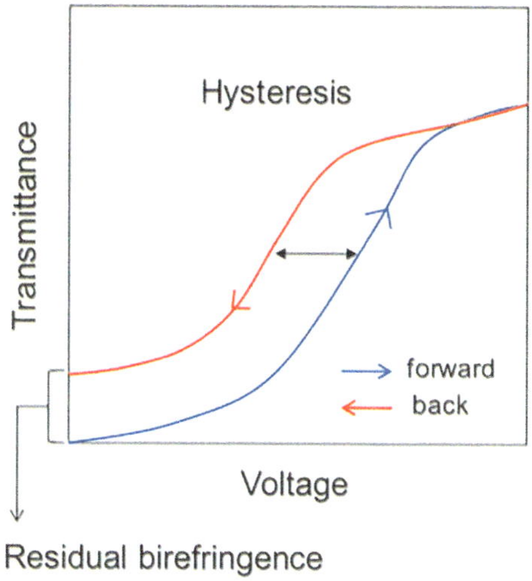

Fig. 4.38 Voltage-dependent transmittance. Reproduced from [9] with permission from the Royal Society of Chemistry

Therein, A denotes a device parameter that exhibits a high degree of dependency on the electrode configuration. K stands for the Kerr constant of the BPLC. A substantial Kerr constant has the capacity to reduce the operating voltage. The configuration of the device plays a pivotal role in the reduction of the operating voltage.

It has been posited that high operating voltage, hysteresis [28], and residual transmittance are critical issues. Figure 4.38 shows hysteresis and residual birefringence in a voltage-transmittance curve. The phenomenon of hysteresis has been demonstrated to exert an influence on the precision of gray-scale control. As delineated in Sect. 4.4.1, when a high electric field is applied, the disclination lines do not revert to their original states. In the event that the electric field is not uniform between the electrodes, a higher field may be present in specific local regions. The higher electric field may induce lattice deformation, thereby resulting in hysteresis.

A plethora of endeavor have been undertaken to ameliorate the issues associated with BPLCD, encompassing the development of novel devices and materials. Of particular significance is the reduction of operating voltage. In the event that the voltage is less than 10 V, the electrostriction effect, which is responsible for causing lattice distortion, would be prohibited. Consequently, the hysteresis would be less than 1% [26].

The most effective approach to reducing an operating voltage is to develop the device structure. The evaluation of electro-optical effects in BPs typically involves the use of a sample positioned between the comb-type interdigitated electrodes operating in an In-Plane Switching (IPS) mode, as depicted in Fig. 4.39.

However, a non-uniform electric field distribution occurs in the IPS cell. This requires a higher applied voltage and induces hysteresis near the electrodes. Two major approaches to improving the device configuration have been undertaken: (1) vertical field switching (VFS) [106] and (2) fringe field switching (FFS) [107, 108]. Figure 4.40a, b show their schematic illustrations. Wu et al. used the same material

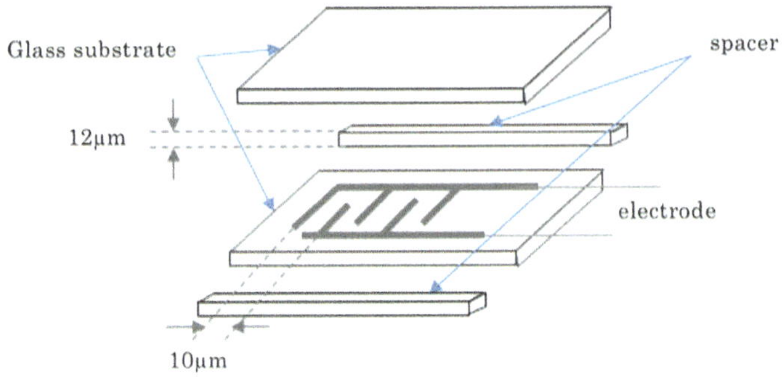

Fig. 4.39 IPS cell geometry for measurements of electro-optical effects in BPs. Reproduced from [9] with permission from the Royal Society of Chemistry

for both IPS and VFS cells and compared their electro-optical properties [106]. The liquid crystal host (JC-BP01M) has a birefringence Δn ~ 0.17 and $\Delta\varepsilon$ ~ 94. After UV stabilization, the blue-phase temperature range is from < 0 °C to approx. 70 °C. Its Kerr constant is approximately 13 nm V^{-2} at 23 °C with a wavelength of 633 nm. Figure 4.41 shows the voltage-dependent transmittance (*VT*) curves measured for the IPS and VFS cells. For the IPS cell, the peak voltage occurs at *V*p of approximately 50 V_{rms}. For the VFS cell, its *V*p is approximately 16 V_{rms}, which is about 3.2 times lower than that of the IPS structure. Hysteresis is defined as the voltage difference (ΔV) between forward and backward scans at half-maximum transmittance. The measured $\Delta V/V$p is approximately 5.8% for the IPS cell, whereas hysteresis-free switching was achieved for the VFS cell. However, the narrow viewing angle is problematic.

From a material design perspective, developing LCs with a large Kerr constant is effective for reducing voltage and achieving high contrast. Recently, Wu et al. demonstrated a BPLC host (JC-BP06N; JNC Petrochemical Corp.) with $\Delta\varepsilon$ ~ 430 and estimated Δn ~ 0.17 [109]. The measured Kerr constant was 33.1 nm V^{-2} in an IPS cell at room temperature, the highest reported value. It was 20.9 nm V^{-2} in a VFS cell at $\lambda = 514$ nm. According to Eq. 4.6, increasing the pitch length is a simple way to increase *K*. However, Bragg reflection must be maintained in the UV region (~350 nm) to achieve a high contrast ratio. LC birefringence is primarily determined by conjugation length. Increasing the conjugation length results in high viscosity and decreased material stability. Therefore, the most effective approach is to increase the dielectric anisotropy. Utilizing a 3.2 μm VFS cell with a BPLC of such a huge $\Delta\varepsilon$, remarkable outcomes were attained, including an operation voltage of approximately 8.4 V_{rms}, a rise time of ~460 μs, and a decay time of ~950 μs at ambident temperature (approximately 22 °C), along with hysteresis of ~0.9%.

The incorporation of nanoparticles [110, 111], nanorods [112], and developing reactive monomers [113] into PSBPLC through doping has been demonstrated to be an effective strategy for reducing driving voltage and mitigating hysteresis. Yang and colleagues presented their findings on low voltage and hysteresis-free blue-phase liquid crystals dispersed by ferroelectric nanoparticles [111]. Ferroelectric

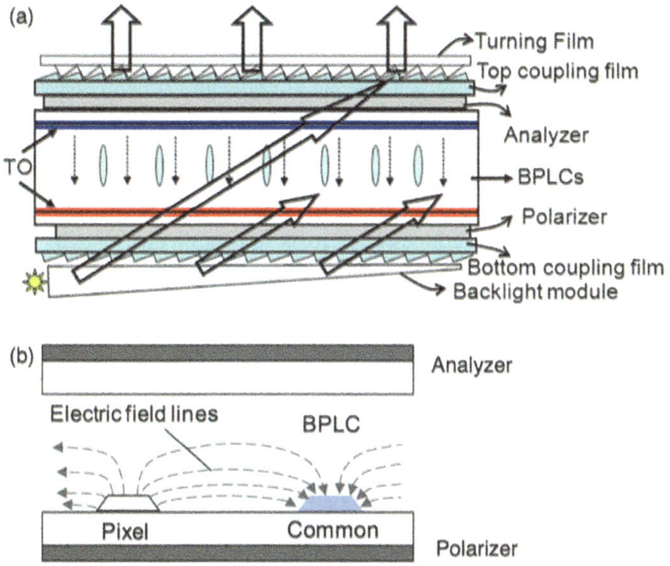

Fig. 4.40 **a** Vertical field switching mode proposed by Wu et al. [106]. Reprinted with permission from [106]. Copyright (2011) AIP publishing. **b** Fringe field switching mode

Fig. 4.41 Measured VT curves and hysteresis of IPS and VFS cells Reprinted with permission from [106]. Copyright (2011) AIP publishing

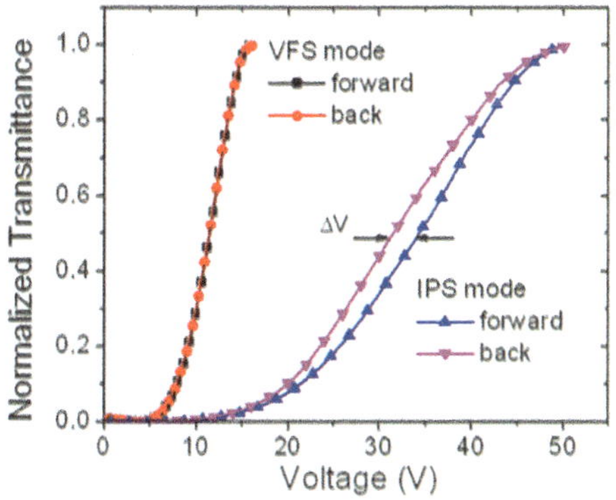

nanoparticles of $BaTiO_3$, with an average diameter of approximately 30 μm, were suspended in a BPI LC composed of low-molecular-mass compounds. BPLC, doped with 0.7 wt% of $BaTiO_3$ NPs, exhibits hysteresis-free switching with a response time of less than 1 ms under a voltage of 42.0 V. The on-state voltage (V_{on}) corresponds to the voltage of peak transmittance. While the incorporation of nanoparticles into BPs has been demonstrated to enhance their physical properties, the application of an electric field has been shown to modify the dispersion of these nanoparticles. It is imperative to assess the long-term stability of nanoparticle-stabilized BPs for application to display devices.

In their seminal work, Lan et al. advanced a novel hypothesis to elucidate the phenomenon of hysteresis in polymer-stabilized BPLC as follows. The phenomenon is influenced by three predominant factors [114]. The primary factor to be considered is the relaxation ability of BPLC. The secondary is steric hindrance of polymer network. The tertiary is the anchoring force between BPLC and polymer network. The achievement of hysteresis-free operation was accompanied through the implementation of dual frequency driving [115]. Chen et al. reported that increasing the domain size of PS-BPLCs reduces hysteresis during electro-optical switching and decreases the driving voltage [116]. A lot of approaches have been proposed to address the hysteresis phenomenon in cubic BPs [117–119]. However, achieving long-term stability remains challenging, as the disclination lines in a cubic BPI exhibit a high degree of sensitivity to the field strength and the history of electric field application, as detailed in Sect. 4.4.1.

As reported by Ozaki et al., the research focused on BP polymer-templated ferroelectric nematic liquid crystal [120]. The researchers achieved the stabilization of the nematic ferroelectric liquid crystal (NFLC) orientation into a BP 3D structure by employing BP polymer networks. The BP polymer-templated NFLC exhibits a cubic BP structure with ferroelectricity. The substantial polarization and dielectric constant exhibited by the NFLCs gave rise to a pronounced electro-optical Kerr effect.

Yang et al. reported an ultrastable cubic BP from molecular synergistic self-assembly of low-molecular-mass liquid crystals [121]. In their work, a series of tailored uniaxial rod-like mesogens that disfavor the formation of blue phases are introduced into a blue phase system comprising biaxial dimeric mesogens. The BP temperature range and the driving voltage of the mixtures are found to be contingent on their compositions. To illustrate this point, we will examine one of the samples below. The compositions are presented in Fig. 4.42. The sample consists of **BPFO*n*** (39.3 wt%), **TPFO*n*** (21.8 wt%), **TTPF** (14.1 wt%), **TP2FTF** (18.8 wt%), **R-5011** (2.0 wt%), and **ISO-(6OBA)$_2$** (4.0 wt%).The phase sequence as follows: Iso 83.5 °C cubic BP −38.8 °C unidentified phase X. The BP temperature range was 122.3 K. The driving voltage was approximately 20 V_{rms} at 20 °C. It was observed that an increase in the concentration of the fluorinated compound **TP2FTF**, which possesses large dielectric anisotropy, led to a significant decrease in the driving voltage. This approach offers a novel and practical methods for the production of cubic BP materials.

The impact of electric field on BPIII has been documented on a limited number of studies [5, 122, 123]. Yang and Crooker's investigation focused on a temperature–electric field phase diagram in a BP mixture, highlighting the occurrence of BPIII. The BPIII transitioned into a new phase (BPE) at $E > 2$ V μm^{-1} [122]. We have previously demonstrated the high-contrast electro-optical switching in BPIII due to electric-field-induced phase transition from BPIII to N for the first time [29]. It has been demonstrated that a well black state is attainable in BPIII without surface treatment, and a homogeneous bright state is achieved in the electric-field-induced N phase [29, 124]. As illustrated in Fig. 4.43, a schematic representation of the electric-field-induced phase transition between BPIII and N is presented. BPIII is characterized by randomly distributed DTCs and appears to be macroscopically isotropic. The

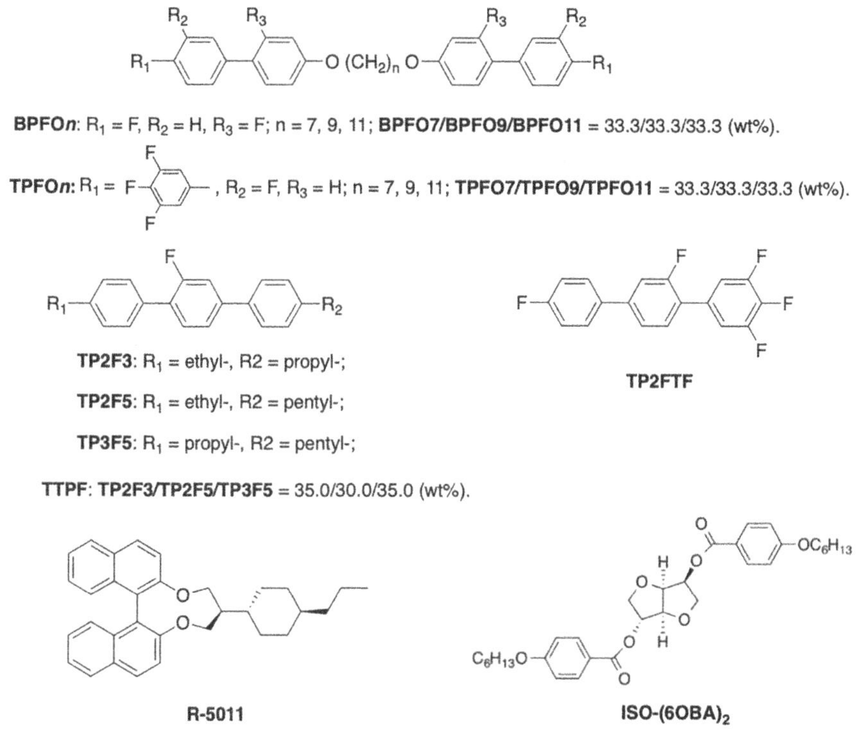

Fig. 4.42 The molecular structures of components for obtaining the low-molecular-mass BP LCs

application of an electric field induces a transformation in the molecular organization, resulting in a change from a double-twisted structure to an anisotropic nematic orientation. Subsequent to the elimination of the electric field, the molecules revert to their original structure, thereby establishing the double-twist organization in BPIII.

T-shaped LCs, H-shaped LCs, and bent-core LCs have been identified as suitable host materials, with the capacity to undergo a transformation into BPIII LC upon the incorporation of a chiral dopant. However, an increase in the viscosity of their BPIII results in slow electro-optical switching. We proposed a ternary system consisting of a rod-like nematic mixture, a BPIII stabilizer, and a chiral dopant [9, 30]. The design concept is illustrated in Fig. 4.44. Physical properties such as the temperature range, dielectric anisotropy, and viscosity are adjusted to facilitate the formation of the nematic mixture. The BP stabilizer exhibits both biaxiality, which stabilizes DTCs, and flexibility, which induces disorder in the system. A rod-like nematic LC mixture doped with a BPIII stabilizer could potentially serve as a host LC, which, upon the addition of a chiral dopant with high twisting power, is converted to BPIII LC.

In accordance with the design concept, a ternary system was formulated, comprising a conventional nematic mixture, a T-shaped BP stabilizer, and a chiral dopant with a large HTP [30]. A host liquid crystal was prepared, consisting of

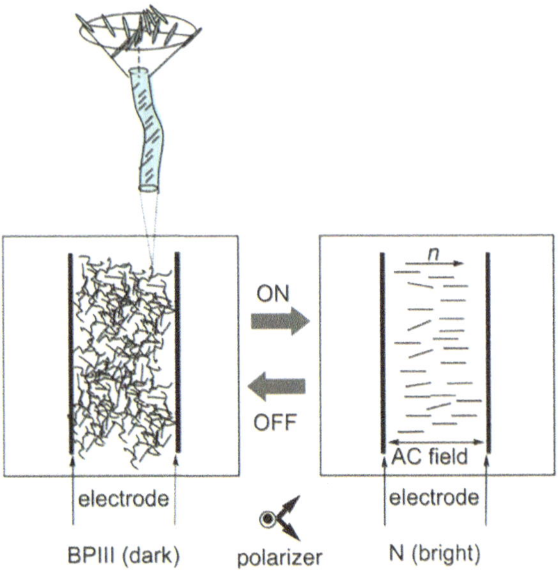

Fig. 4.43 Schematic representation of the electric-field-induced phase transition between BPIII and N. Reprinted with permission from Ref. [125]. Copyright (2008) Society for Information Display

a nematic liquid-crystalline mixture **E7** (95 mol%) and a non-chiral T-shaped compound (5 mol%). The molecular structure of the T-shaped compound is portrayed in Fig. 4.45. **E7** consists of 4-pentyl-4′-cyanobiphenyl (**5CB**, 51 wt%), 4-heptyl-4′-cyanobiphenyl (**6CB**, 25 wt%), 4-octyloxy-4′-cyanobiphenyl (**8OCB**,16 wt%), and 4-pentyl-4″-cyanoterphenyl (**5CT**, 8 wt%). The host LC was doped with 15 wt% of a chiral compound **ISO-(6OBA)$_2$**. The mixture exhibited a phase sequence of Iso 28.4 °C BPIII 21.2 °C N* upon cooling.

As illustrated in Fig. 4.46, the electric-field dependence of transmittances for the forward and backward processes in the BPIII at 26 °C is demonstrated. In the absence of an electric field, the transmittance was found to be 0%. It was 89% with an applied field of 14 V μm^{-1}. The backward curve exhibited a congruence with the forward curve. The device was found to be free from both hysteresis and residual transmittance. The defect structure in BPIII is characterized by a disorder with fluidity, and the DTCs easily rearrange in a liquid-like fashion as detailed in Sect. 4.2. Therefore, BPIII has no original state. Conversely, the 3D lattice structure in a cubic BP is unable to fully relax to its original position during the backward process. The respective rise and decay times with an AC field of 14 V μm^{-1} at 50 Hz were 0.4 and 0.8 ms. Figure 4.47 portrays the front views (Fig. 4.47a) and the oblique views with a viewing angle of approximately 45° (Fig. 4.47b) of the evaluation BPIII cell. The BPIII cell exhibits optical isotropy, which results in wide and symmetrical viewing angles. The electro-optical switching phenomenon is evident in Video S4.1 (see Electronic Supplementary Material).

A comparison was made between the electro-optical properties of PS-BPIII and those of PS-cubic BP [65]. In accordance with the design concept of a BP host LC, a host nematic mixture was prepared consisting of a rod-like nematic mixture (**5CB**

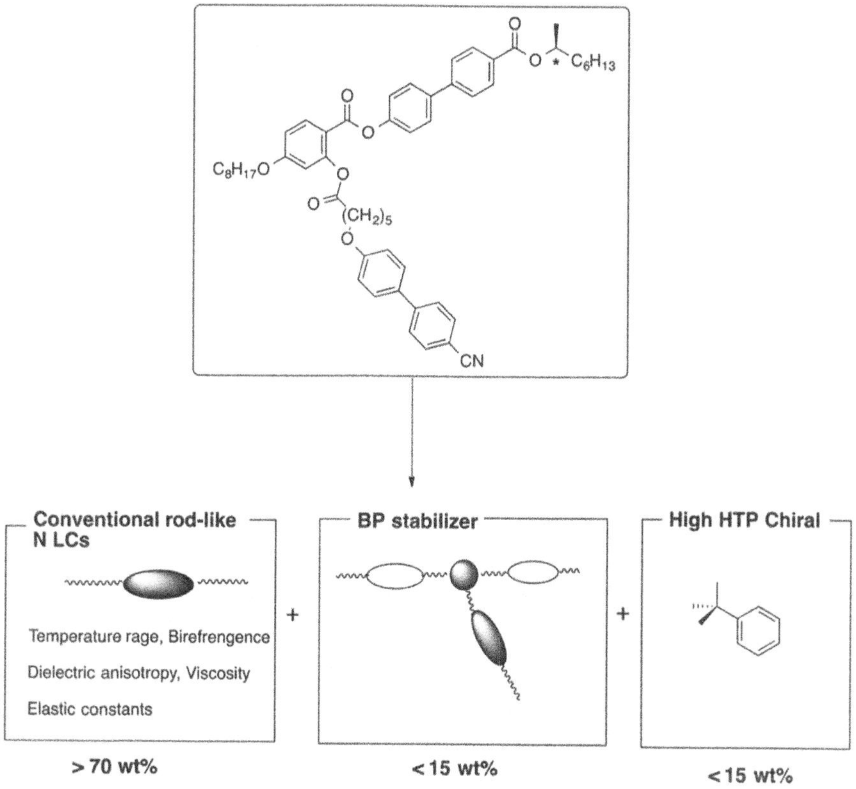

Fig. 4.44 Design concept of the mixing system for a practical BPIII material. Reproduced from [9] with permission from the Royal Society of Chemistry

Fig. 4.45 Molecular structure of the BP stabilizer

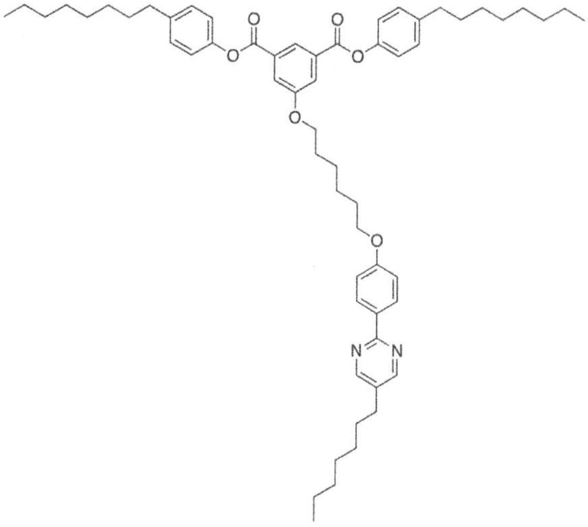

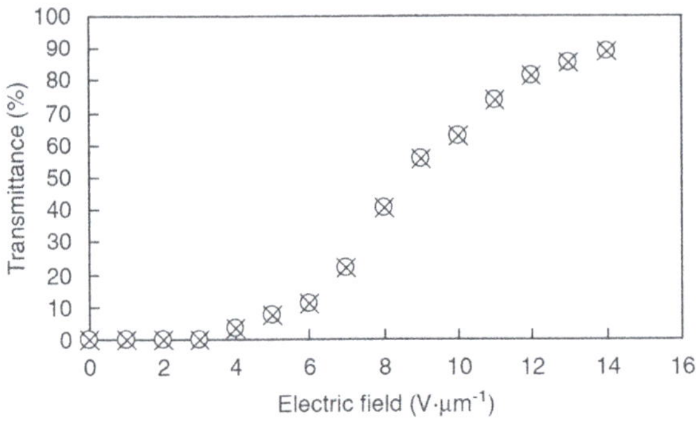

Fig. 4.46 Optical transmittances of the chiral mixture as a function of an AC field at a frequency of 50 Hz in BPIII at 26 °C. The ramping rate was 0.2 s per step. The cell gap was 5 μm Open circles and crosses respectively denote ascending and descending processes. Reprinted with permission from Ref. [30]. Copyright (2011) The Japan Society of Applied Physics

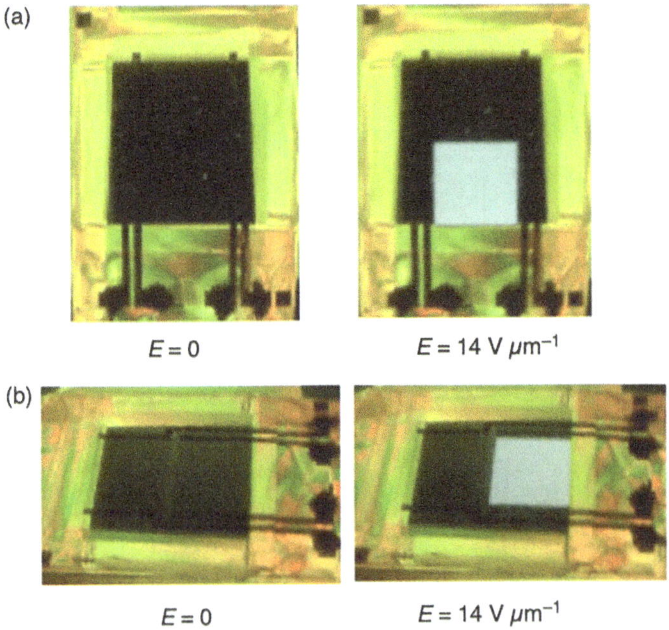

Fig. 4.47 Viewing-angle dependence of transmittance in the BPIII cell. **a** Front and **b** oblique views with a viewing angle of approximately 45°. Reprinted with permission from Ref. [30]. Copyright (2011) The Japan Society of Applied Physics

[50 wt%], **8OCB** [35 wt%], and **5CT** [15 wt%]) and a BP stabilizer (10 wt%). The molecular structure of the BP stabilizer is shown in Fig. 4.48. The host LC doped with a chiral compound **ISO-(6OBA)$_2$** (10 wt%) and that doped with **ISO-(6OBA)$_2$** (15 wt%) were for cubic BP and BPIII, respectively. Subsequently, each BP was polymer-stabilized to yield the corresponding PS-cubic BP and PS-BPIII. Figure 4.49 presents an optical texture of a specimen in the PS-cubic BP without an electric field in the forward process (Fig. 4.49a), that with an electric field of 14 V μm^{-1} (Fig. 4.49b), and that without an electric field in the backward process (Fig. 4.49c). Polymer stabilization has been shown to induce higher residual birefringence in the PS-cubic BP, as illustrated in Fig. 4.49c. In regard to the corresponding PS-BPIII, the transmittance in the absence of an AC field was found to be 0% (Fig. 4.50a). It increased with the increase in the electric field, reaching a saturation value of 84.0% at 14 V μm^{-1} (Fig. 4.50b). Residual transmittance was not detected (Fig. 4.50c). The PS-BPIII demonstrated enhanced higher stability in the presence of an applied electric field in comparison to the PS-cubic BP.

Electro-optical switching in BPIII has been reported to occur through the use of polymer-stabilized materials [14, 50, 125, 126] or a low-molecular-mass system [81, 127]. Gandhi and Chien et al. reported that a polymer network based on BPIII has the capacity to memorize the 3D structure on the original PS-BPIII from which it is nanoengineered. Furthermore, this polymer network has the ability to transfer this structural information to achiral nematic LCs [50]. A comparison was made between the electro-optical behavior of the PS-BPIII and the refilled scaffold. The comparison

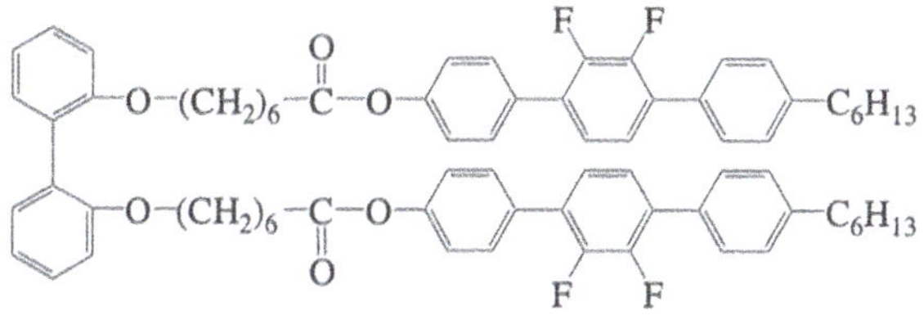

Fig. 4.48 Molecular structure of the U-shaped BP stabilizer

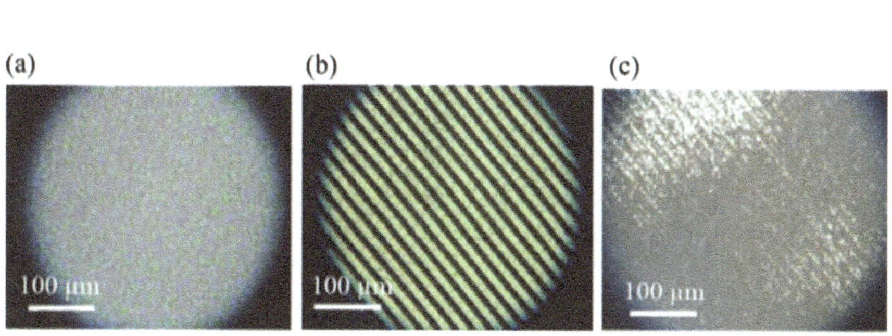

Fig. 4.49 Optical texture of a sample in the PS-cubic BP at 25 C without an electric field in the forward process (**a**), that with an electric field of 14 V μm^{-1} (**b**), and that without an electric field in the backward process (**c**). Reprinted with permission from [65]. Copyright (2015) Taylor & Francis

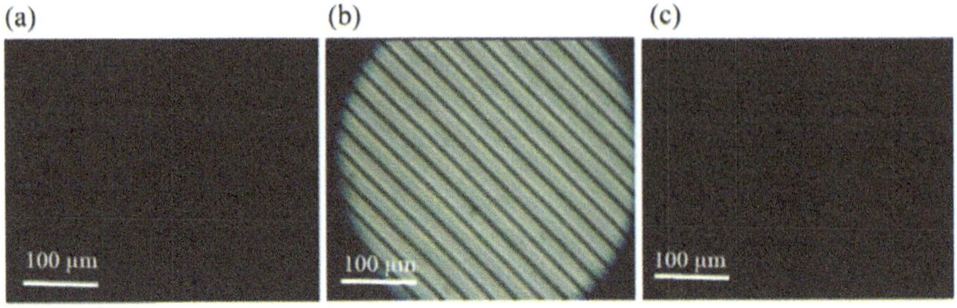

Fig. 4.50 Optical texture of a sample in the PS-BP III at 25 C without an electric field in the forward process (**a**), that with an electric field of 14 V μm^{-1} (**b**) and that without an electric field in the backward process (**c**). Reprinted with permission from [65]. Copyright (2015) Taylor & Francis

reveals that: (1) the driving electric field for both PS-BPIII and the refilled scaffold is approximately 40 V μm^{-1}, (2) the transmittance–electric field curves of the PS-BPIII and the scaffold after refilling **5CB** exhibit a high degree of overlap, and (3) the switch-on and switch-off times of PS-BPIII are both less than 0.2 ms and, remaining in the submillisecond range after refilling the network with **5CB**. These findings demonstrate the EO-memory effects of the polymer network (see Chap. 5).

4.4.3 Photonic Crystals

Three-dimensional (3D) photonic crystals are materials that have garnered significant attention due to their potential for versatile manipulation of light [128–130]. The size of the BP lattice is comparable to the wavelength of visible light, thereby classifying BP as a 3D photonic crystal. The size of the BP unit cell determines the wavelength of reflected light. The relationship between the reflected wavelength, λ, and unit cell size, *a*, of BPs is described by the relationship:

$$\lambda = 2na/\sqrt{h^2 + k^2 + l^2} \tag{4.12}$$

where *n* is the refractive index and (*hkl*) are the Miller indices describing the lattice orientation of the BP. BPs have the capacity to reflect light within the visible spectrum and manifest a range of colors, spanning from blue to red, as illustrated in Fig. 4.51 [12, 131]. BPLCs as photonic crystals have been the focus of extensive research [19, 132, 133]. However, BPs are typically polycrystalline. The various colors of the BPI texture in Fig. 4.2a correspond to different lattice orientations. The majority of applications utilize polycrystalline or nearly amorphous materials. Lattice-aligned BPs demonstrate superior performance in comparison to polycrystalline materials in LCD devices [134]. For sophisticated photonic applications, the utilization of a sufficiently large monocrystalline photonic crystal is of paramount importance.

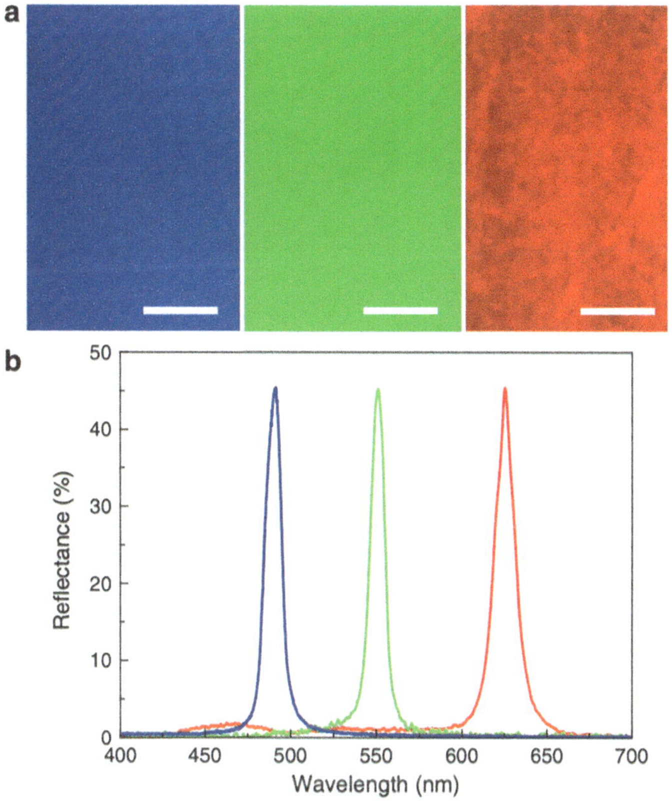

Fig. 4.51 Blue, green, and red reflections from large BPI single crystals. **a** Microscope images of three different BP single crystals (*scale bars*, 200 μm) and **b** respective reflection spectra for linearly polarized probe light [131]

Chen and Hou et al., developed a gradient-temperature technique that enables the fabrication of 3D photonic crystals based on monocrystalline BPs. These crystals can be grown to lateral dimensions of ~1 cm and thickness of ~100 μm [131]. Large and uniform BPI single crystals exhibit high reflectivity and narrow bandwidth (Fig. 4.52). The reflection bandwidth of the monocrystal BPI was found to be significantly narrow than that of a commercially available polycrystal. As illustrated in Fig. 4.53a, the schematic depiction of a gradient-temperature scanning (GTS) system is presented. A micrometer step monitor was utilized to facilitate the transition of the sample between stages, thereby ensuring a regulated shift in the temperature gradient. Figure 4.53b displays an obtained BPI monocrystal of a mixture within a 100 μm-thick cell. The mixture was composed of 32.0 wt% of nematic **HTW114200-500** (from HCCH), 33.0 wt% of **5CB**, and 35.0 wt% of **R-811**. It showed a phase sequence of Iso 35.6 ~ 34.5 °C BPII 34.0 ~ 33.9 °C BPI 29.8 °C N* upon cooling. Large BPLCs prepared by the GTS process have the capacity to adapt to different spectral regimes across the visible spectrum with red, green, and blue reflections

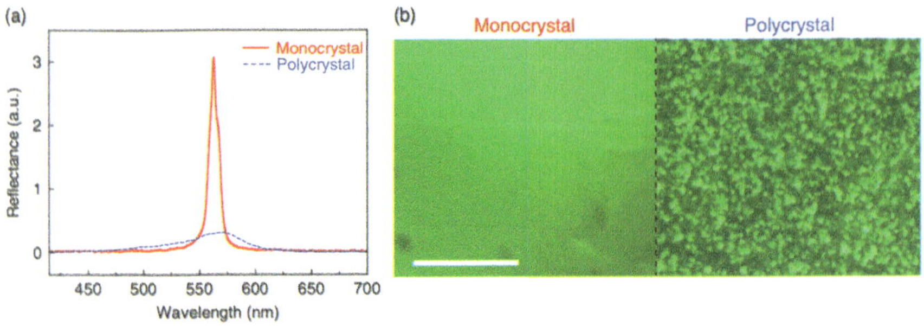

Fig. 4.52 **a** Reflection spectra and **b** microscope images of the BPI crystals (*scale bar*, 300 μm) [131]

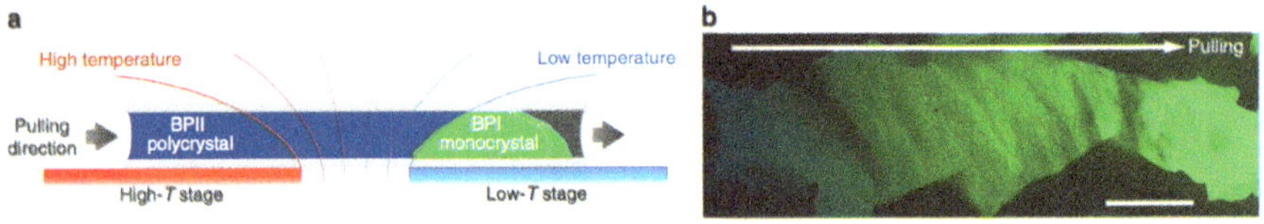

Fig. 4.53 **a** Schematic depiction of the GTS system. **b** Microscope image of a 3 mm-long BPI single crystal fabricated by the GTS system (*scale bar*, 500 μm) [131]

(Fig. 4.51). It has been demonstrated that a blue shift of the photonic bandgap is accompanied by an increase in the concentration of the chiral agent.

The authors fabricated polymer-stabilized versions of these BP single crystals. The LC mixture used for the polymer stabilization was composed of 54.8 wt% of **HTW114200-050**, 36.5 wt% of **R-811**, 4.0 wt% of **RM257**, 4.0 wt% of 2-ethylhexyl acrylate, and 0.7 wt% of a photo initiator. It exhibited a phase sequence of Iso 38.1 °C BPII 34.5 °C BPI 27.0 °C N* upon cooling. The PSBP exhibited the Iso to BPI phase transition at approximately 37.0 °C, and it demonstrated robust stability in the BPI even when the ambient temperature decreased to 13.0 °C. Figure 4.54a demonstrates that these large monodomain PSBP crystals demonstrate a substantially broader temperature range (at least 25 K) in comparison to their polymer-free counterparts (typically ~1–4 K). Additionally, the temperature sensitivity of the photonic bandgap has also been suppressed, measuring $d\lambda/dT \approx 0.16$ nm °C^{-1}. Figure 4.54c reveals that the photonic bandgap can be linearly sifted with increasing electric field from green to red. A dynamic tuning range exceeding 100 nm has been attained with only ~1.3 V μm^{-1}. The lattice distortion is fully reversible, and hysteresis between increasing and decreasing field strength is almost negligible.

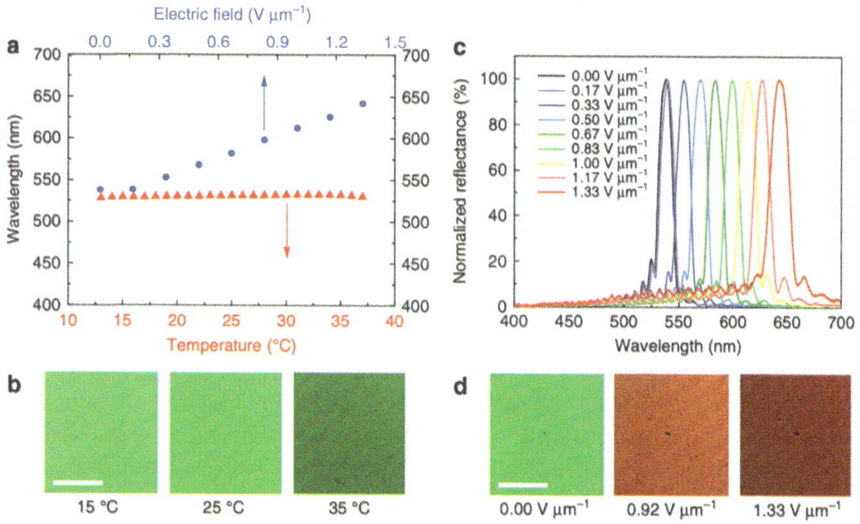

Fig. 4.54 Temperature-invariant and electrically tunable bandgap of large polymer-stabilized BPI monocrystal. **a** Temperature and electric field dependences of the peak reflection wavelength. **b** Microscopic images at different temperatures (*scale bar*, 500 μm). **c** Reflection spectra at different direct current field strengths **d** Microscope images at different field strengths (scale bar, 500 μm) [131]

4.5 Concluding Remarks

Optically isotropic blue phases differ significantly from conventional calamitic liquid-crystalline phases with birefringence. It has been observed that both BPI and BPII are composed of 3D cubic structures with lattice parameters that approximate the visible wavelengths. Conversely, the structure of BPIII has remained ambiguous for an extended period. Recent simulation studies on BPIII and the SEM measurements of the polymer network derived from BPIII reveal that BPIII is a "spaghetti"-like tangled arrangement of DTCs with special dynamics. The defect structure in BPIII is characterized by a disorder with fluidity, thereby the facile rearrangement of DTCs in a liquid-like fashion. The presence of defects that are unfavorable for calamitic phases is a prerequisite for the appearance of the blue phase. Introducing chirality to certain achiral liquid-crystalline phases, such as the N and SmC phases, results in the phase converting to its corresponding chiral version, which possesses a single helical axis. However, it is notable that the blue phases do not exhibit a corresponding achiral phase. The disparity in driving force of phase formation between BPs and the conventional LCs necessitates a novel concept for material design. Two approaches have been identified for widening the BP temperature range. The initial step involves the formation of a defect-free state through the utilization of polymers or nanoparticles. The employment of polymer stabilization techniques has enabled the BP to demonstrate a practical temperature range [15]. The alternative approach involves the molecular design of BP LCs, with the objective of stabilizing DTC structures. It is proposed that a decrease in molecular anisotropy can lead to a reduction in K_{33} and the production of helical twisting power in multiple directions. The U-shaped and T-shaped oligomers have been shown to stabilize the BPs. The liquid-crystalline supermolecules, characterized by molecular biaxiality, have the potential to induce a dynamic origin of chirality. The chiral competition effect has been demonstrated to destabilize the formation of a single twist structure, thereby producing a blue phase. The molecular biaxiality and flexibility have been identified as critical factors in stabilizing BPIII. In addition, there have been documented several methodologies for obtaining practical low-molecular-weight BP liquid crystal compositions. For instance, a cybotactic N LC doped with a chiral additive [78] and a conventional rod-like N LC mixed with a BP stabilizer (fluoro-substituted odd-numbered dimer [120] or a T-shaped trimer [30]) can be used. The BPIII demonstrated submillisecond response, high contrast, and wide-viewing angle at room temperature without surface treatment and an optical compensation film. It is free from both hysteresis and residual transmittance. This is attributed to the inherent dynamics of BPIII, which precludes the existence of an original state. In contrast to cubic BPs, BPIII does not necessitate the alignment of the lattice structure to augment the domain size. The high driving voltage remains a significant challenge, but its mitigation is possible through advancements in material and device technologies. Amorphous BPIII, which exhibits the same symmetry as an isotropic liquid presents not only fundamental interests but also a superior potential for use in next-generation displays. The cubic blue phases demonstrate stimulus responsiveness. The polymer-stabilized large BP monocrystal exhibit remarkable dynamic tunability across a broad color spectrum under direct current electric fields, thereby underscoring the potential for the realization of versatile, reconfigurable, and thermodynamically stable photonic crystals. Blue phases are currently offering the platforms for application as next-generation displays or tunable photonic crystals.

References

1. Stegemeyer, H., Bergmann, K.: Experimental results and problems concerning "Blue Phases." In: Helfrich, W., Heppke, G. (eds.) Liquid Crystals of One- and Two-Dimensional Order, pp. 161–175. Springer, Berlin/Heidelberg, Germany (1980)
2. Stegemeyer, H., Blümel, T., Hiltrop, K., Onusseit, H., Porsch, F.: Thermodynamic, structural and morphological studies on liquid-crystalline blue phases. Liq. Cryst. **1**, 3–28 (1986)
3. Wright, D.C., Mermin, N.D.: Crystalline liquids: the blue phases. Rev. Mod. Phys. **61**, 385–432 (1989)
4. Crooker, P.P.: The blue phases. Liq. Cryst. **5**, 751–775 (1989)
5. Kitzerow, H.-S.: The effect of electric fields on blue phases. Mol. Cryst. Liq. Cryst. **202**, 51–83 (1991)
6. Crooker, P.P.: Blue phases. In: Kitzerow, H.-S., Bahr, C. (eds.) Chirality in Liquid Crystals, pp. 186–222. Springer, New York, NY, USA (2001)
7. Oswald, P., Pieranski, P.: Nematic and Cholesteric Liquid Crystals, pp. 493–552. Taylor & Francis, Boca, Raton FL (2005)

8. Kikuchi, H.: Liquid crystalline blue phases. Struct. Bond. **128**, 99–117 (2008)
9. Yoshizawa, A.: Material design for blue phase liquid crystals and their electro-optical effects. RSC Adv. **3**, 25475–25497 (2013)
10. Asiqur Rahman, M.D., Mohd Said, D., Balamurugan, S.: Blue phase liquid crystal: strategies for phase stabilization and device development. Sci. Tschnol. Adv. Mater. **16**, 033501 (2015)
11. Gleeson, H.F., Miller, R.J., Tian, L., Görtz, V., Goodby, J.W.: Liquid crystal blue phases: stability, field effects and alignment. Liq. Cryst. **42**, 760–771 (2015)
12. Bagchi, K., Emersic, T., Martínez-González, J., de Pablo, J.J., Nealey, P.F.: Functional soft materials from blue phase liquid crystals. Sci. Adv. **9**, 9393 (2023)
13. Dubois-Violette, E., Pansu, B.: Frustration and related topology of blue phases. Mol. Cryst. Liq. Crys. **165**, 151–182 (1988)
14. Hirose, T., Yoshizawa, A.: Odd-even effects of an asymmetric dimer on the double-twist structure in an amorphous blue phase. J. Mater. Chem. C **4**, 8565–8574 (2016)
15. Kikuchi, H., Yokota, M., Hisakado, Y., Yang, H., Kajiyama, T.: Polymer-stabilized liquid crystal blue phases. Nat. Mater. **1**, 64–68 (2002)
16. Coles, H.J., Pivnenko, M.N.: Liqui crystal 'blue phases' with a wide temperature range. Nature **436**, 997–1000 (2005)
17. He, W., Pan, G., Yang, Z., Zhao, D., Niu, G., Huang, W., Yuan, X., Guo, J., Cao, H., Yang, H.: Wide blue phase range in a hydrogen-bonded self-assembled complex of chiral fluoro-substituted benzoic acid and pyrimidine derivative. Adv. Mater. **21**, 2050–2053 (2009)
18. Wang, J., Shi, Y., Yang, K., Wei, J., Guo, J.: Stabilization and optical switching of liquid crystal blue phase doped with azobenzene-based bent-shaped hydrogen-bonded assemblies. RSC Adv. **5**, 67357–67364 (2015)
19. Cao, W., Munoz, A., Palffy-Muhoray, P., Taheri, B.: Lasing in a three-dimensional photonic crystal of the liquid crystal blue phase II. Nat. Mater. **1**, 111–113 (2002)
20. Yokoyama, S., Mashiko, S., Kikuchi, H., Uchida, K., Nagamura, T.: Laser emission from a polymer-stabilized liquid-crystalline blue phase. Adv. Mater. **18**, 48–51 (2006)
21. Chen, H.-Y., Chiou, J.-Y., Yang, K.-X.: Reversible and fast shift in reduction band of a cubic blue phase in a vertical electric field. Appl. Phys. Lett. **99**, 181119 (2011)
22. Chen, H.-Y., Lai, J.-L., Chan, C.-C., Tseng, C.-H.: Fast tunable reflection in amorphous blue phase III liquid crystal. J. Appl. Phys. **113**, 123103 (2013)
23. Hu, W., Sun, J., Wang, Q., Zhang, L., Yuan, X., Chen, F., Li, K., Miao, Z., Yang, D., Yu, H.: Humidity-responsive blue phase liquid-crystalline film with reconfigurable and tailored visual signals. Adv. Funct. Mater. **30**, 2004610 (2020)
24. Yang, Y., Kim, Y.-K., Wang, X., Tsuei, M., Abbott, N.L.: Structural and optical response of polymer-stabilized blue phase liquid crystal films to volatile organic compounds. ACS Appl. Mater. Interfaces **12**, 42099–42108 (2020)
25. Rao, L., Yan, J., Wu, S.-T.: Prospects of emerging polymer-stabilized blue-phase liquid-crystal displays. J. Soc. Inf. Disp. **18**, 954–959 (2010)
26. Yan, L., Rao, L., Jiao, M., Li, Y., Cheng, H.-C., Wu, S.-T.: Polymer-stabilized optically isotropic liquid crystals for next-generation display and photonics. J. Mater. Chem. **21**, 7870–7877 (2011)
27. Lee, H., Park, H.-J., Kwon, O.-J., Yun, S.J., Park, J.H., Hong, S., Shin, S.-T.: The world's first blue phase liquid crystal display. Soc. Inform. Disp. Symp. Dig. Tech. Pap. **42**, 121–124 (2011)
28. Chen, K.-M., Gauza, S., Xianyu, H., Wu, S.-T.: Hysteresis effects in blue-phase liquid crystals. J. Disp. Technol. **6**, 318–322 (2010)
29. Sato, M., Yoshizawa, A.: Electro-optical switching in a blue phase III exhibited by a chiral liquid crystal oligomer. Adv. Mater. **19**, 4145–4148 (2007)
30. Yoshizawa, A., Kamiyama, M., Hirose, T.: Amorphous blue phase III exhibiting submillisecond response and hysteresis-free switching at room temperature. Appl. Phys. Express **4**, 101701 (2011)
31. Yang, Y., Wang, L., Yang, H., Li, Q.: 3D Chiral photonic nanostructures based on blue-phase liquid crystals. Small Sci. **1**, 2100007 (2021)

32. Cladis, P.E., Garel, T., Pieranski, P.: Kossel diagrams show electric-field-induced cubic-tetragonal structural transition in frustrated liquid-crystal blue phases. Phys. Rev. Lett. **57**, 2841–2844 (1986)
33. Johnson, D.L., Flack, J.H., Crooker, P.P.: Structure and properties of the cholesteric blue phases. Phys. Rev. Lett. **45**, 641–644 (1980)
34. Higashiguchi, K., Yasui, K., Kikuchi, H.: Direct observation of polymer-stabilized blue phase I structure with confocal laser scanning microscope. J. Am. Chem. Soc. **130**, 6326–6327 (2008)
35. Marcus, M.: Quasicrystalline behaviour and phase transition in cholesteric "blue" phase. J. Phys. France **42**, 61–70 (1981)
36. Meiboom, S., Sammon, M.: Structure of the blue phase of a cholesteric liquid crystal. Phys. Rev. Lett. **44**, 882–885 (1980)
37. Meiboom, S., Sammon, M.: Blue phases of cholesteryl nonanoate. Phys. Rev. A **24**, 468–475 (1981)
38. Thoen, J.: Adiabatic scanning calorimetric results for the blue phases of cholesteryl nonanoate. Phys. Rev. A **37**, 1754–1759 (1988)
39. Kleiman, R.N., Bishop, D.J., Pindak, R., Taborek, P.: Shear modulus and specific heat of the liquid-crystal blue phases. Phys. Rev. Lett. **53**, 2137–2140 (1984)
40. Zasadzinski, J.A.N., Meiboom, S., Sammon, M.J., Berreman, D.W.: Freeze-fracture electron-microscope observations of the blue phase III. Phys. Rev. Lett. **57**, 364–367 (1996)
41. Finn, P.L., Cladis, P.E.: Cholesteric blue phases in mixtures and in an electric field. Mol. Cryst. Liq. Cryst. **84**, 159–192 (1982)
42. Hornreich, R.M., Kugler, M., Shtrikman, S.: Localized instabilities and the order-disorder transition in cholesteric liquid crystals. Phys. Rev. Lett. **48**, 1404–1407 (1982)
43. Collings, P.J.: Optical rotatory dispersion measurements in the third cholesteric blue phase. Phys. Rev. A 1990–1993 (1983)
44. Hornreich, R.M., Shtrikman, S.: Broken icosahedral symmetry: a quasicrystalline structure for cholesteric blue phase III. Phys. Rev. Lett. **56**, 1723–1726 (1986)
45. Henrich, O., Stratford, K., Cates, M.E., Marenduzzo, D.: Structure of blue phase III of cholesteric liquid crystals. Phys. Rev. Lett. **106**, 107801 (2011)
46. Paul, T., Saha, J.: Origin and structure of liquid crystalline blue phase III. Sci. Rep. **10**, 16016 (2020)
47. Pišljar, J., Ghosh, S., Turlapati, S., Rao, N.V.S., Škarabot, M., et al.: Blue phase III: topological fluid of skyrmions. Phys. Rev. X **12**, 011003 (2022)
48. Castles, F., Day, F.V., Morris, S.M., Ko, D.-H., Gardiner, D.J., Qasim, M.M., Nosheen, S., Hands, P.J.W., Choi, S.S., Coles, H.J., et al.: Blue-phase templated fabrication of three-dimensional nanostructures for photonic applications. Nat. Mater. **11**, 599–603 (2012)
49. Chen, H.-Y., Wu, C.-K., Chen, F.-C., Chen, C.-S.: Investigation on morphology of polymer structure in polymer-stabilized blue phase. Liq. Cryst. **43**, 1351–1358 (2016)
50. Gandhi, S.S., Kim, M.S., Hwang, J.-Y., Chien, L.-C.: Electro-optical memory of a nanoengineered amorphous blue-phase-III polymer scaffold. Adv. Mater. **28**, 8998–9005 (2016)
51. Yoshizawa, A., Kagami, T., Kato, K., Abe, K., Hamada, K., Uemura, T., Yamaguchi, M., Sagisaka, M.: Helical polymerization using a polymer network derived from a blue phase as a template for chirality transfer. Liq. Cryst. **50**, 2268–2279 (2023)
52. Yoshizawa, A., Iwamochi, H., Segawa, S., Sato, M.: The role of a liquid crystal oligomer in stabilizing blue phases. Liq. Cryst. **34**, 1039–1044 (2007)
53. Nishiyama, I., Yoshizawa, A.: Design of blue phase materials. In: Goodby, J.W., Collings, P.J., Kato, T., Tschierske, C., Gleeson, H.F., Raynes, P. (eds.) Handbook of Liquid Crystals, vol. 3, 2nd edn, , pp. 557–586. Wiley-VCH Verlag GmbH, Weinheim, Germany (2014)
54. Alexander, G.P., Yeomans, J.M.: Stabilizing the blue phases. Phys. Rev. E **74**, 061706 (2006)
55. Meiboom, S., Sammon, M., Brinkman, W.F.: Lattice of disclinations: the structure of the blue phases of cholesteric liquid crystals. Phys. Rev. A **27**, 438–454 (1983)
56. Grebel, H., Hornreich, R.M., Shtrikman, S.: Landau theory of cholesteric blue phases. Phys. Rev. A **28**, 1114–1138 (1983)

57. Alexander, G.P., Yeomans, J.M.: Numerical results for the blue phases. Liq. Cryst. **36**, 1215–1227 (2009)
58. Fukuda, J.: Stabilization of blue phases by the variation of elastic constants. Phys. Rev. E **85**, 020701(R) (2012)
59. Meiboom, S., Sethna, J.P., Anderson, P.W., Brinkman, W.F.: Theory of the blue phase of cholesteric liquid crystals. Phys. Rev. Lett. **46**, 1216–1219 (1981)
60. Ravnik, M., Fukuda, J., Yeomans, J.M., Žumer, S.: Confining blue phase colloids to thin layers. Soft Matter **7**, 10144–10150 (2011)
61. Fukuda, J.: Stabilization of a blue phase by a guest component: an approach based on a Landue-de Gennes theory. Phys. Rev. E **82**, 061702 (2010)
62. Fukuda, J.: Stability of cholesteric blue phases in the presence of a guest component. Phys. Rev. E **86**, 041704 (2012)
63. Kitzerow, H.-S., Schmid, H., Ranft, A., Heppke, G., Hikmet, R.A.M., Lub, J.: Observation of blue phases in chiral networks. Liq. Cryst. **14**, 911–916 (1993)
64. Hisakado, Y., Kikuchi, H., Nagamura, T., Kajiyama, T.: Adv. Mater. **17**, 96–98 (2005)
65. Hirose, T., Yoshizawa, A.: Comparison of electro-optical switching between polymer-stabilized cubic and amorphous blue phases. Liq. Cryst. **42**, 1290–1297 (2015)
66. Kim, M.S., Chien, L.-C.: Topology-mediated electro-optical behaviour of a wide-temperature liquid crystalline amorphous blue phase. Soft Matter **11**, 8013–8018 (2015)
67. Yoshida, H., Tanaka, Y., Kawamoto, K., Kubo, H., Tsuda, T., Fujii, A., Kuwabata, S., Kikuchi, H., Ozaki, M.: Nanoparticles-stabilized cholesteric blue phases. Appl. Phys. Express **2**, 121501 (2009)
68. Karatairi, E., Rozic, B., Kutnjak, Z., Tzitzios, V., Nounesis, G., Cordoyiannis, G., Thoen, J., Glorieux, C., Kralj, S.: Nanoparticle-induced widening of the temperature range of liquid-crystalline blue phases. Phys. Rev. E **81**, 041703 (2010)
69. Blatch, A.E., Fletcher, I.D., Luckhurst, G.R.: Symmetric and non-symmetric liquid crystal dimers with branched terminal alkyl chains: racemic and chiral. J. Mater. Chem. **7**, 9–17 (1997)
70. Nakata, M., Takanishi, Y., Watanabe, J., Takezoe, H.: Blue phases induced by doping chiral nematic liquid crystals with nonchiral molecules. Phys. Rev. E **68**, 041710 (2003)
71. Hur, S.-T., Gim, M.-J., Yoo, H.-J., Choi, S.-W., Takezoe, H.: Investigation for correlation between elastic constant and thermal stability of liquid crystalline blue phase I. Soft Matter **7**, 8800–8803 (2011)
72. Zheng, Z., Shen, D., Huang, P.: Wide blue phase range of chiral nematic liquid crystal doped with bent-shaped molecules. New J. Phys. **12**, 113018 (2010)
73. Wang, H., Zheng, Z., Shen, D.: Blue phase liquid crystals induced by bent-shaped molecules based on 1,3,4-oxadiazole derivatives. Liq. Cryst. **39**, 99–103 (2012)
74. Wang, L., Yu, L., Xiao, X., Wang, Z., Yang, P., He, W., Yang, H.: Effects of 1,3,4-oxadiazoles with different rigid cores on the thermal and electro-optical performances of liquid crystalline blue phases. Liq. Cryst. **39**, 629–638 (2012)
75. Wang, L., He, W., Xiao, X., Yang, Q., Li, B., yang, P., Yang, H.: Wide blue phase range and electro-optical performances of liquid crystalline composites doped with thiophene-based mesogens. J. Mater. Chem. **22**, 2383–2386 (2012)
76. Khan, R.K.: Blue phase III from bent-core liquid crystals: unveiling thermal stability & unique traits—A comprehensive review. ChemPhysChem **25**, e202400779 (2024)
77. Yelamaggad, C.V., Shashikala, S., Liao, G., Rao, D.S.S., Prasad, S.K., Li, Q., Jakli, A.: Blue phase, smectic fluids and unprecedented sequences in liquid crystal dimers. Chem. Mater. **18**, 6100–6102 (2006)
78. Taushanoff, S., Le, K.V., Williams, J., Twieg, R.J., Sadashiva, B.K., Takezoe, H., Jakli, A.: Stable amorphous blue phase of bent-core nematic liquid crystals doped with a chiral material. J. Mater. Chem. **20**, 5893–5898 (2010)
79. Chiang, I.-H., Long, C.-J., Lin, H.-C., Chuang, W.-T., Lee, J.-J., Lin, H.-C.: Broad ranges and fast responses of single-component blue-phase liquid crystals containing banana-shaped 1,3,4-oxadiazole cores. ACS Appl. Mater. Interfaces **6**, 228–235 (2014)

80. Cordoyiannis, G., Lavric, M., Trcek, M., Tzitzios, V., Lelidis, I., Nounesis, G., Daniel, M., Kutnjak, Z.: Quantum dot-driven stabilization of liquid-crystalline blue phases. Front. Phys. **8**, 315 (2020)
81. Khatun, N., Sridurai, V., Csorba, K.F., Nair, G.G.: Topological defects stabilized by a soft twist-bend dimer and quantum dots lead to a wide thermal range and ultra-fast electro-optic response in a liquid crystalline amorphous blue phase. J. Mater. Chem. C **11**, 9686–9694 (2023)
82. Baghla, A., Sahai, M., Yadav, N., Gupta, S.P., Punjani, V., manjuladevi, V., Vij, J.K., Pal, S.K.: Unusual polar ordering and room-temperature blue phase stabilization in tetrafluorinated bent-shaped mesogens. Chem. Sci. **16**, 8002–8013 (2025)
83. Li, B., He, W., Wang, L., Xiao, X., Yang, H.: Effect of lateral fluoro substituents of rodlike tolanecyano mesogens on blue phase temperature ranges. Soft Matter **9**, 1172–1177 (2013)
84. Kishikawa, K., Itoh, H., Akiyama, S., Kobayashi, T., Kohmoto, S.: Stabilization of the blue phases of simple rodlike monoester compounds by addition of their achiral homologues. J. Mater. Chem. **22**, 8484–8491 (2012)
85. Guo, J., Shi, Y., Han, X., Jin, O., Wei, J., Yang, H.J.: Stabilizing blue phases of a simple cyanobiphenyl compound by addition of achiral mesogen monomer with a branched end group and chiral hydrogen-bonded assemblies. J. Mater. Chem. C **1**, 947–957 (2013)
86. Rokuhone, J., Yoshizawa, A.: An unusual phase sequence of iso liq-blue phase-smectic A for novel binaphthyl mesogenic derivatives. J. Mater. Chem. **15**, 275–279 (2005)
87. Kogawa, Y., Yoshizawa, A.: Chiral effects of blue phase stabilization of a binaphthyl derivative. Liq. Cryst. **38**, 303–307 (2011)
88. Jeong, H.-C., Aya, S., Kang, S., Araoka, F., Ishikawa, K., Takezoe, H.: Are chiral dopants with higher twisting power advantageous to induce wider temperature range of the blue phases? Liq. Cryst. **40**, 951–958 (2013)
89. Yoshizawa, A., Kogawa, Y., Kobayashi, K., Takanishi, Y., Yamamoto, J.: A binaphthyl derivative with a wide temperature range of a blue phase. J. Mater. Chem. **19**, 5759–5764 (2009)
90. Yoshizawa, A.: Nanostructured assemblies of liquid-crystalline supermolecules: from display to medicine. Liq. Cryst. **46**, 1950–1972 (2019)
91. Tanaka, M., Yoshizawa, A.: U-shaped oligomers with a molecular biaxiality stabilizing blue phase. J. Mater. Chem. C **1**, 315–320 (2013)
92. Kogawa, Y., Hirose, T., Yoshizawa, A.: Biphenyl derivative stabilizing blue phases. J. Mater. Chem. **21**, 19132–19137 (2011)
93. Yoshizawa, A., Sato, M., Rokunohe, J.: A blue phase observed for a novel chiral compound possessing molecular biaxiality. J. Mater. Chem. **15**, 3285–3290 (2005)
94. Marik, M., Mukherjee, A., Jana, D., Yoshizawa, A., Chaudhuri, B.K.: Dielectric spectroscopy of T-shaped blue-phase-III liquid crystal. Phys. Rev. E **38**, 012502 (2013)
95. Sayama, S., Yoshizawa, A.: Achiral H-shaped liquid crystals exhibiting an electric-field-induced chiral nematic phase. J. Mater. Chem. C **43**, 8865–8868 (2019)
96. Titibocchi, A., Gonnella, G., Marenduzzo, D., Orlandini, E.: Switching dynamics in cholesteric blue phases. Soft Matter **7**, 3295–3306 (2011)
97. Fukuda, J., Žumer, S.: Field-induced dynamics and structures in a cholesteric-blue-phase cell. Phys. Rev. E **87**, 042506 (2013)
98. Rao, L., Ge, Z., Gauza, S., Chen, K.-M., Wu, S.-T.: Emerging liquid crystal displays based on the Kerr effect. Mol. Cryts. Liq. Cryst. **527**, 186–198 (2010)
99. Gerber, P.R.: Electro-optical effects of a small-pitch blue-phase system. Mol. Cryst. Liq. Cryst. **116**, 197–206 (1985)
100. Singh, U., Keyes, P.H.: Measurement of the Kerr effect in cholesteric blue phases. Liq. Cryst. **8**, 851–860 (1990)
101. Porsch, F., Stegemeyer, H., Hiltrop, K.: Electric field-induced birefringence in liquid-crystalline blue phases. Z. Naturforsch. A. **39a**, 475–480 (1984)
102. Heppke, G., Jerome, B., Kitzerow, H.-S., Pieranski, P.: Electrostriction of the cholesteric blue phases BPI and BPII in mixtures with positive dielectric anisotropy. J. Phys. France **50**, 2991–2998 (1991)

103. Stegemeyer, H., Porsch, F.: Electric field effect on phase transitions in liquid-crystalline blue-phase systems. Phys. Rev. A **30**, 3369–3371 (1984)
104. Glesson, H.F., Coles, H.J.: Dynamic properties of blue-phase mixtures. Liq. Cryst. **5**, 917–926 (1989)
105. Li, Y., Chen, Y., Yan, J., Liu, Y., Cui, J., Wang, Q., Wu, S.-T.: Polymer-stabilized blue phase liquid crystal with a negative Kerr constant. Opt. Mater. Express **2**, 1135–1140 (2012)
106. Cheng, H.-C., Yan, J., Ishinabe, T., Wu, S.-T.: Vertical field switching for blue-phase liquid crystal devices. Appl. Phys. Lett. **98**, 261102 (2011)
107. Lee, S.H., Lee, S.L., Kim, H.Y.: Electro-optic characteristics and switching principle of a nematic liquid crystal cell controlled by fringe-field switching. Appl. Phys. Lett. **73**, 2881–2883 (1998)
108. Lee, S.H., Bhattacharya, S.S., Jin, H.S., Jeong, K.-U.: Devices and materials for high-performance mobile liquid crystal displays. J. Mater. Chem. **22**, 11893–11903 (2012)
109. Chen, Y., Xu, D., Wu, S.-T., Yamamoto, S., Haseba, Y.: A low voltage and submillisecond-response polymer-stabilized blue phase liquid crystal. Appl. Phys. Lett. **102**, 141116 (2013)
110. Wang, L., He, W., Xiao, X., Meng, F., Zhang, Y., Yang, P., Wang, L., Xiao, J., Yang, H., Lu, Y.: Hysteresis-free blue phase liquid-crystal-stabilized by ZnS nanoparticles. Small **8**, 2189–2193 (2012)
111. Wang, L., He, W., Xiao, X., Wang, M., Wang, M., Yang, P., Zhou, Z., Yang, H., Yu, H., Lu, Y.: Low voltage and hysteresis-free blue phase liquid crystal dispersed by ferroelectric nanoparticles. J. Mater. Chem. **22**, 19629–19633 (2012)
112. Wong, J.-M., Hwang, J.-Y., Chien, L.-C.: Electrically reconfigurable and thermally sensitive optical properties of gold nanorods disperses liquid crystal blue phases. Soft Matter **7**, 7956–70959 (2011)
113. Hussain, Z., Danner, D., Masutani, A., Hollfelder, N., Pleis, F., Nelles, G., Kilickiran, P.: Investigation of reactive acrylic monomers for their effect on the temperature range and operating voltage of polymer-stabilized optically isotropic liquid crystal blue phases. Liq. Cryst. **39**, 1345–1357 (2012)
114. Lan, Y.-F., Tsai, C.-Y., Lu, J.-K., Sugiura, N.: Mechanism of hysteresis in polymer-network stabilized blue phase liquid crystal. Polymer **54**, 1876–1879 (2013)
115. Chen, Y., Wu, S.-T.: Recent advances on polymer-stabilized blue phase liquid crystal materials and devices. J. Appl. Polym. Sci. **131**, 40556 (2014)
116. Chen, H.-S., Lin, Y.-H., Wu, C.-H., Chen, M., Hsu, H.-K.: Hysteresis-free polymer-stabilized blue phase liquid crystals using thermal recycles. Opt. Mater. Express **8**, 1149–1155 (2012)
117. Yoshizawa, D., Okumura, Y., Yamamoto, J., Kikuchi, H.: Decreasing the operating voltage of a polymer-stabilized blue phase based on intermolecular affinity. Polymer J. **5**, 667–673 (2019)
118. Bo, S., Chen, B., Zhu, D., Feng, S., Chen, Z.: Low-voltage and faster-response polymer-stabilized blue-phase liquid crystals achieved a new organosilicone monomer. Soft Matter **18**, 8188–8193 (2022)
119. Zhao, Y., Lu, W., Yin, S., Ge, S., Ma, H., Sun, Y.: Polymer-stabilized blue phase liquid crystal devices with ultra-low hysteresis and driving stability. J. Mol. Liq. **396**, 123963 (2024)
120. Nakajima, K., Kamifuji, H., Nakase, M., Nishi, K., Kikuchi, H., Ozaki, M.: Blue phase-polymer-templated ferroelectric nematic liquid crystal. ACS Appl. Mater. Interfaces **16**, 66552–66559 (2024)
121. Hu, W., Wang, L., Wang, M., Zhong, T., Wang, Q., Zhang, L., Chen, F., Li, K., Miao, Z., Yang, D., Yang, H.: Ultrastable liquid crystalline blue phase from molecular synergistic self-assembly. Nat. Commun. **12**, 1440 (2021)
122. Yang, D.K., Crooker, P.P.: Blue phase III of chiral liquid crystals in an electric field. Phys. Rev. A **37**, 4001 (1988)
123. Kitzerow, H.-S., Crooker, P.P., Heppke, G.: Line shapes of field-induced blue-phase-III selective reflections. Phys. Rev. Lett. **67**, 2151–2154 (1991)
124. Yoshizawa, A.: Molecular design for stabilizing a blue phase III and electro-optical switching in the blue phase. J. Soc. Inf. Disp. **16**, 1189–1194 (2008)

125. Chen, H.-Y., Lu, S.-F., Hsieh, Y.-C.: Unusual electro-optical behavior in a wide-temperature BPIII cell. Opt. Express **21**, 9774–9779 (2013)
126. Chen, H.-Y., Lu, S.-F., Wu, P.-H., Wang, C.-S.: Transflective BPIII mode with no internal reflector. Liq. Cryst. **44**, 473–478 (2017)
127. Khan, R.K., Mohiuddin, G., Begum, N., Turlapati, S., Namdiraju, R.V.S., Debbarma, B.K., Ghosh, S.: Extremely high Kerr constant and low operating voltage in a stable room-temperature blue phase III derived from three-ring-based bent-core molecules. ACS Appl. Mater. Interfaces **14**, 42628–42634 (2022)
128. Joannopoulos, J.D., Villeneuve, P.R., Fan, S.: Photonic crystals: putting a new twist on light. Nature **386**, 143–149 (1997)
129. Aresenault, A., Fournier-Bidoz, S., Ozin, G.A., et al.: Towards the synthetic all-optical computer: science fiction or reality? J. Mater. Chem. **14**, 781–794 (2004)
130. von Freymann, G., Ledermann, A., Thiel, M., Staude, I., Essig, S., Busch, K., Wegner, M.: Three-dimensional nanostructures for photonics. Adv. Func. Mater. **20**, 1038–1052 (2010)
131. Chen, C.-W., Bunning, T.J., Khoo, I.-C., Lin, T.-H., et al.: Large three-dimensional photonic crystals based on monocrystalline liquid crystal blue phases. Nat. Commun. **8**, 727 (2017)
132. Hur, S.-T., Lee, B.R., Gim, M.-J., Park, K.-W., Song, M.H., Choi, S.-W.: Liquid-crystalline blue phase laser with widely tunable wavelength. Adv. Mater. **25**, 3002–3006 (2013)
133. Castles, F., Morris, S.M., Coles, H.J., et al.: Stretchable liquid-crystal blue-phase gels. Nat. Mater. **13**, 817–821 (2014)
134. Nayek, P., Jeong, H., Park, H.R., et al.: Tailoring monodomain in blue phase liquid crystal by surface pinning effect. Appl. Phys. Express **5**, 051701 (2012)
135. Yoshizawa, A.: Amorphous blue phase III: structure, materials, and properties. Materials **17**, 1291 (2024)

Chapter 5
Supramolecular Chirality Arising from Spontaneous Mirror Symmetry Breaking: Its Amplification and Transfer

Abstract The creation of homochiral materials from achiral molecules has garnered considerable significant interest. The underlying mechanisms are of significance not only for the production of useful chiral materials, such as pharmaceuticals, but also for the inference of the emergence of homochirality in nature. The spontaneous mirror symmetry breaking has the capacity to produce chirality in an achiral system. Investigations have been conducted into asymmetric autocatalytic reactions that do not require a chiral source, as well as other methods that utilize external stimuli, such as mechanical rotation, circularly polarized light, and asymmetric reaction fields. Recent findings have revealed that mirror symmetry breaking occurs in liquid crystals. This chapter elucidates the formation of supramolecular chiral materials by the transfer of chirality arising from the mirror symmetry breaking in achiral liquid-crystalline trimers. The symmetry-breaking process leads to the formation of conglomerates, where chiral domains with opposite twist senses are present in a non-equal population. The dominant homochiral domains undergo amplification to generate a homochiral template. The supramolecular chirality is transferred to a polymer film during the template polymerization of achiral monomers. Finally, we discuss the implications of this mechanism in relation to the occurrence of homochirality in life.

Keywords Chirality · Mirror symmetry breaking · Chiral transfer · Liquid crystals · Helical polymers

5.1 Introduction

Chirality is seen in everywhere from molecular to macroscopic levels in nature. For example, from small amino acid molecules to the biological macromolecule deoxyribonucleic acid. Only *L*-amino acids exist in nature, and the question why only one

---

Supplementary Information The online version contains supplementary material available at https://doi.org/10.1007/978-981-92-2524-8_5.

A. Yoshizawa, *Liquid-Crystalline Supermolecules*, The Materials Research Society Series, https://doi.org/10.1007/978-981-92-2524-8_5

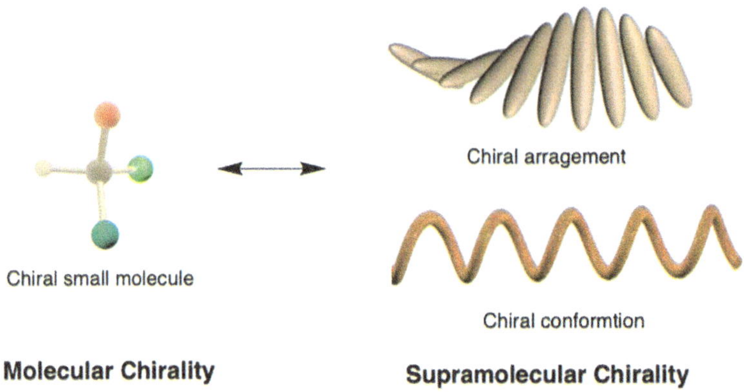

Fig. 5.1 Schematic illustration of chirality from molecular to supramolecular levels. Reproduced from [7]

enantiomer is present remains unknown. Chiral materials play an important role in various fields, such as medicine and optical materials [1]. The efficient production of the chiral materials is in high demand. Generally, producing the chiral materials requires a chiral source, such as chiral building blocks or chiral catalysts [1]. Enantioselective synthesis using asymmetric catalysts in particular has been employed for the production of many useful chiral substances [2]. Recently, the formation of chiral architectures by the self-organization of achiral small molecules has been intensively investigated [3–7]. We use the term “supramolecular chirality” to describe the chirality [7]. Figure 5.1 shows the schematic illustration of chirality from molecular to supramolecular levels. The supramolecular chirality usually appears in the chiral arrangement of molecules and the helical polymer. The appearance of chirality in an achiral world is important for understanding the emergence of homochirality in nature [8, 9]. We will show how mirror symmetry breaks in achiral systems with the following examples.

The first example shows asymmetric autocatalysis inducing the molecular chirality [10–15]. Soai et al. discovered the amplification of enantiomer excess (ee) during the asymmetric autocatalysis of 2-alkynylpyrimidine-alkanol after the addition of i-Pr_2Zn to 2-alkynylpyrimidine-5-carbaldehyde (Fig. 5.2) [10, 16]. Pyrimidine alkanol with low ee (ca 5×10^{-5}% ee) acts as an asymmetric autocatalyst with amplification of ee to afford itself with high ee (>99.5%). Furthermore, Soai et al. reported that optically active (S)- or (R)-pyrimidyl alkanol was generated from the reaction between 2-alkynylpyrimidine-5-carbaldehyde and i-Pr_2Zn without adding chiral compounds (Fig. 5.3) [16, 17]. The aldehyde **1** is treated with i-Pr_2Zn in a mixed ether–toluene solvent to give the racemic reaction intermediate. However, the racemate does not statistically contain exactly the same number of each enantiomer. Therefore, an initial bias in ee resulting from a fluctuation in the racemate would be enhanced to produce an enantiomeric product by the asymmetric autocatalysis [16]. The absolute configurations of the pyrimidyl alkanol show an approximate stochastic distribution. This autocatalytic reaction is spontaneous absolute asymmetric synthesis, inducing chirality at a molecular level. Because it is an autocatalytic reaction, there are limitations on the products obtained from this reaction.

Next, we present some examples of supramolecular chiral induction in an achiral system under an external stimulus. Enantiomeric crystallization was observed under stirring conditions in $NaClO_3$ [18] and 1,1′-binaphthyl melt [19, 20]. Recently, Ishi et al. reported a rotary-evaporation-induced enantioselective aggregation of achiral phthalocyanines (**H_2Pc**) [21]. The chiral films were obtained on the bottom of the flask. The **H_2P_C** molecules form face-to-face type aggregates (H-aggregates) showing helicity based on the rotation of the constituent molecules around Z_H axis (Fig. 5.4). Interestingly, the handedness of the chirality depends on the rotational direction of the evaporator. The H-aggregates of **H_2P_C** prepared with anticlockwise (or clockwise) were twisted to the right-hand (or the left-hand) around the Z_H axis. The enantioselective aggregation of **H_2Pc** can be explained by the helicity of the fluid motion. In the case of anticlockwise rotation of the evaporation flask, the

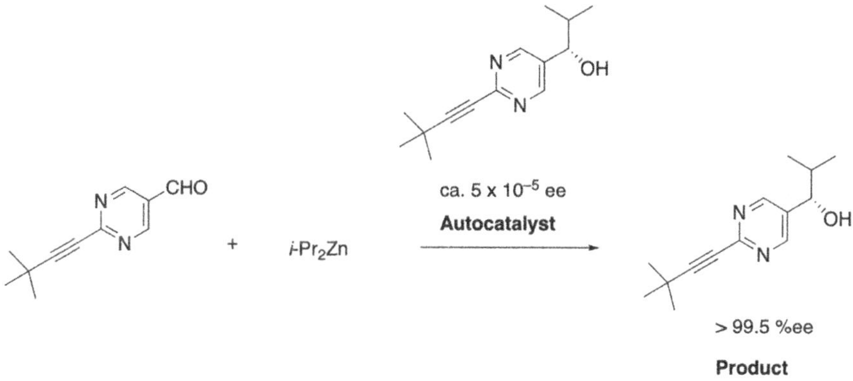

Fig. 5.2 Asymmetric autocatalysis of pyrimidyl alkanol

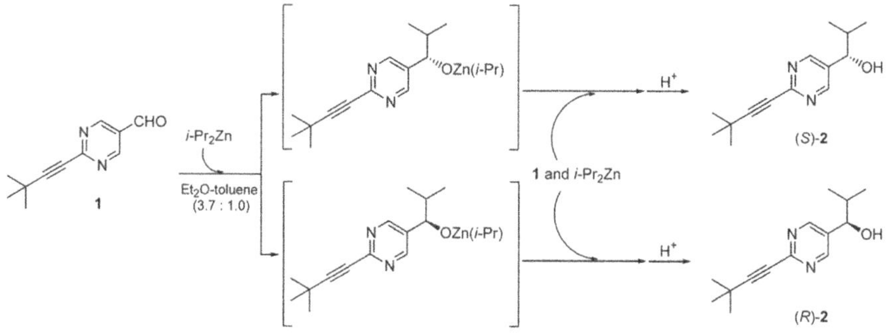

Fig. 5.3 Asymmetric synthesis of pyrimidyl alkanol without adding chiral substances. Reprinted with permission from [17]. Copyright 2003 Elsevier Science Ltd.

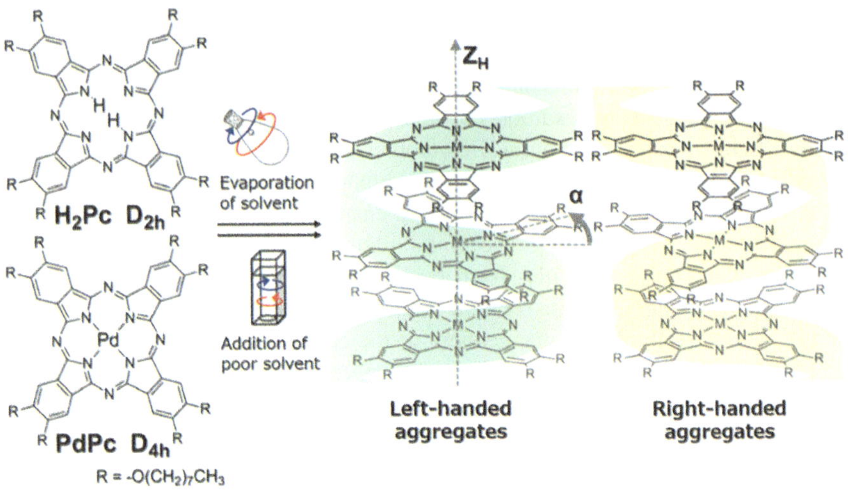

Fig. 5.4 Molecular structures of phthalocyanine (H_2Pc) and chiral H-aggregates of phthalocyanines prepared by mechanical rotations. Blue and red arrows show clockwise and anticlockwise rotations, respectively. Reprinted with permission from [21]. Copyright 2019 WILEY-VCH Verlag GmbH & Co. KGaA, Weinheim

right-handed helicity of the fluid motion is transferred to the H-aggregates of $\mathbf{H_2Pc}$ to form the right-handed helicity. Fluid dynamics, resulting from the macroscopic mechanical rotation, selectively induces one of two enantiomers in the nanoscale supramolecular aggregation.

Tejedor et al. reported that irradiating circularly polarized light (CPL) induced a helical organization in an achiral nematic glassy phase of azobenzene polymer P100 (Fig. 5.5a) [22]. CPL is a chiral electromagnetic radiation. The chiral exciting light induces an asymmetric photoisomerization of an azobenzene group in the side chain. The photo-induced chiral component could act as a chiral dopant for the nematic glassy phase, inducing a supramolecular chiral organization as the chiral nematic version. The helical handedness can be switched using orthogonal CPL. The occurrence of the supramolecular chirality is explained by the chirality transfer from CPL to the azopolymer P100 through the asymmetric photoisomerization (Fig. 5.5b) [22]. The question remains as to whether the chiral component originates from *E*- or *Z*-azobenzene. This will be discussed in Sect. 5.4.

The spontaneous mirror symmetry breaking has been observed in liquid-crystalline systems of achiral molecules [23–40]. It produces not a homochiral state but chiral conglomerates consisting of domains with opposite handedness. For example, we describe helical nanofilament phases (HNF phases, also assigned as B4 phases). They exhibit the mirror symmetry braking and have a characteristic helical network [33]. Figure 5.6a shows the molecular structure of achiral bent-core molecule **P-9-OPIMB** exhibiting the HNF phase [33]. Figure 5.6b shows the optical textures of **P-9-OPIMB** mixed with a rod-like liquid crystal in the HNF phase between crossed and uncrossed polarizers [41]. The texture under crossed polarizers

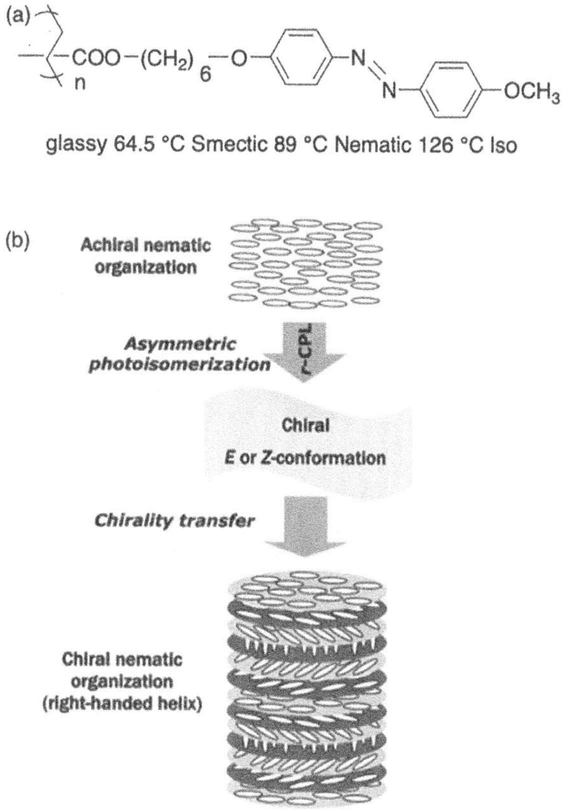

Fig. 5.5 **a** Structure and thermal properties of the side-chain liquid crystal azopolymer **P100**. **b** Proposed model for the photoinduction of chirality in azo materials. Reprinted with permission from [22]. Copyright 2007 WILEY-VCH Verlag GmbH & Co. KGaA, Weinheim

was nearly dark, suggesting that it is optically isotropic. By observing the sample under slightly uncrossed polarizers, the texture was split into darker and brighter domains. By uncrossing the polarizers in opposite directions by the same angle, the darker and brighter domains were exchanged. Thus, the two domains have optical activity with opposite handedness. Figure 5.6c shows a freeze-fracture transmission electron microscopy (FF-TEM) image of the HNF phase of **P-9-OPIMB** and the structural model of the helical nanofilaments [42]. There are two proposed mechanisms for the occurrence of the supramolecular chirality: layer chirality [43] and conformational chirality [25]. We will discuss the mechanisms later. It should be noted that the symmetry breaking in liquid crystals cannot achieve a homochiral phase.

We then describe the supramolecular chirality of helical polymers. The synthesis of chiral polymers from achiral small molecules or racemates has been reported [3, 6, 44–49]. Among them, the supramolecular chirality occurs during the polymerization of achiral monomers under the influence of noncovalent interactions, such as molecular chirality of a catalyst, CPL, and chiral solvents [6]. Akagi et al. reported the synthesis of helical polyacetylene in chiral nematic liquid crystals [50,

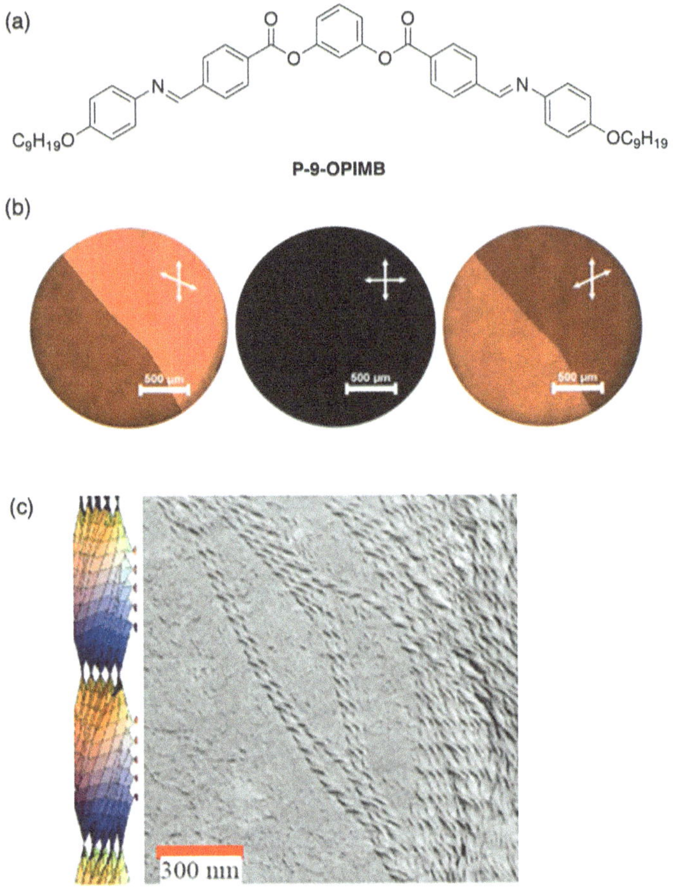

Fig. 5.6 **a** Molecular structure of ahiral bent-core liquid crystal **P-9-OPIMB**. **b** Polarized optical textures of the HNF phase of **P-9-OPIMB** blended with the rod-like liquid crystal between uncrossed and crossed polarizers. The figure was reproduced from [41]. **c** FF-TEM image of the helical nanofilamants of **P-9-OPIMB** with their structural model. Reprinted with permission from [42]. Copyright 2015American Chemical Society

51]. The chiral nematic LC is used as an asymmetric solvent. The synthetic procedure is as follows. The euimolecular mixture of the nematic liquid crystals **PCH302** and **PCH304** was mixed with a chiral dopant such as a binaphthyl derivative or a rod-like chiral compound. The molecular chirality of the chiral dopant transforms the nematic (N) phase to the chiral nematic (N*) phase with the supramolecular helical ordering. Ziegler-Natta catalyst, Ti(O-*n*-Bu)$_4$-AlEt$_3$, was dissolved into the N*-LC to produce the asymmetric reaction field for the acetylene polymerization (Fig. 5.7). Left-handed and right-handed helical polyacetylene chains were formed in right-handed and left-handed chiral nematic liquid crystals, respectively. The supramolecular chirality of the N*-LC transfers to the helical polyacetylene in the course of polymerization of

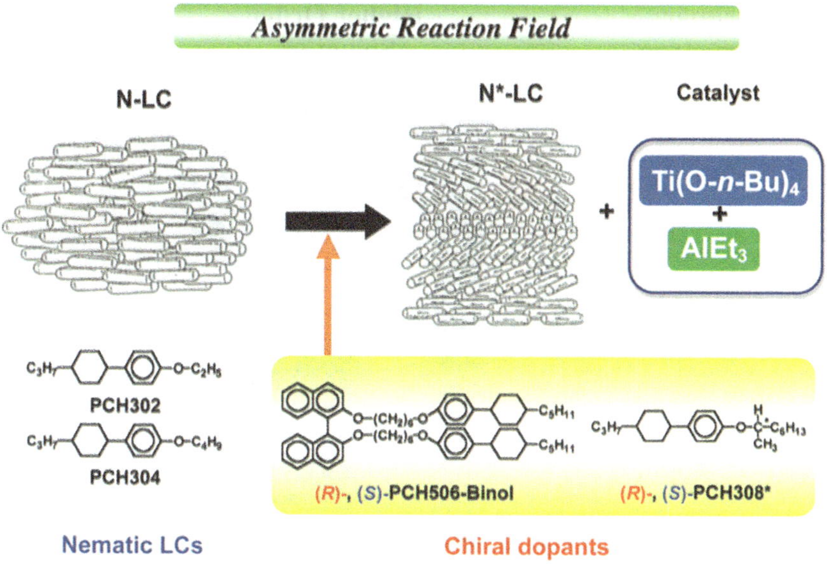

Fig. 5.7 Asymmetric reaction field for the formation of helical polyacetylene. Reprinted with permission from [51]. Copyright 2009 American Chemical Society

achiral acetylene monomers in the asymmetric solvent. However, the helical directions of the helical polyacetylene are opposite to those of the corresponding N*-LCs. The primary helical chains are bundled to form helical fibrils. The bundles of fibrils further form the spiral morphology as shown in Fig. 5.8.

As described above, external asymmetric stimuli, such as asymmetric rotational directions, CPL, and asymmetric solvent, induce the supramolecular chirality to produce the homochiral macrostructures. On the other hand, the spontaneous mirror symmetry breaking in achiral liquid crystals only gives the chiral conglomerates. This chapter describes how the spuramolecular chirality emerging from achiral liquid crystals is transferred to a homochiral helical polymer.

5.2 Design Concept of a Dynamic Chiral Origin

Figure 2.10 in Chap. 2 shows the design concept of the liquid-crystalline supermoleculs. This chapter presents how the desigend liquid-crystalline supermolecules induce the supramolecular chirality. Some linear-shaped dimers affects the interlayer correlation in the smectic phases as described in Chap. 2. We already reported a linear chiral dimesognic molecule in which two phenylpyrimidine cores are connected via optically active 3-methyladipic acid [52]. It has a chiral center in the spacer. Although the molecule itself does not have a large helical twisting power, the mixtures with a host smectic liquid crystal exhibited two kinds of twist grain boundary (TGB) phases

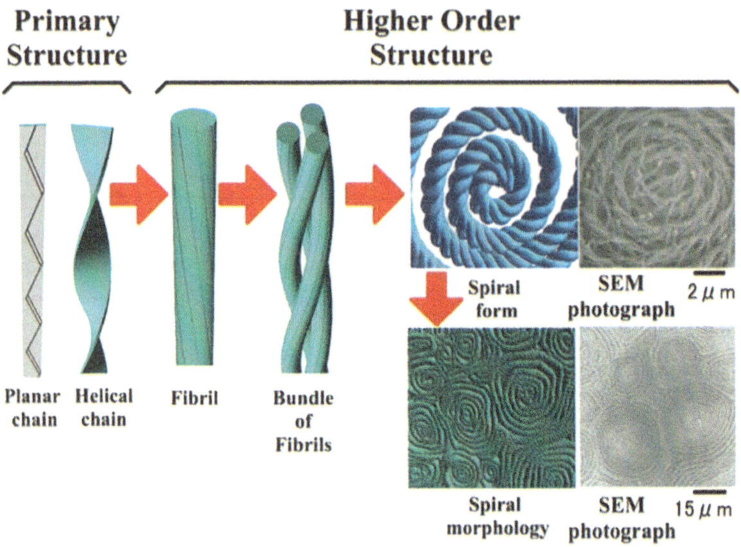

Fig. 5.8 Hierarchical helical structures from primary to high order in helical polyacetylene. Reprinted with permission from [51]. Copyright 2009 American Chemical Society

(Fig. 5.9) [52]. The appearance of the TGB phases of the mixture can be interpretd in terms of intercalated chirality as follows. The intercalated dimer enhances the interlayer interactions between the adjacent layers of the mixture. The twisting power originated from the chiral space desires to form a spiral along the direction perpnedicular to the layer normal. The competition between the interlayer interaction and the twisting force the helical ordering of the smectic blocks.

If a liquid-crystalline molecule has an order resembling a pseudo-layered structure and can form a twist conformation, the spontaneously mirror symmetry breaking might occur in the achira liquid crystal phase (Fig. 5.10). The deigned liquid-crystalline supermolecules do not have a molecular chirality but can produce the dynamic chiral origin in the molecular aggregates.

5.3 Spontaneous Mirror Symmetry Breaking in Achiral Flexible Liquid Crystal Trimers

Since the discovery of liquid crystals, molecular chirality had been considered essential for the emergence of chiral LCs and ferroelectric LCs. Niori et al. discovered ferroelectric switching in achiral bent-core molecule **P-n-OPIMB** in 1996 [23], and then, Sekine et al. suggested the emergence of chirality in the fluid smectic B2 phase [24]. These discoveries brought a breakthrough to research on polar ordering and/or chirality in liquid crystals [53, 54]. Two mechanisms were poposed for the

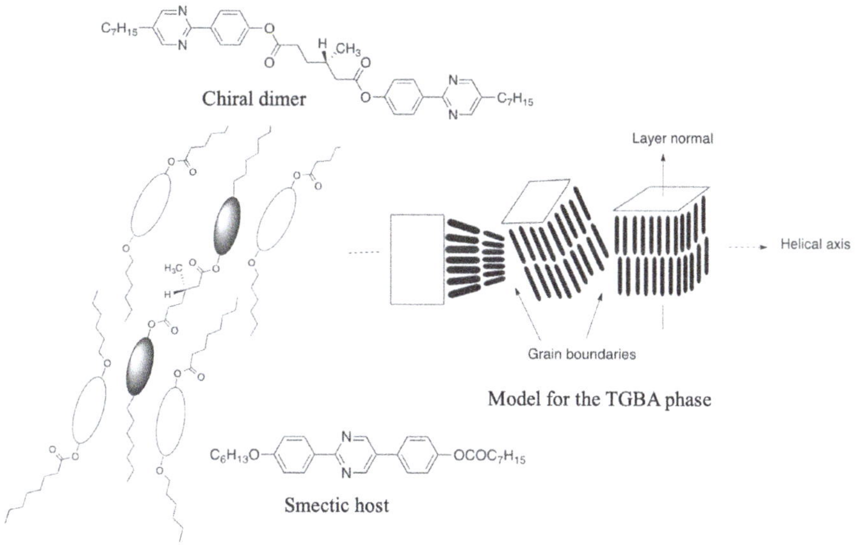

Fig. 5.9 Intercaleted chirality model for the foematio of the TGBA phase

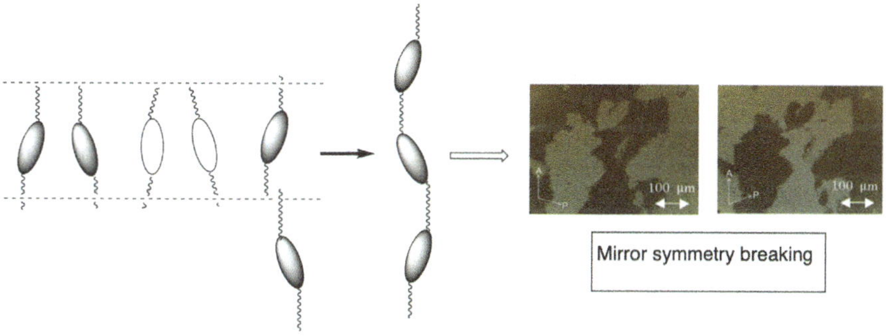

Fig. 5.10 Design concept of the dynamic chiral origin

origin of chirality in achiral bent-core liquid crystals. One of them is layer chirality. Figure 5.11a shows two bent-core moleculs tileted in opposite directions relative to the layer normal. For the molecule on the left, the convex direction and tilt direction of the molecule are clockwise when they are rotated around the layer's normal direction. On the other hand, those of the molecule on the right are anticlockwise. The other one is conformational chirality (Fig. 5.11b). The bent-core molecules have twisted side wings, forming a twist conformation.

Optically isotropic layered phases such as liquid-crystalline sponge phases [32], helical nanofilament phases [33], and helical nanocrystalline (HNC) phases [34], exhibit characteristic three-dimensional structures. They appear as chiral conglomerates by the spontaneous mirror symmetry breaking. Figure 5.12 shows their schematic

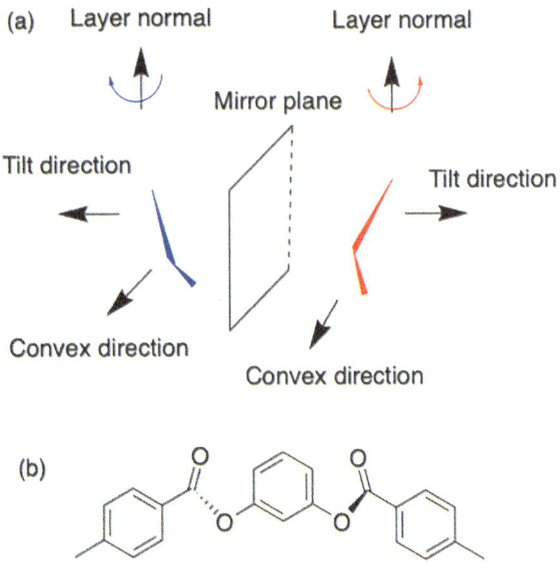

Fig. 5.11 **a** Layer chirality [43] and **b** conformational chirality [25]

models. The tilt organization of bent-core molecules in the layers drives the twisting of those layers to produce the saddle splay curvature, which can relieve the stress in the layer structures. Compounds exhibiting those phases possess a rigid bent-core structure, with a few exceptions. Biphenyl-based symmetric dimers with an odd number of carbon atoms in the methylene spacer exhibited the HNF phase [55]. The inherent twist of the rigid bent-core as shown in Fig. 5.11b is essential for the formation of these phases. When the twisted molecules overlap in each layer, nano-ordered smectic spirals are formed to induce the supramolecular chirality. The symmetry breaking has also been observed in achiral LCs without layer ordering [37–39, 56–62]. Cestari et al. the twist–bend nematic (N_{TB}) phase in 1,7-bis(4-cyanobiphenyl-4′-yl)heptane in 2011 [37]. The N_{TB} phase has been observed in liquid crystal dimers with an odd number of methylene spacers, as well as in bent-core liquid crystals. Dozov predicted that bent-shaped molecules can exist the N_{TB} phase [57]. The bent shape is a key for the formation of the N_{TB} phase. Figure 5.13 shows how the bent-shaped molecules form the N_{TB} phase. They cannot fill the space in the nematic phase. Some molecules rise to form helical structures. On the other hand, Ivic et al. suggested that not only the geometry shape of the molecule but also the twist conformation contributes to the helical ordering in the N_{TB} phase [63]. The molecular long axis is tilted with respect to the helical axis in the N_{TB} phase. Figure 5.14 shows the molecular structures of typical bent-core molecules and the flexible dimer.

We reported the spontaneous mirror symmetry breaking in achiral flexible liquid crystal trimers [65]. It was observed in trimer **I-(7,7)**. Figure 5.15 shows the molecular structure and the optical textures of the chiral conglomerates. The texture between crossed polarizers was nearly dark. Chiral domains with opposite twist senses were observed the POM textures between slightly uncrossed polarizers. Therefore, we

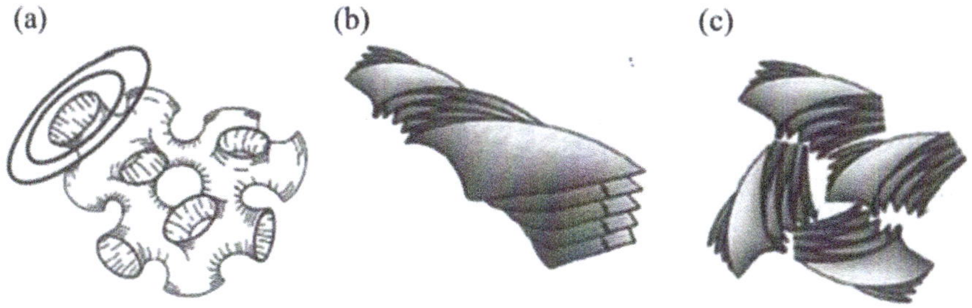

Fig. 5.12 **a** Sponge phase, **b** helical nanofilament phase, and **c** helical nanocrystalline phase. Reproduced from [64]

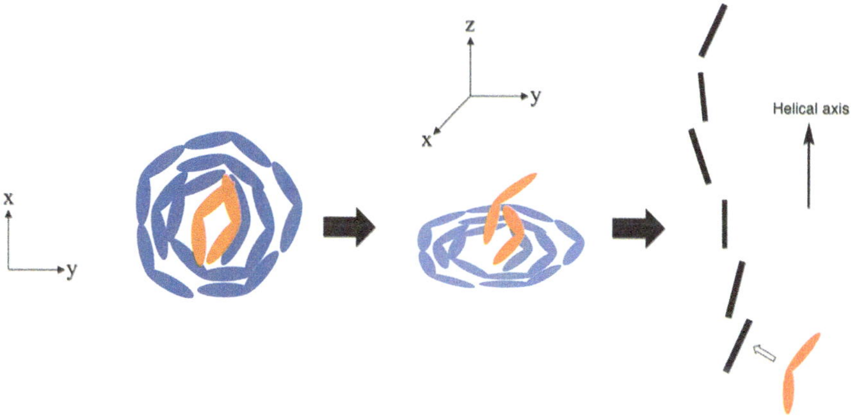

Fig. 5.13 Schematic illustration for the formation of the N_{TB} phase

call this state as dark chiral conglomerate (DC) phase. Figure 5.16 shows the DSC thermograms. Trimer **I-(7,7)** exhibited a phase sequence of Iso–N–DC on cooling. The N–DC phase transition accompanied a large entropy change ($\Delta S/R$) of 10.4. Therefore, the DC phase is not a liquid crystal but a soft crystal. On heating a virgin sample, two crystalline phases existed below the N phase. No optical activity was detected in either the crystalline phases or the enantiotropic N phase.

The DC phase has a layer structure with a periodicity of 56.5 Å and the correlation length for the periodicity is 350 Å, which corresponds to about six layers. Figure 5.17 shows MOPAC models for the trimer with an extended conformation (Fig. 5.17a) and that with a twisted conformation (Fig. 5.17b) and their estimated molecular lengths. These indicate that the trimer molecules form a twisted conformation in the layered DC phase. The DC phase may form nano-ordered smectic spirals.

Figure 5.18 shows the phase transition temperatures for a homolog series of trimers **I-(*n*,*m*)**. Trimers **I-(7,7)**, **I-(7,9)**, and **I-(9,9)** exhibited the DC phase. They possess longer spacers with an odd number of methylene units. They have larger flexibility than the others due to their longer spacers. If we assume that the spacers of the

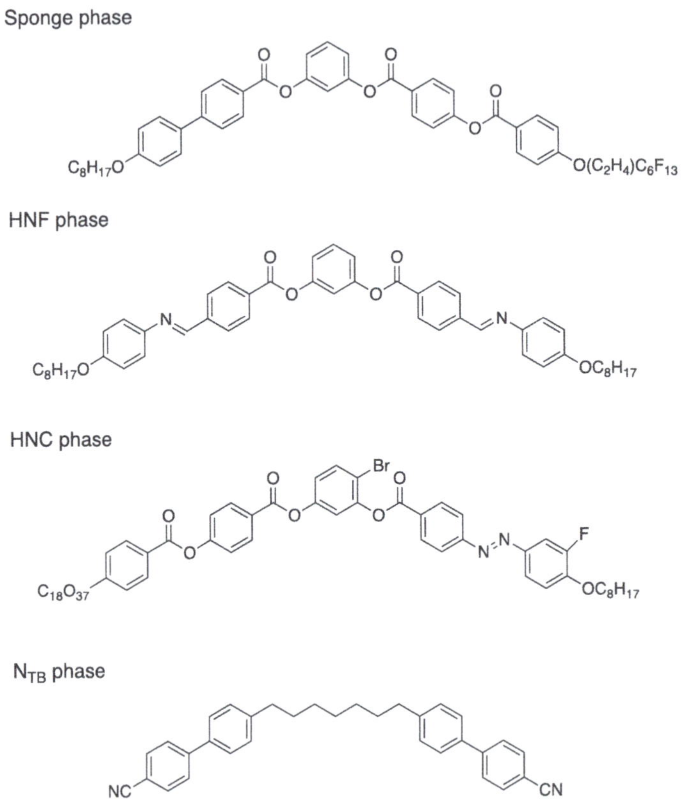

Fig. 5.14 Bent-shaped LCs exhibiting the layered chiral conglomerate phases and the flexible dimer exhibiting the twist-bend nematic phase. Sponge phase [32]; HNF phase [33]; HNC phase [34]; N_{TB} phase [37]

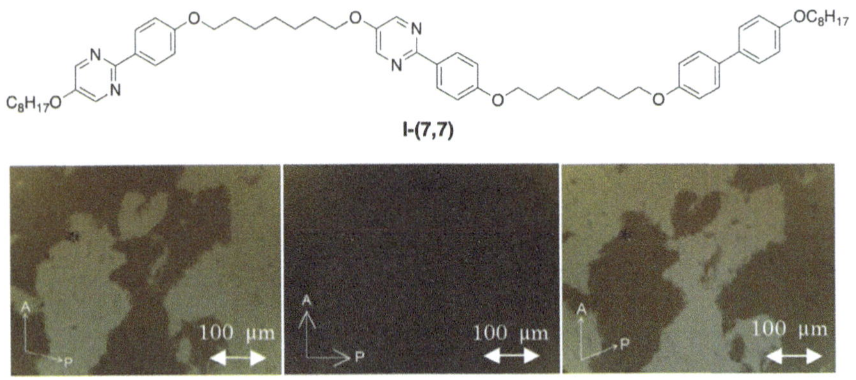

Fig. 5.15 Molecular structure and polarized optical textures of trimer **I-(7,7)** on a glass slide with a cover glass in the DC phase at 130 °C under uncrossed and crossed polarizers. Reproduced from [65] with permission from the Royal Society of Chemistry

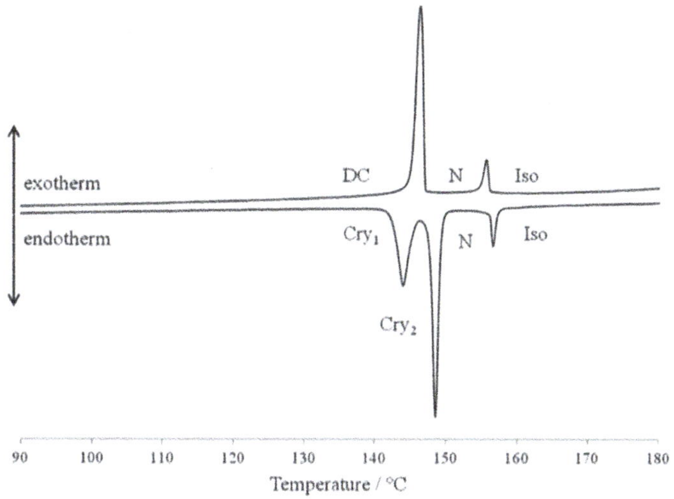

Fig. 5.16 DSC thermogram of trimer **I-(7,7)**. The rate of cooling and heating was 5 °C min^{-1}. Reproduced from [65] with permission from the Royal Society of Chemistry

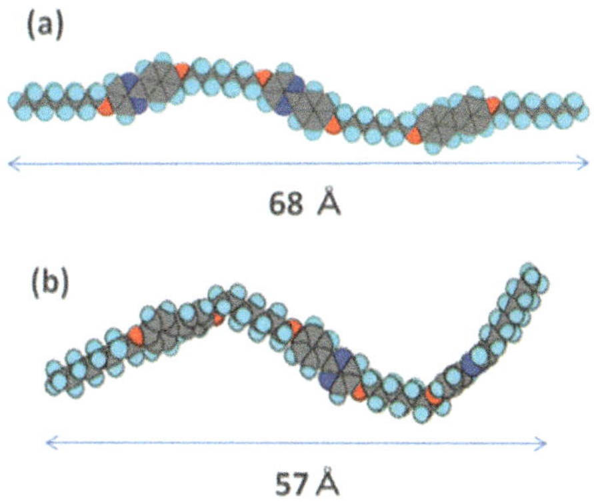

Fig. 5.17 MOPAC models for trimer **I-(7,7)** with **a** an extended conformation and **b** a twisted conformation. Reproduced from [65] with permission from the Royal Society of Chemistry

trimers adopt an all-trans conformation, the odd-numbered trimers form a zigzag shape. Therefore, they may also form twisted structures. The trimer forms an achiral conformer in the nematic phase. The intermolecular core-core interactions in the soft-crystalline DC phase cause the trimer to adopt a metastable twisted conformer. The chiral synchronization of the transiently chiral molecule forms the nano-ordered smectic spirals. Figure 5.19 shows a schematic model how the supramolecular chirality appears in the DC phase. The achiral trimer produces a dynamic chiral origin.

We observed the surface structures of the trimers exhibiting the DC phase by using atomic force microscopy (AFM) [66]. The results are shown in Fig. 5.20. Trimer **I-(7,7)** showed a disordered surface relief exhibiting dimples with 40 ± 20 nm diameter,

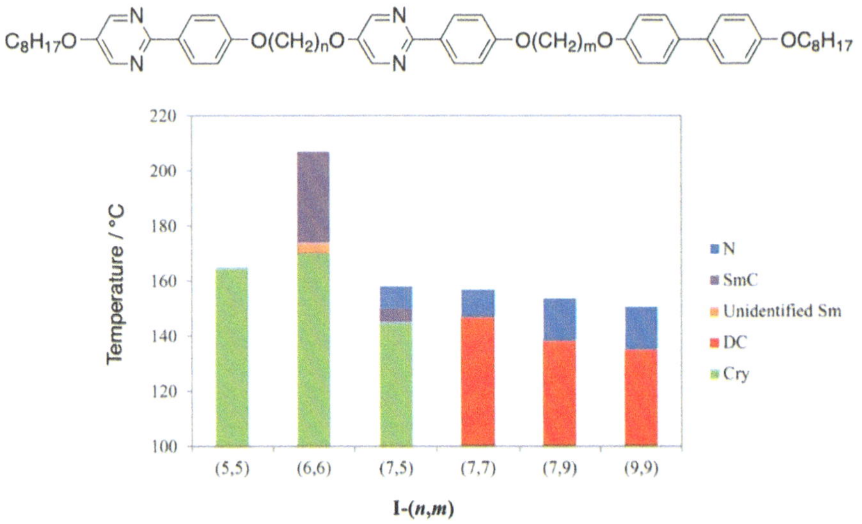

Fig. 5.18 Phase diagram on cooling for a homologue series of trimers **I-(*n*,*m*)**. Reproduced from [65] with permission from the Royal Society of Chemistry

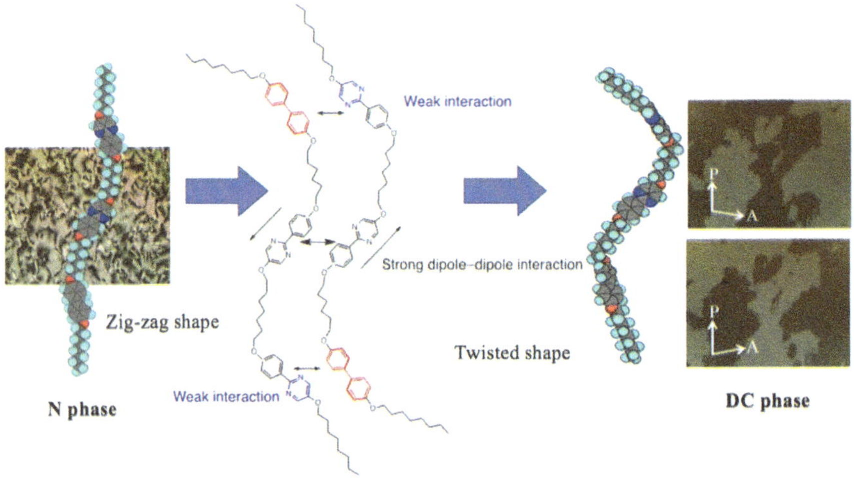

Fig. 5.19 Schematic model for the N–DC phase transition of trimer **I-(7,7)**

whereas trimer **I-(9,9)** showed a well-ordered surface relief exhibiting dimples with 95 ± 15 nm diameter.

Then we investigated both the surface and inside structures of trimer **I-(9,9)** by using field-emission scanning electron micrograph (FE-SEM). The air/liquid crystal interface shows bicontinuous networks (Fig. 5.21a). As shown in its higher magnification (Fig. 5.21b), the networks accompanying hexagonal voids are organized. The

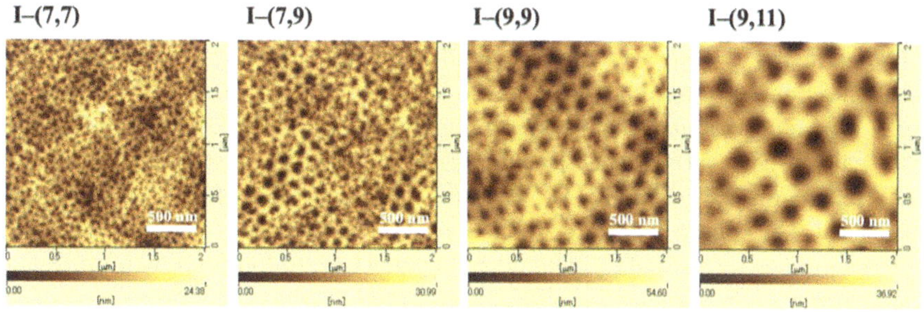

Fig. 5.20 AFM images of trimers **I-(7,7)**, **I-(7,9)**, **I-(9,9)**, and **I-(9,11)** in the DC phases at room temperature. Each image shows a microstructure at the air/liquid crystal interface of a droplet sample on an untreated glass substrate. Reproduced from [66] with permission from the Royal Society of Chemistry

network consists of bicontinuous triply branchings, and defects exist in the boundaries between adjacent branches. Helically disordered layer structures may exist in each unit block. On the other hand, the cross-section area shows a sponge-like structure (Fig. 5.21c). In lipid–water systems, a sponge phase is regarded as a disordered version of the bicontinuous cubic phases [67]. We explain the mechanism of the formation of the DC phase as follows. Upon cooling to the DC phase, layer ordering occurs and then the layers are helically deformed at the surface. On the other hand, both the layer formation and the helical ordering of trimer molecules occur simultaneously in the bulk. The competition between those orderings induces the layer deformation to produce the disordered version of the bicontinuous cubic structure.

Szepke et al. prepared a DC thin film of **I-(9,9)** and investigated the surface structure and optical effects [68]. The DC film was prepared as follows. A toluene solution of **I-(9,9)** was slowly drop-cast to a standard glass slide kept at 70 °C. 10 μm silica spacers were placed near edges of the glass. The material was covered with a second glass, then quickly heated above the melting point, cooled down to 120 °C and at that temperature frozen in liquid nitrogen. AFM topography image of the surface structure revealed that the relative depths of cavities, ~6 nm, remained consistent. Figure 5.22

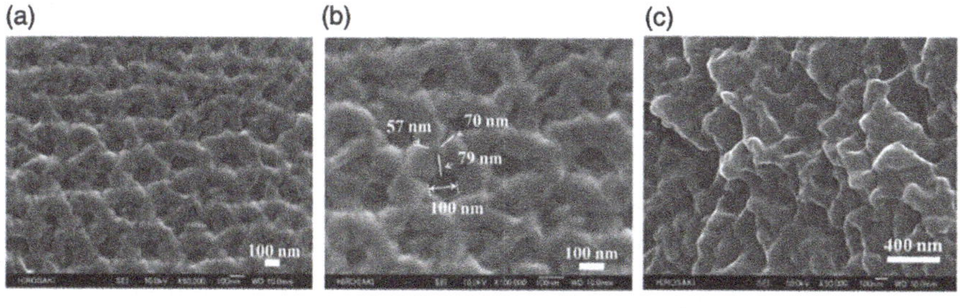

Fig. 5.21 **a** FE-SEM image of the surface structure of **I-(9,9)**. **b** Higher magnification of the surface structure of (**a**), and **c** that of the cross-section area. Reproduced from [66] with permission from the Royal Society of Chemistry

shows the UV-vis and CD spectra of the DC thin film. Mirror spectra showing bands with maxima at ~345 nm were obtained for the domains with the opposite optical rotation. They demonstrated that the DC phase can serve as a chiral template for directing the self-assembly of gold nanoparticles into 3D spiral-like structures (Fig. 5.23). A gold nanoparticle surface was functionalized with liquid–crystal-like ligand (**L1**) and decanethiol (**L2**) (Fig. 5.23a). Spherical gold nanoparticles with 4 nm diameter were used. The composites were prepared by mixing solutions of the functionalized gold nanospheres (Au4) and liquid crystal trimer **I-(9,9)**, followed by drop-casting the mixture and heat annealing of the film. Figure 5.23b shows the TEM image of gold nanospheres assembled in the **I-(9,9)** matrix. The nanoparticles followed curved paths forming nested, spiral-like structures within the cavities of the organic matrix. The nanoparticles follow the twisted layer morphology of the DC phase and form the helix. The DC phase proves to be a good host for gold nanoparticles, allowing for their co-assembly into structures with a 3D spiral-like symmetry. The DC phase consists of chiral conglomerates, therefore, homochiral assemblies cannot be obtained.

Interestingly, symmetric trimers **II-*n*** exhibited the DC phase with a large homochiral domain [69]. Their cooling phase transition temperatures are shown in Fig. 5.24. Similar to trimer **I-(*n*,*m*)**, an odd–even effect of the methylene number in the spacer is observed in the appearance of the DC phase. However, it does not depend on the spacer length. Furthermore, all the odd-numbered trimers showed the direct Iso–DC transition. Trimer **II-9** exhibited two homochiral droplets with opposite handedness in the DC phase as shown in Fig. 5.25. The DC phase has a layer structure with a periodicity length of 43.1 Å and the correlation length is about 700 Å. Considering its molecular length of 67 Å, the DC phase forms am intercalated structure as shown in Fig. 5.26.

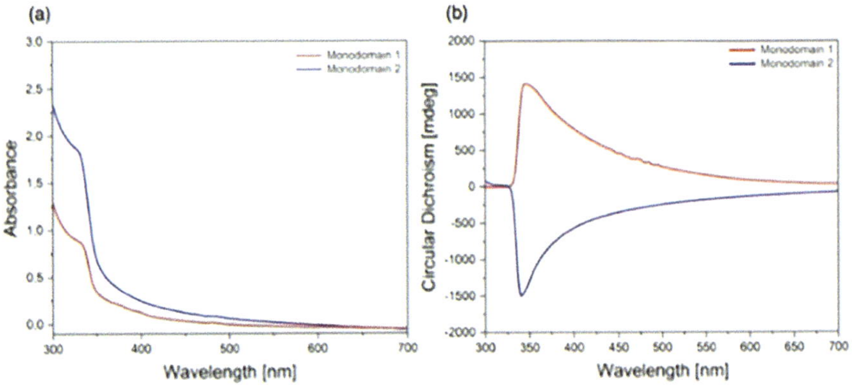

Fig. 5.22 CD measurements of the DC film. Measured spots were picked with POM for measured area to contain a majority of domains of one handedness to assess the qualitative optical activity of a sample. **a** Absorbance spectra and **b** circular dichroism spectra. Reprinted with permission from [68]. Copyright 2024 AIP publishing

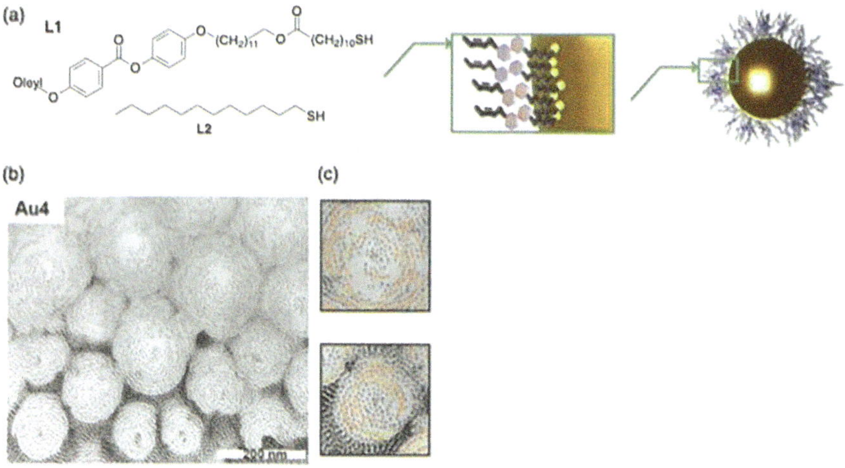

Fig. 5.23 I3D spiral-like assembly of gold nanoparticles inside oof the DC phase of **I-(9,9)**. **a** Molecular structures of **L1** and **L2** with an illustration of the functionalized gold nanosphere. **b** TEM image of 4 nm diameter gold nanospheres (Au4) assembled in the **I-(9,9)** matrix. **c** A close-up of two single cavities with Au4 assembled in a spiral-like manner; curved nanoparticle paths are falsely colored for eye guidance. Reprinted with permission from [68]. Copyright (2024) AIP publishing

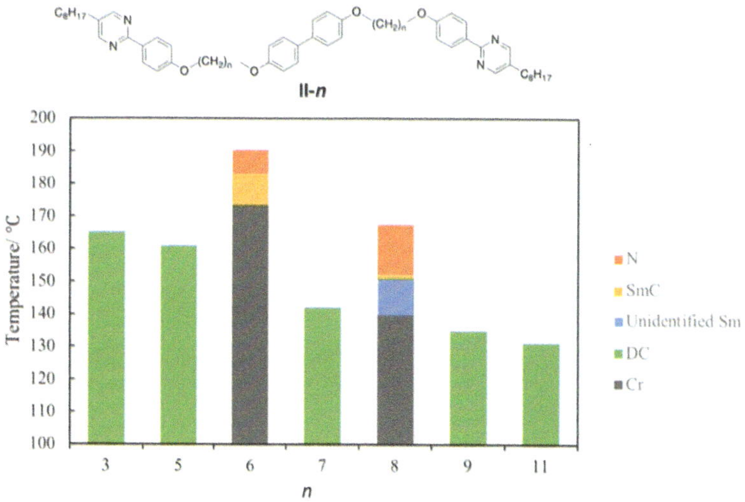

Fig. 5.24 Phase transition diagram on cooling of trimers **II-*n***. Reproduced from [69] with permission from the Royal Society of Chemistry

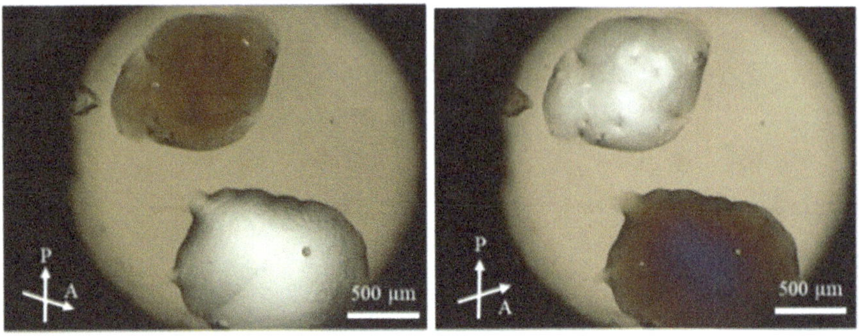

Fig. 5.25 Polarized optical textures of trimer **II-9** in the DC phase under uncrossed polarizers. Reproduced from [69] with permission from the Royal Society of Chemistry

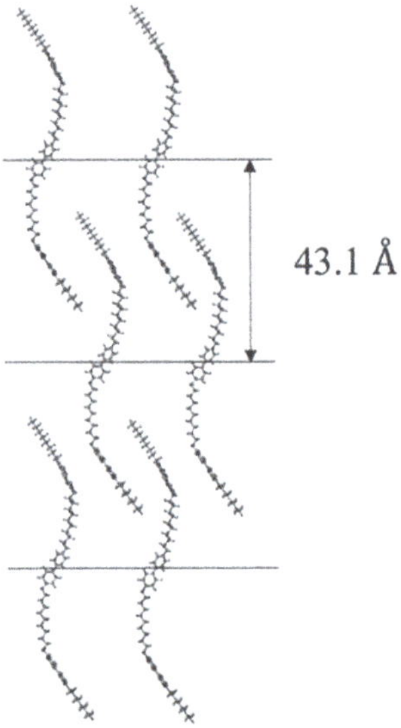

Fig. 5.26 Molecular organization model for trimer **II-9** in the DC phase. Reproduced from [69] with permission from the Royal Society of Chemistry

The DC phase of other odd-numbered trimers also exhibited an intercalated structure, but their nanostructures differed depending on the spacer length. Trimer **II-5** showed hollow cylinders whereas trimers **II-7**, **II-9**, and **II-11** formed toroidal pits at the surface (Fig. 5.27). On the other hand, all three of these trimers formed disordered structures in their cross-section areas. Trimer **II-3** showed cylinders at the surface (Fig. 5.28a) but formed double-winding cylindrical tubes in the cross-section area. (Fig. 5.28b) A spiral-like periodicity can be seen for the tube (see Fig. 5.28c).

We infer the formation of a large homochiral droplet as follows. At the initial stage, ether a right-handed helical structure or a left-handed helical structure appears.

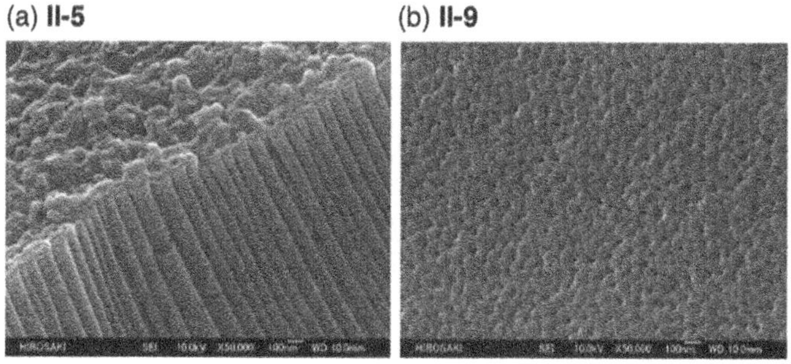

Fig. 5.27 **a** FE-SEM image of the nanostructure at the air/trimer interface of trimer **II-5** and **b** that of trimer **II-9**. The FE-SEM measurements were carried out at room temperature. Reproduced from [69] with permission from the Royal Society of Chemistry

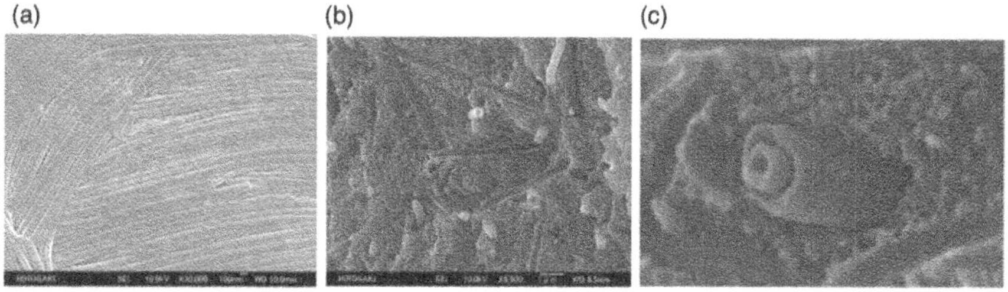

Fig. 5.28 **a** FE-SEM image of the microstructure at the surface of trimer **II-3**, **b** that of the cross-section area, and **c** an enlarged image of the central part of **b** for clarity. The SEM measurements were carried out at room temperature. Reproduced from [69] with permission from the Royal Society of Chemistry

It depends on the statistics. The appeared chirality is synchronized to produce the large homochiral area through the cooperative formation of the intercalated layer structure. Next, we discuss the effect of the spacer length of the odd-numbered trimers on the formation of nanostructure in their DC phases (Fig. 5.29). In the DC phase of trimer **II-3**, the interlayer correlation is strong. The helicity of the metastable twisted conformer is synchronized along the layer normal to produce the spiral-like cylinders. On the other hand, in the DC phase of trimer **II-9**, the dynamic molecular helicity propagates within the layer to form the helical toroidal bits.

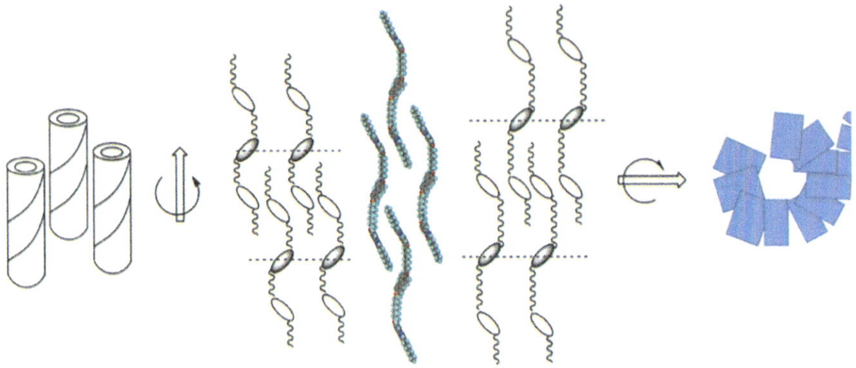

Fig. 5.29 Molecular organization model for the formation of the supramolecular architectures. The arrows show their helical axes. Reproduced from [69] with permission from the Royal Society of Chemistry

5.4 External-Field-Induced Mirror Symmetry Breaking in Liquid Crystals

5.4.1 Photofield

External stimuli such as light or electric fields can also cause the breaking of mirror symmetry in liquid crystals. The photo-induced isomerization of an azobenzene unit causes a remarkable change in the physical properties of liquid crystals. Azobenzene derivatives have been used for applications in photomemory [70, 71], photoalignment [72–77], or photomobile [78, 79]. Light-driven supramolecular chirality has attracted considerable interests. Circular polarized light (CPL) is usually used for the irradiation light [80]. CPL is a type of special polarization where the electric field vector of light rotates at a constant speed relative to the direction of propagation, with its leading edge tracing a circular path. Figure 5.30 shows a schematic illustration of the electric field vectors of a traveling circularly polarized electromagnetic wave. This wave is right-handed/clockwise circularly polarized as defined from the point of view of the source.

Nikolova et al. reported the inducing of large circular anisotropy in films of liquid crystalline side-chain azobenzene polyesters by illumination with CPL [81]. The twist sense of the induced chirality is the same as that of the light polarization. Iftime et al. reported that irradiation with CPL on a film of an achiral azobenzene liquid crystalline polymer (**p4MAN**) produces chirality. And the chirality is reversibly switched by modulation of the CPL (left or right) [82]. Above Tg, **p4MAN** films exhibited a smectic A phase. The irradiation with CPL (l = 514 nm) destroys the smectic layers by forming cis isomers. The cis isomers relax back to the trans isomers, leading to a new state. By the irradiation, the achiral films become chiral. Opposite CPL produced enantiomeric supramolecular structures. Their CD spectra are complete

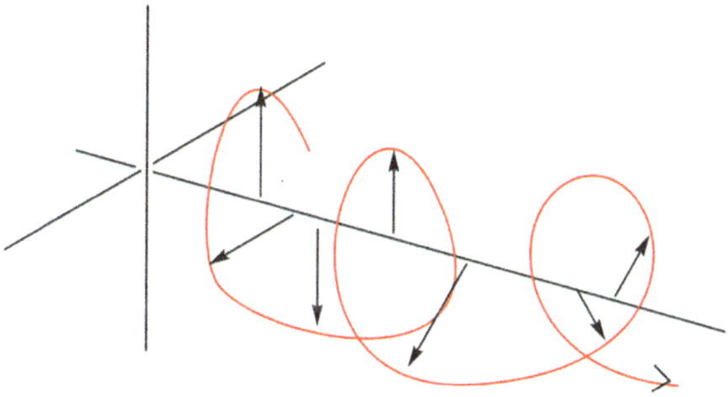

Fig. 5.30 Schematic illustration of right-handed CPL

mirror images of each other. The authors proposed the formation of a TGB-like phase for the photo-induced supramolecular structures (Fig. 5.31). The polymer backbones exist mainly in the planes defining the smectic layers, but can also go between the various smectic grains, thus illustrating the potential for chirality for the main chain. The helical axis is along the film thickness, and the azobenzene units are parallel to the film plane. The helix inversion with the opposite CPL was observed for columnar mesomorphic assembles of propeller-like hydrogen-bonded complexes [83]. A lot of research has been conducted on azobenzene-containing polymer system [84]. On the other hand, photo-induced isothermal phase transitions have been observed for chiral conglomerates composed of achiral liquid-crystalline molecules. Some examples are shown in Fig. 5.32 [85–88]. Almost all the photo-induced isothermal phase transitions are chiral to achiral phase transitions, except for the SmA to Iso_1* transition [87]. In this case mirror symmetry breaking takes place at the transition from a higher to a lower order phase, which is contrary to common knowledge.

As described above, the photo-induced isomerization of an azobenzene unit can induce the supramolecular chirality in higher ordered molecular assemblies. In Sect. 5.1, the question was raised as to whether the chiral component originates from *E*- or *Z*-azobenzene [22]. Non-planar twisted conformations of both *E*- and *Z*-azobenzene units have been reported [89, 90]. We designed a binary system in which achiral azobenzene liquid crystal trimers intercalate with the adjacent layers composed of host smectic molecules [91]. Figure 5.33 shows the molecular structures and transition properties of trimer **III-*n*** (n = 7–11) and **Host LC**. All the trimers except trimer **III-7** showed a nematic phase. Trimer **III-7** showed no liquid crystallinity. Pronounced odd–even effects are seen for the isotropic liquid–N phase transition temperatures and the enthalpy changes of the trimers as the parity of the methylene number in the spacers is varied. Therefore, the trimers with odd-numbered spacers have a zig-zag shape whereas those with even-numbered spacers have a linear shape. Figure 5.34 shows MOPAC models for trimers **III-8** and **III-9**. **Host LC** exhibited N and smectic C (SmC) phases with wide temperatures.

We observed the UV-vis spectra of trimers **III-8** and **III-9** in a chloroform solution (2×10^{-5} M) during UV light irradiation with a wavelength of 365 nm as a function of the irradiation intensity (Fig. 5.35). For both trimers, as increasing the irradiation intensity, an absorbance at 360 nm associated with the π–π* transition of the *trans*

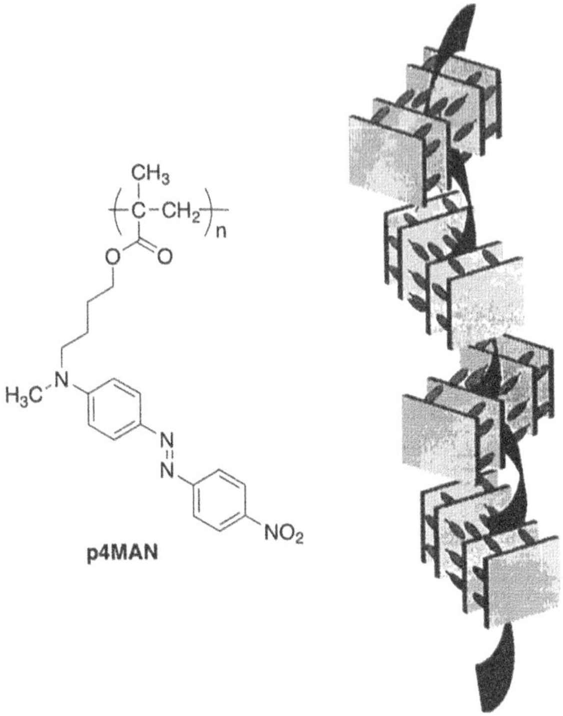

Fig. 5.31 The molecular structure of **p4MAN** and the TGB-like model for the formation of the photo-induced chiral supramolecular structures. Reprinted with permission from [82]. Copyright 2000 American Chemical Society

form decreased whereas an absorbance at 450 nm associated with n–π* transition of the *cis* form increased. These results revealed that the trans–cis photoisomerization of the azobenzene unit in both trimers occurs in dilute solutions.

We prepared a binary mixture of trimer **III-11** (20 wt%) and **Host LC** (80 wt%), and observed the phase transition behavior of the mixture on a glass slide with a cover glass with 365 nm UV irradiation at a power of 20 mW cm^{-2}. On cooling from the isotropic liquid, the N phase appeared. The molecules were homeotropically aligned. Then the N phase changed to the SmC phase, which exhibited chiral conglomerates with bright grain boundaries (Fig. 5.36). On the other hand, the mixture of trimer **III-10** (20 wt%) with **Host LC** (80 wt%) also showed chiral conglomerates but did not accompany such boundaries (Fig. 5.37). Trimer **III-9** induced similar chiral conglomerates in the mixture with **Host LC** as observed for trimer **III-11**. However, trimers **III-7** and **III-8** showed neither chiral nature nor grain boundaries in their mixtures with **Host LC**.

Why did timers **III-9**, **III-10**, and **III-11** induce the supramolecular chirality in their mixtures with **Host LC**? Those trimers themselves did not exhibit a chiral nature in their nematic phase by photo irradiation. The photo-induced *cis*-isomer

HNF to Iso

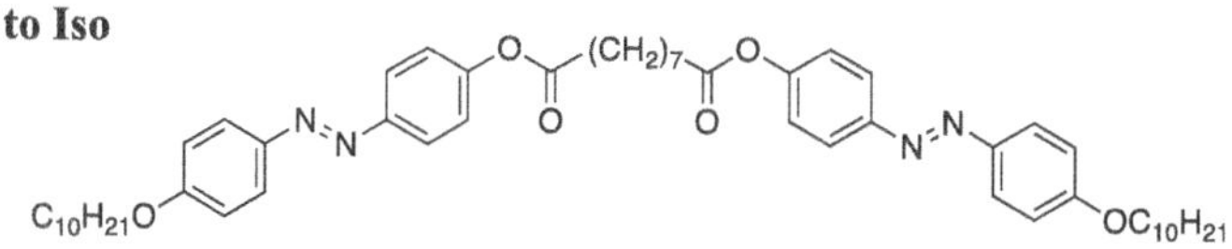

N_{TB} to N

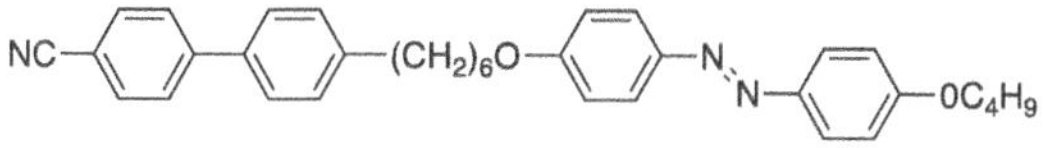

SmA to chiral isotropic (Iso_1*)

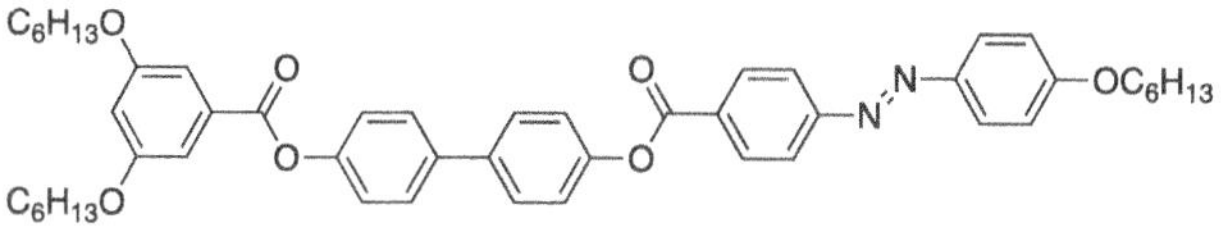

DC to N

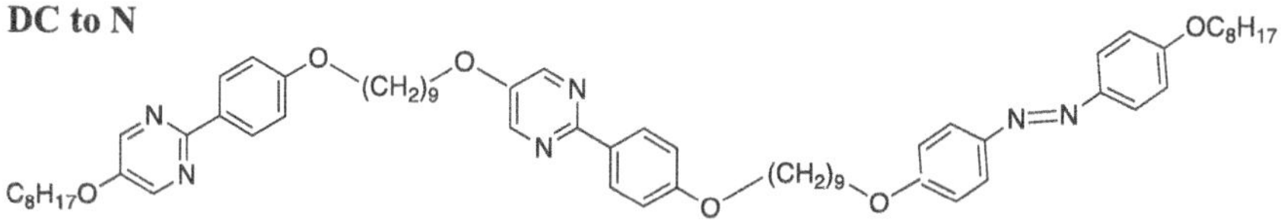

Fig. 5.32 Achiral molecules exhibiting photo-induced isothermal phase transitions. HNF to Iso [85]; N_{TB} to N [86]; SmA to Iso1*[87]; DC to N [88]

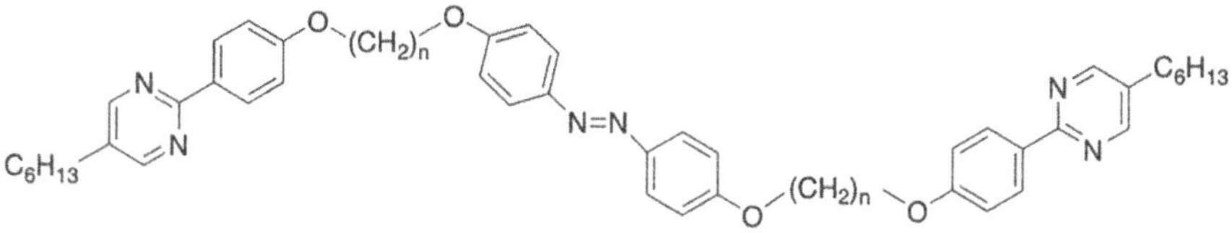

III-*n*

III-7: Iso 163.9 (87) Cry mp 168.2
III-8: Iso 184.8 (17) N 169.8 (49) Cry1 137.8 (31) Cry2 mp 172.5
III-9: Iso 150.8 (4.6) N 147.5 (83) Cry mp 149.6
III-10: Iso 168.4 (1.7) N 152.6 (37) Cry 1 145.3 (51) Cry 2 mp 156.6
III-11: Iso 145.5 (7.1) N 139.5 (93) Cry mp 142.4

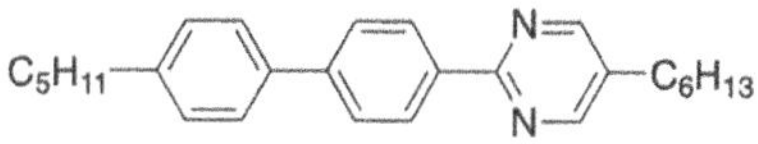

Host LC

Iso 162.7 (0.84) N 80.7 (0.30) SmC 41.1 (6.3) Cry

Fig. 5.33 Molecular structures, phase transition temperatures (°C), and enthalpies (kJ mol^{-1}, in parentheses) of transition for trimer **III-*n*** and **Host LC**

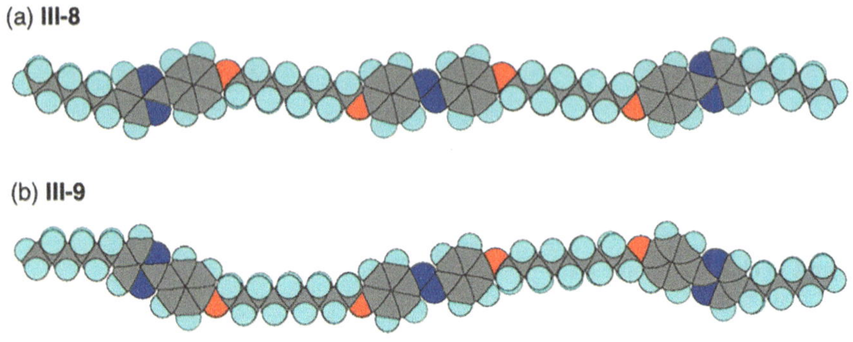

Fig. 5.34 MOPAC models for **a** *trans*-**III-8** and **b** *trans*-**III-9**. Reproduced from [91] with permission from the Royal Society of Chemistry

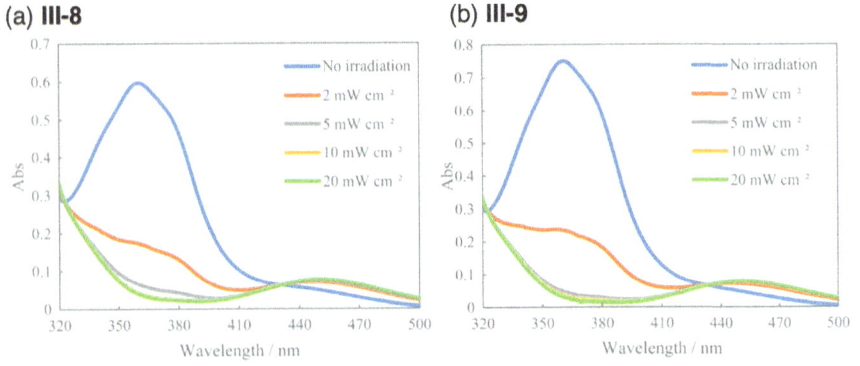

Fig. 5.35 UV-vis spectra of **a** trimer **III-8** and **b** trimer **III-9** in chloroform solutions (2.5×10^{-5} M). After the sample was exposed to UV light of wavelength 365 nm at room temperature for 10 s, the UV–vis spectra were recorded. Reproduced from [91] with permission from the Royal Society of Chemistry

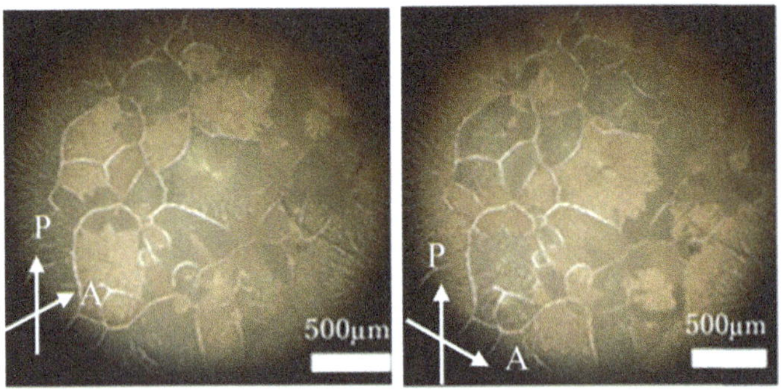

Fig. 5.36 Optical textures of a mixture of trimer **III-11** (20 wt%) and **Host LC** (80 wt%) cooling from the isotropic liquid with 365 nm UV irradiation at a power of 20 mW cm^{-2} between uncrossed polarizers. Reproduced from [91] with permission from the Royal Society of Chemistry

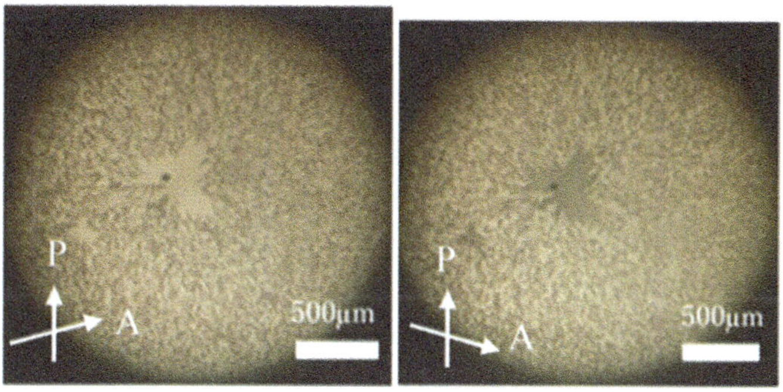

Fig. 5.37 Optical textures of a mixture of trimer **III-10** (20 wt%) and **Host LC** (80 wt%) cooling from the isotropic liquid under the same condition. Reproduced from [91] with permission from the Royal Society of Chemistry

of trimer **III-*n*** ($n = 9$, 10, and 11) intercalates with adjacent layers composed of **Host LC** molecules. Figure 5.38 shows MOPAC models for trimers **III-10** and **III-11**. Both azobenzene units form a twist-bent structure. Therefore, the twist-bent deformation of the azobenzene unit can induce the neighboring molecules to form the twist-bend orientation through core–core interactions as shown in Fig. 5.39. This twist orientation is synchronized in the layer normal via interlayer interactions to produce the supramolecular chirality. For the mixture of trimer **III-10** possessing even-numbered spacers with Host LC, the director is twisted continuously (Fig. 5.39a). With respect to trimer **III-11** possessing odd-numbered spacers, the molecules themselves are thought to have one twist-bent core and two flexible bent structures. The odd-numbered trimer molecules have a desire to induce bend deformation of the director of the surrounding molecules, destabilizing the layer structure to produce the grain boundaries (Fig. 5.39b). With respect to trimers **III-7** and **III-8**, they cannot sufficiently intercalate with the adjacent layer due to their sorter spacers. The photo-isomerization of *trans*-azobenzene produces *cis*-azobenzene with a twist-bent structure. The transiently molecular chirality is stabilized and amplified in the intercalated system to produce the supramolecular chirality. We can say that *Z*-azobenzene potentially act as the origin of chirality at a molecular level.

5.4.2 Electric Field

There are a few reports on electric-field-induced mirror symmetry breaking in liquid crystals. One of the reports is a study on a bent-core LC under an AC field [92]. Bent-core liquid crystals exhibit various characteristic phases so called as banana phases. The most studied phase is B2 phase. According to the interlayer correlations of polarization and tilt direction, four states are possible for the B2 phase as shown in Fig. 5.40 [53] Subscripts S and A for C stand for synclinic and anticlinic

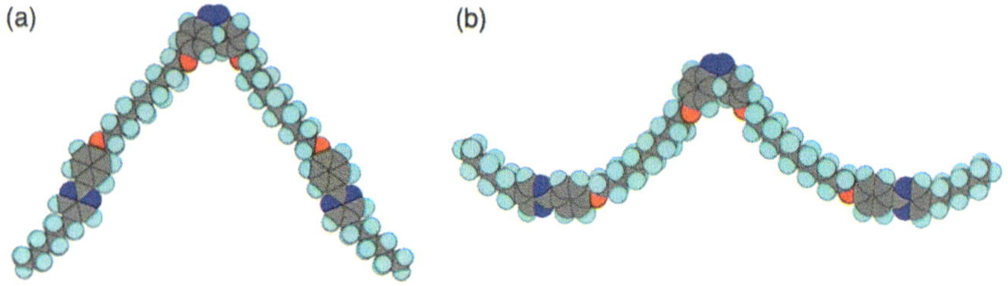

Fig. 5.38 MOPAC models for *cis*-isomers of the trimers: **a III-10** and **b III-11**. Reproduced from [91] with permission from the Royal Society of Chemistry

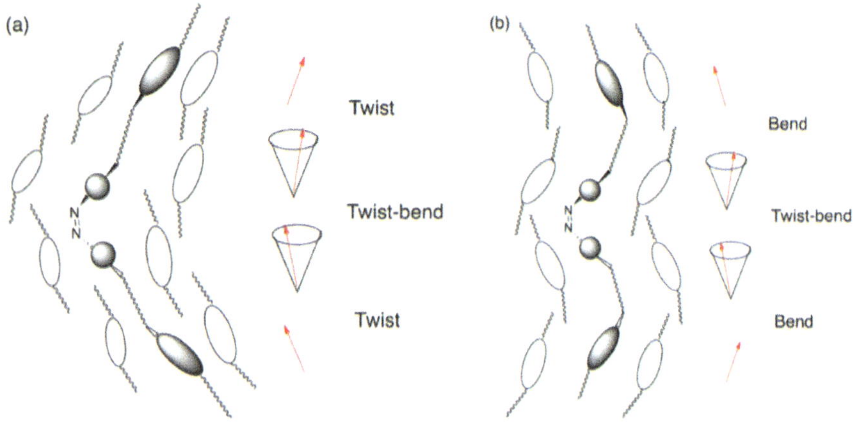

Fig. 5.39 **a** Schematic model for the formation of chiral conglomerates for a mixture of photo-induced *cis*-isomer **III-10** (20 wt%) and **Host LC** (80 wt%). **b** Model for the formation of chiral conglomerates accompanying grain boundaries for a mixture of photo-induced *cis*-isomer **III-11** (20 wt%) and **Host LC** (80 wt%). Reproduced from [91] with permission from the Royal Society of Chemistry

between adjacent layers. Subscripts F and A stand for ferroelectric and antiferroelectric. Allows show the polar direction. Blue- and red-colored molecules specify right-handed and left-handed layer chiralities, respectively (see Fig. 5.9a). The layer is homochiral in SmC_SP_F and SmC_AP_A. In these phases, two homochiral domains can exist in equal probability. On the other hand, SmC_AP_F and SmC_SP_A are racemic.

The bent-shaped 4-cyanoresorcinol liquid crystal shown in Fig. 5.41 exhibited the synclinic and antiferroelectric SmC_sP_A phase. When an AC field was applied on the bent-core LC in the SmC_sP_A phase and then the field was switching off, the heliconical $Sm(CP)^{hel}$ phase was induced [92]. Lehmann et al. explained the field-induced helix formation as follows (also see Fig. 5.41). The electric field induces SmC_SP_A to SmC_SP_F transition by rotation around the long axis. After the electric filed is removed, a helix can develop in the field-induced chiral SmC_SP_F structure by

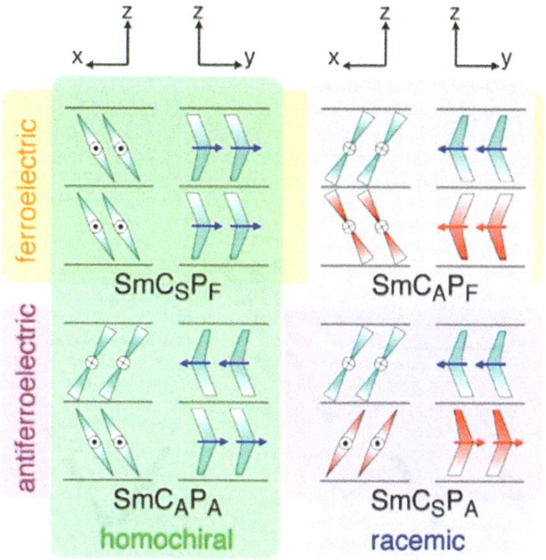

Fig. 5.40 Four possible structures in B2 phase. Reprinted with permission from [53]. Copyright (2006) The Japan Society of Applied Physics

a helical distribution of the polar layers. Thus, the system escapes from macroscopic polarization.

The other report is a study on electric-filed-induced helix formation in a nematic phase of an achiral H-shaped trimer [93]. Figure 5.42 shows the molecular structure and transition properties of the H-shaped trimer. It exhibited a monotropic N phase.

Figure 5.43 shows the electric-field-induced textures of the trimer with different electric fields. Before applying the electric field, the molecules were aligned uniformly parallel to the rubbing direction in the N phase When applying an AC field of 0.8 V μm^{-1}, the texture changed to gray. The nematic director is not perpendicular to the top and bottom substrates. When the electric field was removed, the gray texture transformed back into the original nematic texture. On the other hand, after applying an AC field of 6.0 μm^{-1} and then removing it, the texture turned completely dark and subsequently changed to orange. When rotating the sample, the texture remained almost unchanged. These results can be explained as follows. When the trimer molecules are aligned perpendicular to the substrates by applying an electric field, they form a chiral nematic phase rather than the original nematic phase after the electric field is removed.

Figure 5.44 shows the temporal evolution of the optical texture after the electric field is removed. Before applying the electric field, the molecules were aligned uniformly parallel to the rubbing direction in the N phase (Fig. 5.44a). When applying an AC field of 12 Vμm^{-1}, the texture was dark. After 2.5 s following the removal of the electric field, the nematic texture appeared but it differed from that observed before the electric filed was applied (Fig. 5.44b). The light green color changed to

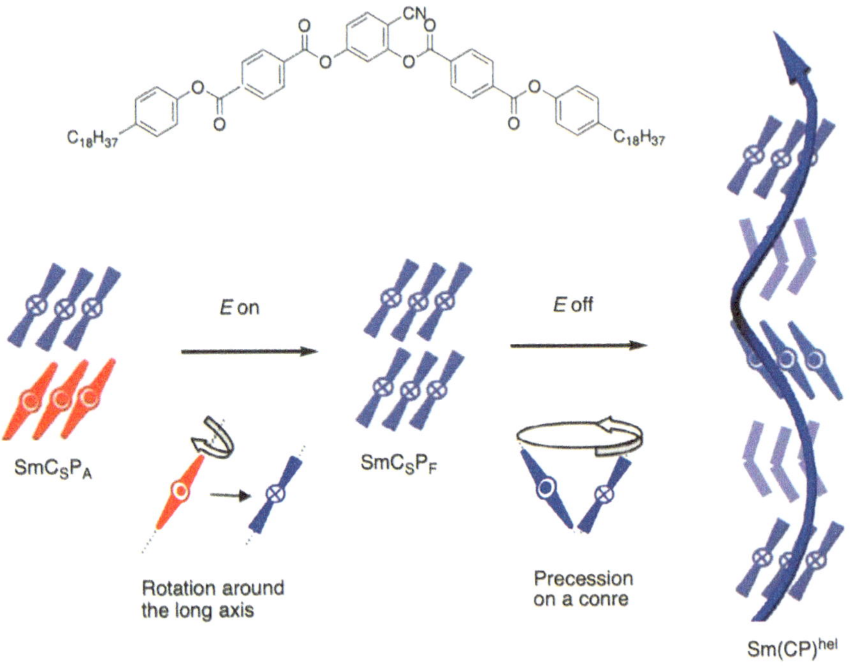

Fig. 5.41 Model for the electric-field-induced SmC$_s$P$_A$ to Sm(CP)^{hel} transition. Reproduced from [92]

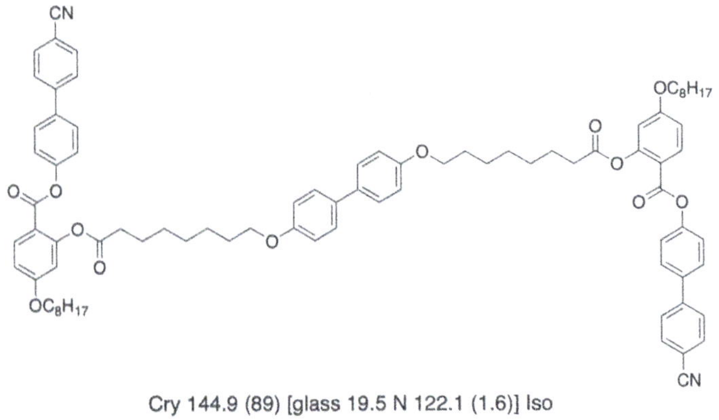

Fig. 5.42 Molecular structure, phase transition temperatures (°C), and enthalpies (kJ mol^{-1}, in parentheses) of transition for the achiral H-shaped trimer. Square brackets show monotropic transitions

Fig. 5.43 Electric-field induced textures of the trimer in the N phase at T_{IN}–30 °C with **a** $E = \pm 0.8\ \text{V}\ \mu\text{m}^{-1}$ and **b** $E = \pm 6.0\ \text{V}\ \mu\text{m}^{-1}$. The compound was confined in a 5 μm homogeneously aligned cell. The rubbing directions on top and bottom glass substrates are parallel. Black bars indicate the rubbing direction. Reproduced from [93] with permission from the Royal Society of Chemistry

orange (Fig. 5.40d) through the mottled texture consisting of gray and dark domains (Fig. 5.44c). The orange texture was split into red and dark yellow domains between uncrossed polarizers (Fig. 5.44e, f). Chiral domains with opposite handedness coexisted in the N* phase. Applying a DC field on the sample did not induce the N* phase. When we used an ITO cell without rubbing treatment instead of the rubbing cell, the electric-field-induced N* phase was not observed.

How does the supramolecular chirality appear? Based on the MOPAC model for the H-shaped trimer, it forms a rod-like shape **A** keeping the alkyl chains outside (Fig. 5.45a). Applying an AC field above the threshold voltage on the sample, the director reorients perpendicular to the substrate. During the process, the molecules change the conformation from **A** to a rod-like shape **B** in which the terminal cyano units are oriented toward the top and bottom electrodes (Fig. 5.45b). Upon removing the electric field, the trimer molecules inside the sample are twisted during the reorganization process, whereas the trimer molecules near the surface form a rod-like shape **A** and reorient parallel to the rubbing direction (Fig. 5.45c). The twisted shape **C** returns to the rod-like shape **A** to drive helical interactions among molecules, producing a helical molecular alignment (Fig. 5.45d). The coupling of the on/off switching of the electric field and the surface anchoring is responsible for breaking the mirror symmetry in the nematic phase of the flexible H-shaped trimer.

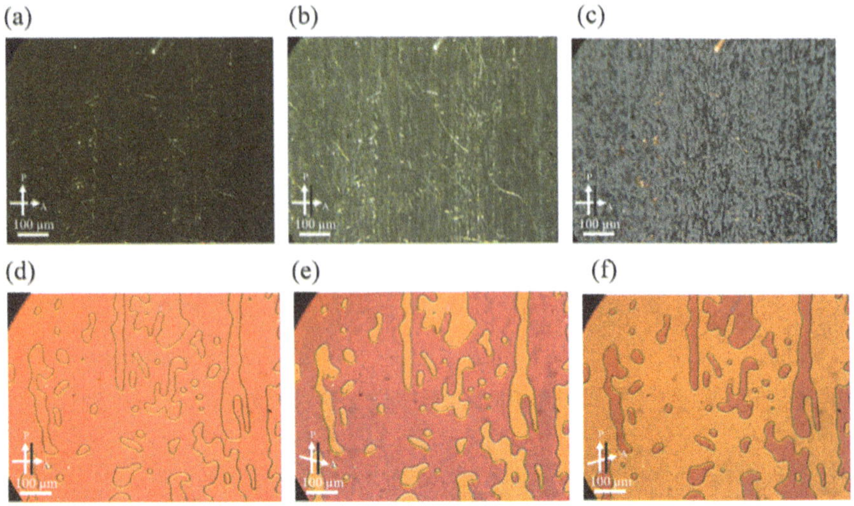

Fig. 5.44 Optical textures of the trimer after the removal of an AC field of 12 V μm^{-1}. The sample was confined in a 5 μm homogeneously aligned cell. **a** Texture before applying the electric field. **b** Texture at 2.5 s after the removal of the electric field. **c** Texture at 11 s after the removal of the electric field. **d**–**f** Textures at 17 s after the removal of the electric field between crossed and uncrossed polarizers. Reproduced from [93] with permission from the Royal Society of Chemistry

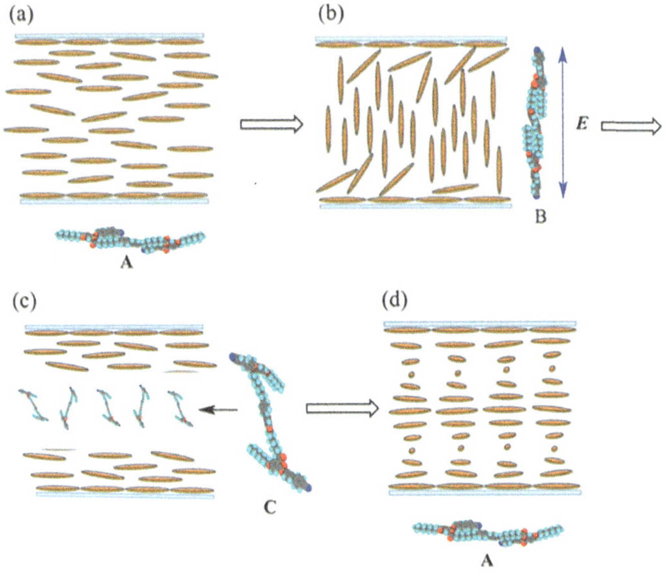

Fig. 5.45 Model for the electric-field-induced phase transition. **a** Molecular orientation without an AC field, **b** molecular reorientation with an AC field, **c** that upon removing the AC field, and **d** molecular orientation producing the helical alignment. Reproduced from [93] with permission from the Royal Society of Chemistry

5.5 The Formation of Helical Polymers Without a Chiral Component

5.5.1 Chirality Transfer from the Blue Phase

Significant achievements have been made in the transfer and amplification of chirality within chiral supramolecular nanomaterials [94]. The supramolecular chirality transfer process with hierarchical self-assembly plays a vital role in the fabrication of multiscale chiral structures [7]. Template effects have been applied to the chirality transfer [1, 95–97]. Blue phases act as a template for the chirality transfer [98] (please see Chap. 4). Photopolymerization of monomers in BPI forms a polymer network along the defects. After removing the BPLC, injecting an achiral nematic liquid crystal into the space results in the emergence of BPI templated by the polymer network.

Helical polymers were prepared from achiral monomers using a polymer network derived from an amorphous blue phase III (BPIII) [99]. A BP mixture of **T-shaped LC**, **S-** or **R-811**, a reactive monomer mixture (**RM257** and **C12A**), and a photo-initiator **DMAP** were used. Their molecular structures are shown in Fig. 5.46. The BP mixture was irradiated with UV light in the BPIII to give the polymer-stabilized BPIII (PS-BPIII). The PS-BPIII material was immersed in toluene to remove the low-molar-mass materials to give the polymer network. It exhhiited a strong CD signal at about 305 nm originated from the supramolecular chirality (Fig. 5.47). The optical sense of the polymer network was the same as that of the corresponding PS-BPIII. Furthermore, several weak peaks exist in the wavelength region below 300 nm due to the twist conformation of bifunctional monomer **RM257** (Fig. 5.48). The chiral field in the BPIII induces the axial chirality of the mesogenic core of **RM257** during the polymer stabilization to form the homochiral polymer network. The supramolecular chirality is transferred from the BPIII to the polymer network.

Then, the photopolymerization of achiral monomers was performed with the BPIII polymer network as a template. The polymer network was derived from the BP mixture containing of **R-811**. The monomer mixture consisted of 4-[(6-acryloyloxy)hexyloxy]-4′-cyanobiphenyl (**AHCB**, 14 wt%), 1,1,1-tris(acryloyloxymethyl)propane (**TMPTA**, 81 wt%), and a photo-initiator (5.0 wt%) (Fig. 5.49). The polymer film exhibited a negative CD at around 350 nm (Fig. 5.50). The optical sense of the film was the same as that of the polymer network. According to the SEM image of the polymer network (please see Fig. 4.10b), it has a porous structure with pore sizes of approximately 100–200 nm. The results are explained as follows. The reactive monomers are thought to be polymerized along the pores. They follow the twisted morphology of the polymer network and produce the helical polymer. The BP polymer network proves to be a good host for achiral monomers, allowing their helical polymerization.

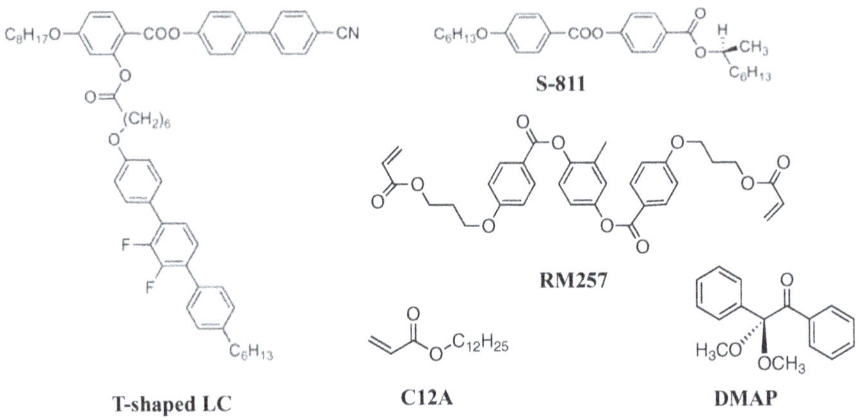

Fig. 5.46 Molecular structures of the BP mixture [99]

Fig. 5.47 CD spectra of the BPIII polymer network derived from the BP mixture containing **S-811** or **R-811**. Reprinted with permission from [99]. Copyright 2023 Taylor and Francis

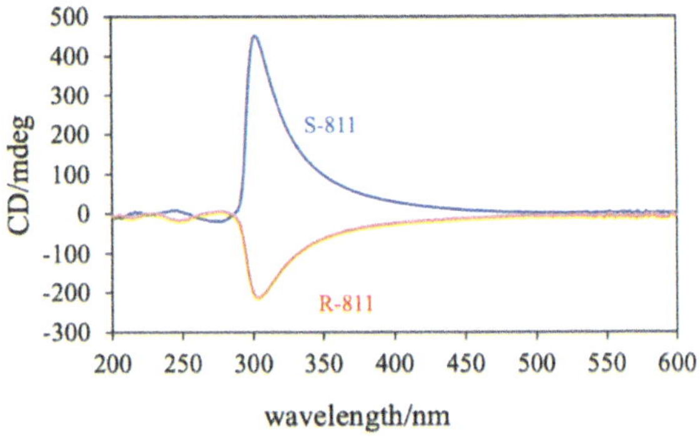

Fig. 5.48 Schematic model of the polymer network

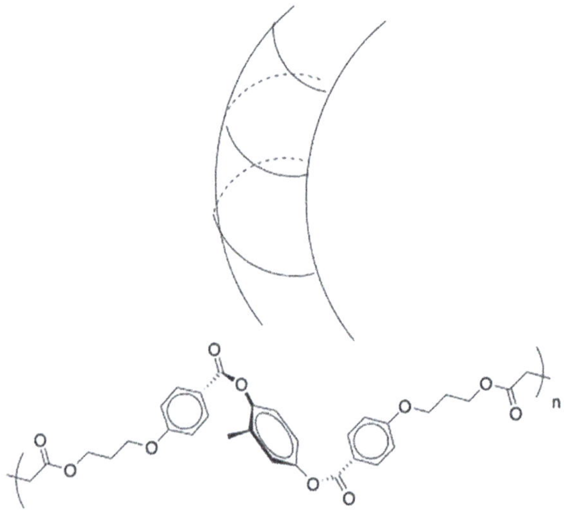

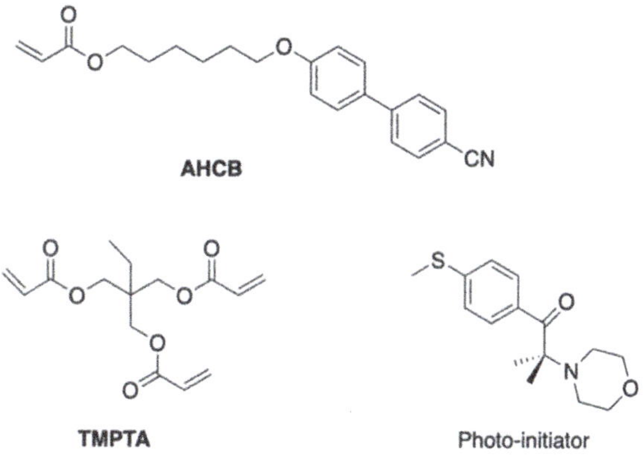

Fig. 5.49 Molecular structures of **AHCB**, **TMPTA**, and a photo-initiator [99]

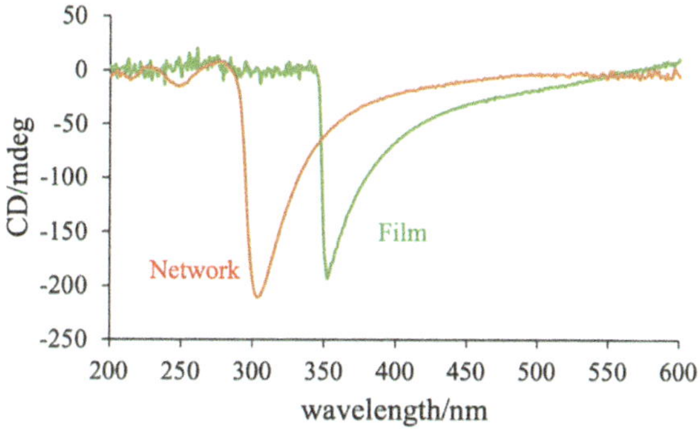

Fig. 5.50 CD spectrum of the BPIII polymer network (red) and that of the polymer film (green). Reprinted with permission from [99]. Copyright 2023 Taylor and Francis

5.5.2 Chirality Transfer from the Optically Isotropic Layered Phase

A helical nanofilament network composed of achiral bent-core molecules was utilized as a template for the fabrication of a chiral nanoporous film [100]. A mixture of a bent-core compound **P7** (49.5 wt%), a photoreactive monomer **LC242** (33.2 wt%), a nematic LC **5CB** (16.3 wt%), and a photoinitiator (1 wt%) was prepared (Fig. 5.51). A cell filled was with the mixture. The phase sequence of the mixture was Iso 140 °C HNF/Iso 120 °C HNF/N 80 °C HNF/Cryst. The coexistence of HNF and N indicates **P7** is in the HNF phase whereas the mixture of **LC242** and **5CB** (RMC) is in the N phase. The cell was irradiated with UV light at 100 °C. This state is the photopolymerized nanosegregated phase (stage 1). Figure 5.52a shows POM textures of the film between crossed and uncrossed polarizers. Two optically active domains with opposite chirality coexist. The sample cell was immersed in an organic solvent to remove the HNF of **P7** and the other low-molecular-mass materials. The photocured polymer was obtained (stage 2). Figure 5.52b shows POM textures of the polymer. Chiral domains are undistinguishable. This is because the absorption band of the polymer is far shorter than visible wavelengths. The chirality was confirmed by detecting very weak CD responses. Figure 5.53a shows the UHR FE-SEM image. Nanosized voids are seen in the image. The morphological chirality is transferred from the HNF network to the photocured polymer, as illustrated in Fig. 5.53b. The chiral nanoporous film has an inverse helical nanofilament structure. Then, the cell was refilled with a commercially available achiral nematic LC mixture (HTW109110-100) (stage 3). Figure 5.52c shows POM textures of the film. The two enantiomeric regions were restored, with the same sign of optical activity as in the photopolymerized nanosegregated phase. The achiral conglomerate phase appeared due to the transfer of chirality from the inverse helical network to achiral nematic LC molecules. The proposed method for obtaining chiral structures without a chiral component is very interesting. However, the problem is that HNF phases usually exhibit as chiral conglomerates. Therefore, no homochiral material can be obtained by using this system.

In this chapter so far, we have discussed that the mirror symmetry breaking introduces the supramolecular chirality in achiral liquid crystal systems. Furthermore, the template surface of the DC material proves to be a good host for the formation of the spiral-like structures. However, the supramolecular chiral phases exist as chiral conglomerates, where two chiral domains with opposite twist senses coexist in a non-equal population. We succeeded in expanding the dominant chiral domain across the entire area [101]. The process is as follows. A mixture of trimer **II-9** (see Fig. 5.24, 70.6 wt%), **RM257** (23.3 wt%), a photo-initiator, and a thermal polymerization inhibitor was heated to the isotropic liquid at 155 °C. Then, it was cooled from the isotropic liquid to the DC phase. The DC phase still consisted of chiral domains with opposite handedness (Fig. 5.54a). The sample cell was exposed to UV irradiation. The photo-polymerization produced the polymer network in the DC phase. The cell was heated again to the isotropic liquid at 200 °C and then cooled to 25 °C. The Iso–DC phase transition occurred. Interestingly, the DC phase was not chiral conglomerates but a homochiral phase (Fig. 5.54b). It is called as the polymer-stabilized DC (PS-DC) phase. Its helical sense cannot be controlled. Figure 5.55 shows POM textures of the PS-DC materials with opposite optical activity and their corresponding CD

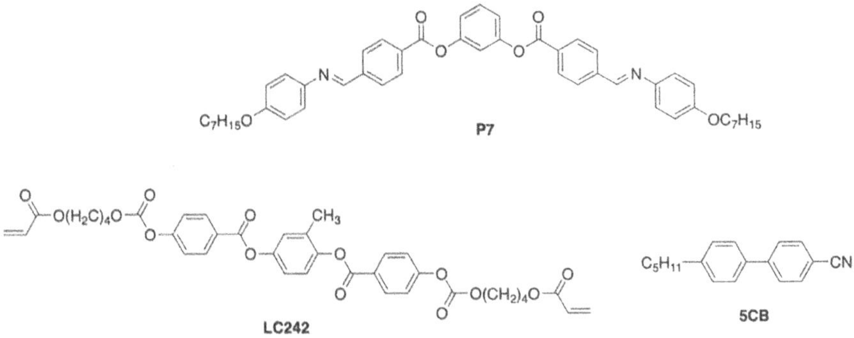

Fig. 5.51 Molecular structures of **P7**, **LC242**, and **5CB**

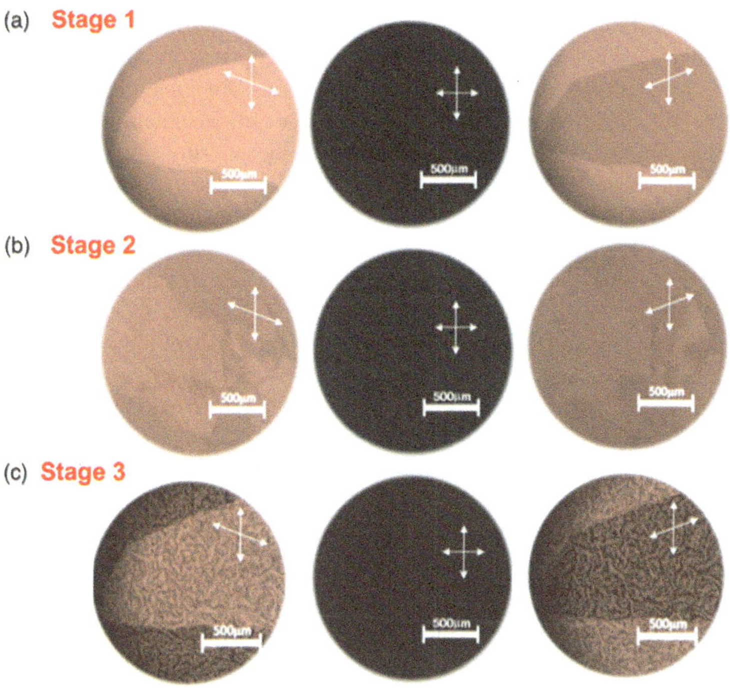

Fig. 5.52 Polarized optical textures of **a** the material in the photopolymerized nanosegregated phase **b** the photocured nanoporous polymer film, and **c** the nanoporous film filling with the achiral nematic LC mixture. The textures were observed under crossed and uncrossed polarizers. Reprinted with permission from [100]. Copyright 2020 American Chemical Society

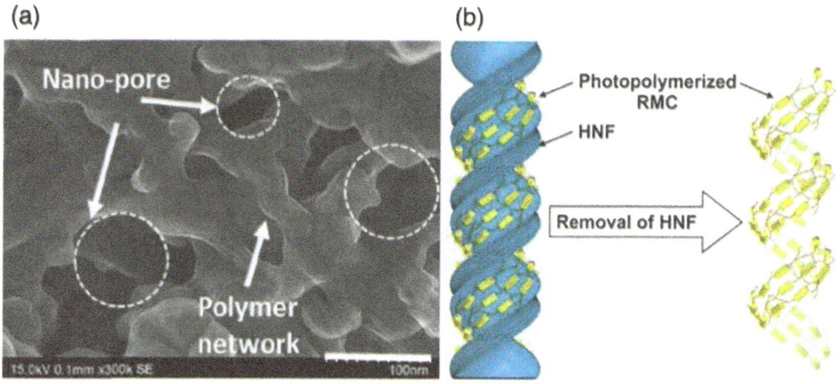

Fig. 5.53 **a** Ultra-high-resolution field-emission scanning electron microscopy (UHR FE-SEM) image of the polymer network coexisting nano-pores. **b** Schematic image of the polymer network and removal of HNF. Reprinted with permission from [100]. Copyright 2020 American Chemical Society

spectra. The CD spectra are mirror images. The chirality of the dominant domain is amplified to the whole area, providing a homochiral material. Very weak signals exist in the wavelength region below 320 nm. Those peaks are thought to have originated from the molecular twist of the mesogenic core of bifunctional monomer **RM257**. The polymer network plays an important role in the chirality amplification.

Let us describe the effect of the presence or absence of the polymer network on the Iso–DC phase transition. The movies of the time-dependent POM textures can be in Video S1 (see Electronic Supplementary Material). In the absence of the polymer network, a few domains with opposite handedness appeared and they were spread independently. The boundary between the chiral domains with opposite sese was linear and angular. On the other hand, in the presence of the polymer network, many small homochiral dots appeared and they were spread simultaneously. The spread speed of the latter was much faster than the former. The helical sense of each chiral domain in the DC phase is memorized in the polymer network during the photopolymerization of the recative monomer **RM275**. The twist units with opposite handedness coexist in non-equal population in the polymer. Then, chiral amplification by 'Majority rule' produces almost a single helical structure in the polymer network. The rule is that even a small excess of majority units could lead to a preference for epimerization of the minority units to the configuration of the majority units, which would lead to a fuller excess of the majority helical sense [45]. The chiral polymer network obtained in this manner is thought to be a seed, which accelerates the chiral synchronization to produce the homochiral state.

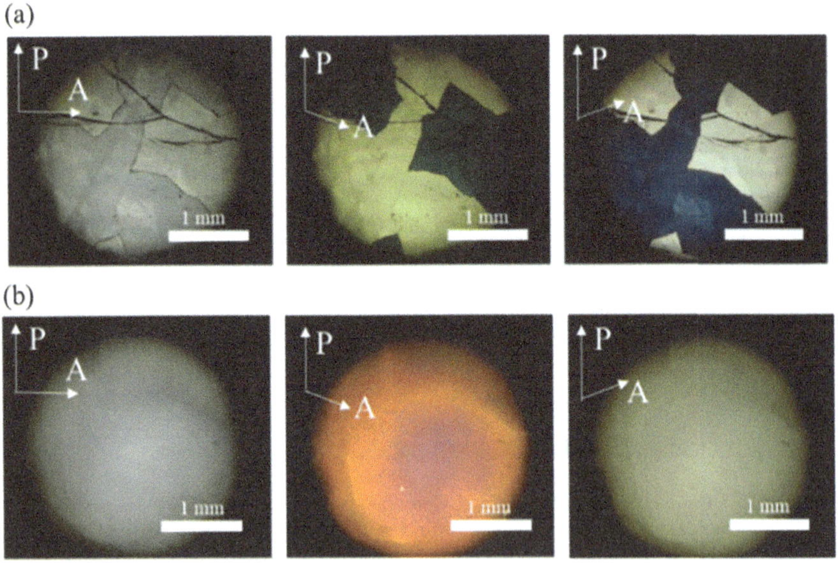

Fig. 5.54 **a** Optical textures of trimer **II-9** in the DC phase and **b** those of the mixture containing trimer **II-9** in the polymer-stabilized DC phase between crossed and uncrossed polarizers at room temperature. Reproduced from [101] with permission from the Royal Society of Chemistry

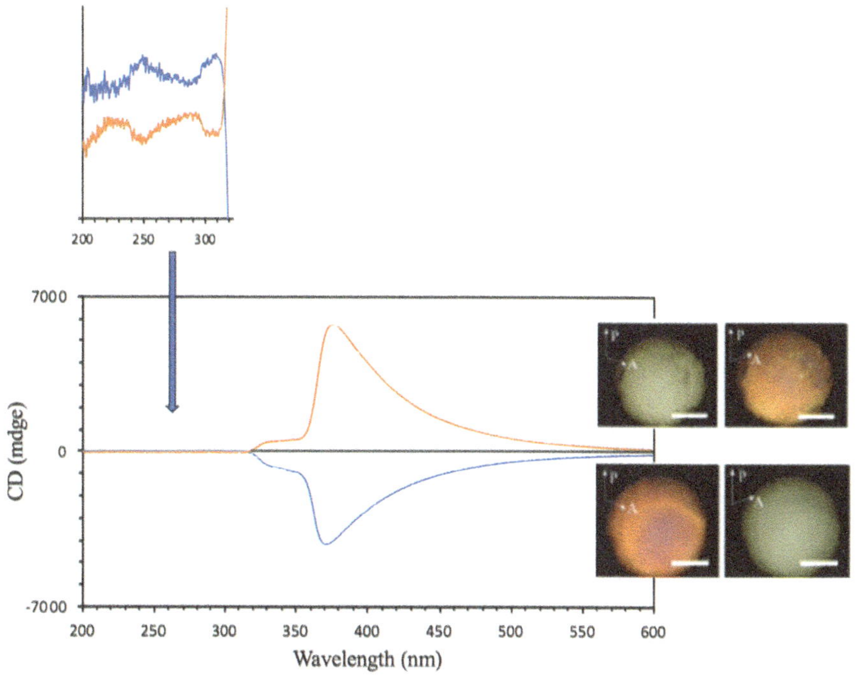

Fig. 5.55 POM textures of the PS-DC materials with opposite optical activity under uncrossed polarizers and the corresponding CD spectra. Reproduced from [101] with permission from the Royal Society of Chemistry

The photo-polymerization of achiral reactive monomers was performed on the surface of the PS-DC material [101]. We used a monomer mixture of benzyl methacrylate (**BM**, 47.5 wt%), 1,1,1-tris(acryloyloxymethyl)propane (**TMPTA**, 47.5 wt%), and a photo-initiator (5.0 wt%). Figure 5.56 shows a schematic illustration of the preparation process of the photo-polymerized film. The mixture was put on the homochiral PS-DC material. The sample cell was exposed to UV irradiation at room temperature. After the photo-polymerization, the cell was immersed in choloroform under ultrasonic condition. The polymer was isolated from the DC template to give a transparent film. A comparison of the FE-SEM images between the PS-DC material and the polymer film reveals that the surface structure of the PS-DC material is reversely printed to the surface of the polymer film.

Figure 5.57 shows CD spectra of the polymer films prepared by using each PS-DC material possessing opposite sense. The POM textures of the PS-DC materials are inserted in the figure. The supramolecular chirality is transferred from each PS-DC material to the corresponding polymer film. Figure 5.58 shows a schematic sketch of the chirality transfer process. The surface exhibiting voids with a mesoscopic size has a chiral environment. The reactive monomer molecules follow the twisted layer morphology of the DC phase and polymerize in the helical manner.

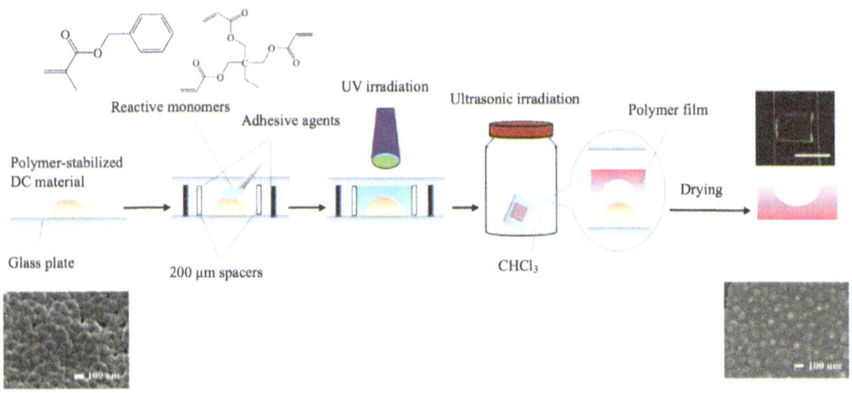

Fig. 5.56 Schematic illustration of the preparation process of the photo-polymerized film with a photograph of the polymer film on a glass plate and FE-SEM images of the surface structures of the PS-DC material and the polymer film. Reproduced from [101] with permission from the Royal Society of Chemistry

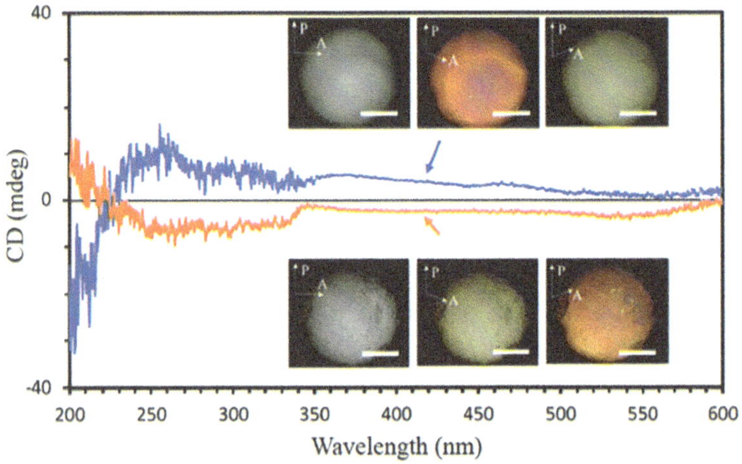

Fig. 5.57 CD spectra of the polymer films prepared by using each PS-DC material. Optical textures of the PS-DC materials are inserted. The white bars represent 1 mm. Reproduced from [101] with permission from the Royal Society of Chemistry

5.6 Concluding Remarks

Figure 5.59 shows our strategy how we obtain chiral materials in an achiral world. The achiral trimer possessing odd-numbered methylene spacers can form a twisted conformer by intermolecular interactions in the soft-crystalline phase. They create a dynamic chiral origin, which triggers the formation of supramolecular chirality in the molecular aggregations. The spontaneous mirror symmetry breaking occurs in this

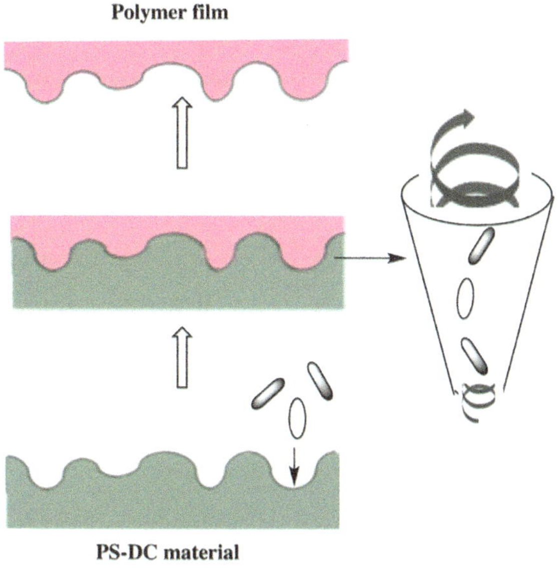

Fig. 5.58 Schematic sketch of the chirality transfer from the PS-DC material to the polymer film. Reproduced from [101] with permission from the Royal Society of Chemistry

manner. However, it produces chiral conglomerates, where two chiral domains with opposite sense exist in non-equal population. The chirality in the dominant domain is memorized in a polymer network. The chiral polymer becomes a seed, accelerates the chiral synchronization to produce the homochiral state. This process is the chirality amplification. The homochiral DC phase stabilized by the polymer network acts as a template for the helical polimerization of achiral reactive monomers. However, the direction of the spaital cannot be controlled.

We have not yet succeeded in transforming the supramolecular chirality to molecular chirality, such as a chiral center. We attempted to perform the reaction of acetone with benzaldehyde in the presence of a catalytic amount of trimer **I-(9,9)** in the DC phase [102]. The aldol product did not exhibit optical activity. This is a future challenge.

Let us consider the emergence of homochirality on Earth. There is a hypothesis that the Earth's movements, i.e., the Earth's right-handed rotation near the moon, alongside with the right-handed Earth's orbital around Sun, is a strong candidate for inducing the chirality [103–107]. At the laboratory level, it has been observed that the direction of mechanical rotation determines that of the chirality. The organic analysis of the samples corrected from the surface of asteroid Ryugu revealed the presence of both *L*- and *D*-amino acids [108]. Furthermore, with respect to the samples from asteroid (101955) Bennu, all chiral non-protein amino acids that could be enantiomerically resolved, including isovaline, norvaline, b-amino-*n*-butyric acid, b-aminoisobutyric acid, and 3-amnnopentanoic acid, were present as racemic or near racemic mixtures [109]. The following scenario is proposed regarding the emergence of homochirality on Earth. The primary Earth was an achiral world. The motion of the Earth induced mirror symmetry breaking, resulting in an excess of one chirality.

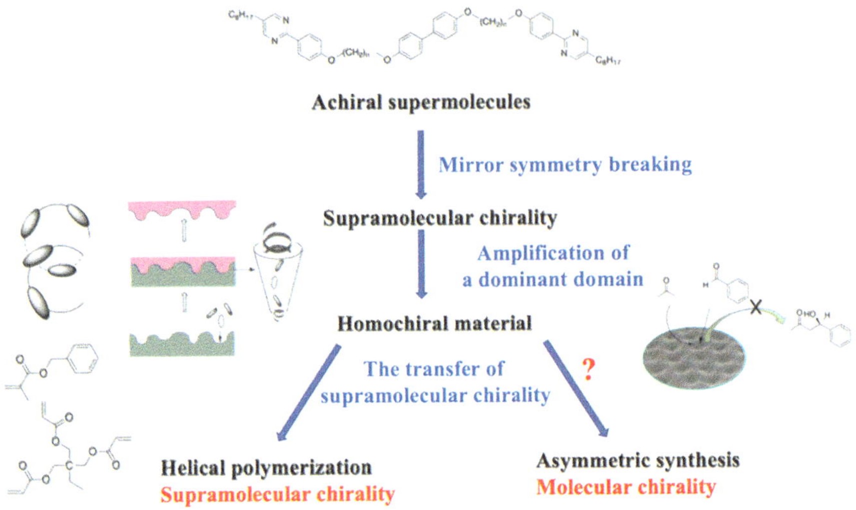

Fig. 5.59 Strategy for producing chiral materials without a chiral component

The dominant chirality was synchronized with an amplification system to emerge the homochiral template. The chirality was transferred to higher-order structures via template reactions of achiral molecules.

This chapter have demonstrated how chirality occurs in an achiral system and how the chirality is transferred to a higher ordered architecture. The methodology describes here provides useful information not only for the production of chiral substances but also for the origin of homochirality in life.

References

1. Niu, X., Zhao, R., Yan, S., Pang, Z., Li, H., Yang, X., Wang, K.: Chiral materials: progress, applications, and prospects. Small **19**, 2303059 (2023)
2. Noyori, R.: Asymmetric catalysis: science and opportunities (Nobel Lecture). Angew. Chem. Int. Ed. **41**, 2008–2022 (2002)
3. Liu, M., Zhang, L., Wang, T.: Supramolecular chirality in self-assembled systems. Chem. Rev. **115**, 7304–7397 (2015)
4. Yashima, E., Ouska, N., Taura, D., Shimomura, K., Ikai, T., Maeda, K.: Supramolecular helical systems: helical assemblies of small molecules, foldamers, and polymers with chiral amplification and their functions. Chem. Rev. **116**, 13752–13990 (2016)
5. Xing, P.Y., Zhao, Y.: Controlling supramolecular chirality in multicomponent self-assembled systems. Acc. Chem. Res. **51**, 2324–2334 (2018)
6. Chang, B., Li, X., Sun, T.: Self-assembled chiral materials from achiral components or racemates. Eur. Polym. J. **118**, 365–381 (2019)
7. Huang, S., Yu, H., Li, Q.: Supramolecular chirality transfer toward chiral aggregation: asymmetric hierarchical self-assembly. Adv. Sci. **8**, 2002132 (2021)
8. Guijarro, A., Yus, M.: The Origin of Chirality in the Molecules of Life. RSC Publishing, Cambridge, UK (2008)

9. Mauksch, M., Wei, S., Freund, M., Zamfir, A., Tsogoeva, S.B.: Spontaneous mirror symmetry breaking in the aldol reaction and its potential relevance in prebiotic chemistry. Orig. Life Evol. Biosph. **40**, 79–91 (2010)
10. Soai, K., Shibata, T., Morioka, H., Choji, K.: Asymmetric autocatalysis and amplification of enantiomeric excess of a chiral molecule. Nature **378**, 767–768 (1995)
11. Huber, L., Trapp, O.: Symmetry breaking by consecutive amplification: efficient paths to homochirality. Orig. Life Evol. Biosph. **52**, 75–91 (2022)
12. Kitamura, M., Okada, S., Suga, S., Noyori, R.: Enantioselective addition of dialkylzincs to aldehydes promoted by chiral amino alcohols. Mechanism and nonlinear effect. J. Am. Chem. Soc. **111**, 4028–4036 (1989)
13. Córdova, A., Engqvist, M., Ibrahem, I., Casas, J., Sundén, H.: Plausible origins of homochirality in the amino acid catalyzed neogenesis of carbohydrates. Chem. Commun. 2047–2049 (2005)
14. Viedma, C.: Chiral symmetry breaking during crystallization: complete chiral purity induced by nonlinear autocatalysis and recycling. Phys. Rev. Lett. **94**, 065504 (2005)
15. Satyanarayana, T., Abraham, S., Kagan, H.B.: Nonlinear effects in asymmetric catalysis. Angew. Chem. Int. Ed. **48**, 456–494 (2009)
16. Soai, K., Kawasaki, T., Matsumoto, A.: Asymmetric autocatalysis of pyrimidyl alkanol and its application to the study on the origin of homochirality. Acc. Chem. Res. **47**, 3643–3654 (2014)
17. Soai, K., Sato, I., Shibata, T., Komiya, S., Hayashi, M., Matsueda, Y., Imamura, H., Hayase, T., Morioka, H., Tabira, H., et al.: Asymmetric synthesis of pyrimidyl alkanol without adding chiral substances by the addition of diisopropylzinc to pyrimidine-5-carbaldehyde in conjunction with asymmetric autocatalysis. Tetrahedron Asymmetry **14**, 185–188 (2003)
18. Kondepudi, D.K., Kaufman, R.J., Singh, N.: Chiral symmetry breaking in sodium chlorate crystallization. Science **250**, 975–976 (1990)
19. Kondepudi, D.K., Laudadio, J., Asakura, K.: Chiral symmetry breaking in stirred crystallization of 1,1-binaphthyl melt. J. Am. Chem. Soc. **121**, 1448–1451 (1999)
20. Kondepudi, D.K., Asakura, K.: Chiral autocatalysis, spontaneous symmetry breaking, and stochastic behavior. Acc. Chem. Res. **34**, 946–954 (2001)
21. Kuroha, M., Nambu, S., Hattori, S., Kitagawa, Y., Niimura, K., Mizuno, Y., Hamba, F., Ishii, K.: Chiral supramolecular nanoarchitectures from macroscopic mechanical rotations: effects on enantioselective aggregation behavior of phthalocyanines. Angew. Chem. Int. Ed. **58**, 18454–18459 (2019)
22. Tejedor, R.M., Oriol, L., Serrano, J.L., Partal Ureña, F., López González, J.J.: Photoinduced chiral nematic organization in an achiral glassy nematic azopolymer. Adv. Func. Mater. **17**, 3486–3492 (2007)
23. Niori, T., Sekine, J., Watanabe, J., Furukawa, T., Takezoe, H.: Distinct ferroelectric smectic liquid crystals consisting of banana shaped achiral molecules. J. Mater. Chem. **6**, 1231–1233 (1996)
24. Sekine, T., Niori, T., Sone, M., Watanabe, J., Choi, S.-W., Takanishi, Y., Takezoe, H.: Origin of helix in achiral banana-shaped molecular systems. Jpn. J. Appl. Phys. **36**, 6455–6463 (1997)
25. Thisayukta, J., Takezoe, H., Watanabe, J.: Study on helical structure of the B4 phase formed from achiral banana-shaped molecule. Jpn. J. Appl. Phys. **40**, 3277–3287 (2001)
26. Le, K.V., Takezoe, H., Araoka, F.: Chiral superstructure mesophases of achiral bent-shaped molecules—Hierarchical chirality amplification and physical properties. Adv. Mater. **29**, 1602737 (2017)
27. Takezoe, H.: Spontaneous achiral symmetry breaking in liquid crystalline phases. Top. Curr. Chem. **318**, 303–330 (2012)
28. Görtz, V., Goodby, J.W.: Enantioselective segregation in achiral nematic liquid crystals. Chem. Commun. 3262–3264 (2005)
29. Kajitani, T., Kohomoto, S., Yamamoto, M., Kishikawa, K.: Spontaneous chiral induction in a cubic phase. Chem. Mater. **17**, 3812–3819 (2005)

30. Dressel, C., Liu, F., Prehm, M., Zeng, X., Ungar, G., Tschierske, C.: Dynamic mirror-symmetry breaking in bicontinuous cubic phases. Angew. Chem. Int. Ed. **53**, 13115–13120 (2014)
31. Roche, C., Sun, H.-J., Prendergast, M.E., Leowanawat, P., Partridge, B.E., Heiney, P.A., Araoka, F., Graf, R., Spiess, H.W., Zeng, X., et al.: Homochiral columns constructed by chiral self-sorting during supramolecular helical organization of hat-shaped molecules. J. Am. Chem. Soc. **136**, 7169–7185 (2014)
32. Hough, L.E., Spannuth, M., Nakata, M., Coleman, D.A., Jones, C.D., Dantlgraber, G., Tschierske, C., Watanabe, J., Körblova, E., Walba, D.M., et al.: Chiral isotropic liquids from achiral molecules. Science **325**, 452–456 (2009)
33. Hough, L.E., Jung, H.T., Krürke, D., Heberling, M.S., Nakata, M., Jones, C.D., Chen, D., Link, D.R., Zasadzinski, J., Heppke, G., et al.: Helical nanofilament phases. Science **325**, 456–460 (2009)
34. Alaasar, M., Prehem, M., Tschierske, C.: Helical nano-crystallite (HNC) phases: chirality synchronization of achiral bent-core mesogens in a new type of dark conglomerates. Chem. Eur. J. **22**, 6583–6597 (2016)
35. Dressel, C., Reppe, T., Prehm, M., Brautzsch, M., Tschierske, C.: Chiral self-sorting and amplification in isotropic liquids of achiral molecules. Nat. Chem. **6**, 971–977 (2014)
36. Tschierske, C., Ungar, G.: Mirror symmetry breaking by chirality synchronization in liquids and liquid crystals of achiral molecules. ChemPhysChem **17**, 9–26 (2016)
37. Cestari, M., Diez-Berart, S., Dunmur, D.A., Ferrarini, M.R., de la Fuente, M.R., Jackson, D.J.B., Lopez, D.O., Luckhurst, G.R., Perez-Jubindo, M.A., Richardson, R.M., et al.: Phase behavior and properties of the liquid-crystal dimer 1,7-bis(4-cyanobiphenyl-4-yl)heptane: a twist-bend nematic liquid crystal. Phys. Rev. E **84**, 031704 (2011)
38. Chen, D., Porada, J.H., Hooper, J.B., Klittnick, A., Shen, Y., Tuchband, M.R., Korblova, E., Bedrov, D., Walba, D.M., Glaser, M.A., et al.: Chiral helicoidal ground state of nanoscale pitch in a nematic liquid crystal of achiral molecular dimers. Proc. Natl. Acad. Sci. U.S.A. **110**, 15931–15936 (2013)
39. Borshch, V., Kim, Y.-K., Xiang, J., Gao, M., Jákli, A., Panov, V.P., Viji, J.K., Imrie, C.T., Tamba, M.G., Mehl, G.H., et al.: Nematic twist-bend phase with nanoscale modulation of molecular orientation. Nat. Commun. **4**, 2635 (2013)
40. Tschierske, C., Dressel, C.: Mirror symmetry breaking in liquids and their impact on the development of homochirality in abiogenesis: Emerging proto-RNA as source of biochirality? Symmetry **12**, 1098 (2020)
41. Kim, J.-Y., Lee, J.-J., Park, Choi, Y.-J., Choi, S.-W.: Control of the induced handedness of helical nanofilaments employing cholesteric liquid crystal fields. Molecules **26**, 6055 (2021)
42. Zhu, C., Wang, C., Young, A., et al.: Probing and controlling liquid crystal helical nanofilaments. Nano Lett. **15**, 3420–3424 (2015)
43. Link, D.R., Natale, G., Shao, R., Maclennan, J.E., Clark, N.A., Körblova, E., Walba, D.M.: Spontaneous formation of macroscopic chiral domains in a fluid smectic phase of achiral molecules. Science **278**, 1924–1927 (1997)
44. Okamoto, Y., Nakano, T.: Asymmetric polymerization. Chem. Rev. **94**, 349–372 (1994)
45. Green, M.M., Park, J.-W., Sato, T., Teramoto, A., Lifson, S., Selinger, R.L.B., Selinger, J.V.: The macromolecular route to chiral amplification. Angew. Chem. Int. Ed. **38**, 3138–3154 (1999)
46. Wu, Z.-Q., Nagai, K., Banno, M., Okoshi, K., Onitsuka, K., Yashima, E.: Enantiomer-selective and helix-sense-selective living block copolymerization of isocyanide enantiomers initiated by single-handed helical poly(phenyl isocyanide)s. J. Am. Chem. Soc. **131**, 6708–6718 (2009)
47. Siriwardane, D.A., Kulikov, O., Rokhlenko, Y., Perananthan, S., Novak, B.M.: Stereocomplexation of helical polycarbodiimides synthesized from achiral monomers bearing isopropyl pendants. Macromolecules **50**, 9162–9172 (2017)
48. Stals, P.J.M., Korevaar, P.A., Gillissen, M.A.J., de Greef, T.F.A., Fitié, C.F.C., Sijbesma, R.P., Palmans, A.R.A., Meijer, E.W.: Symmetry breaking in the self-assembly of partially fluorinated benzene-1,3,5-tricarboxamides. Angew. Chem. Int. Ed. **51**, 11297–11301 (2012)

49. Karunakaran, S.C., Cafferty, B.J., Weigert-Munoz, A., Schuster, G.B., Hud, N.V.: Spontaneous symmetry breaking in the formation of supramolecular polymers: implications for the origin of biological homochirality. Angew. Chem. Int. Ed. **58**, 1453–1457 (2019)
50. Akagi, K., Piao, G., Kaneko, S., Sakamaki, K., Shirakawa, H., Kyotani, M.: Helical polyacetylene synthesized with a chiral nematic reaction field. Science **282**, 1683–1686 (1998)
51. Akagi, K.: Helical polyacetylene: asymmetric polymerization in a chiral liquid-crystal field. Chem. Rev. **109**, 5354–5401 (2009)
52. Yoshizawa, A., Nishiyama, I.: Mesophasic helical structures with high twisting power in optically active 3-methyladipic acid bis esters. J. Mater. Chem. **4**, 449–456 (1994)
53. Takezoe, H., Takanishi, Y.: Bent-core liquid crystals: their mysterious and attractive world. Jpn. J. Appl. Phys. **45**, 597–625 (2006)
54. Eremin, A., Jakli, A.: Bent-shape liquid crystals–from molecular bend to layer splay and chirality. Soft Matter **9**, 615–637 (2013)
55. Białecka-Florjanczyk, E., Sledzinska, I., Górecka, E., Przedmojski, J.: Odd–even effect in biphenyl-based symmetrical dimers with methylene spacer—Evidence of the B4 phase. Liq. Cryst. **35**, 401–406 (2008)
56. Meyer, R.B.: Structural problems in liquid crystal physics. In: Balian, R., Weill, G. (eds.) Molecular Fluids; Volume XXV-1973 of Les Houches Summer School in Theoretical, Physics, pp. 273–373. Gordon and Breach, New York, NY, USA (1976)
57. Dozov, I.: On the spontaneous symmetry breaking in the mesophases of achiral banana-shaped molecules. Eur. Lett. **56**, 247–253 (2001)
58. Henderson, P.A., Imrie, C.T.: Methylene-linked liquid crystal dimers and the twist-bend nematic phase. Liq. Cryst. **38**, 1407–1414 (2011)
59. Mandle, R.J.: The dependency of twist-bend nematic liquid crystals on molecular structure: a progression from dimers to trimers, oligomers and polymers. Soft Matter **12**, 7883–7901 (2016)
60. Walker, R., Pociecha, D., Storey, J.M.D., Górecka, E., Imrie, C.T.: The chiral twist-bend nematic phase (N^*_{TB}). Chem. Eur. J. **25**, 13329–13335 (2019)
61. Walker, R.: The twist-bend phases: structure–property relationships, chirality and hydrogen-bonding. Liq. Cryst. Today **29**, 2–14 (2020)
62. Dunmur, D.: Anatomy of a discovery: the twist-bend nematic phase. Crystals **12**, 309 (2022)
63. Ivsic, T., Vinkovic, M., Baumeister, U., Mikleusevic, A., Lesac, A.: Milestone in the N_{TB} phase investigation and beyond: direct insight into molecular self-assembly. Soft Matter **10**, 9334–9342 (2014)
64. Alaasar, M., Prehm, M., Tschierske, C.: Mirror symmetry breaking in fluorinated bent-core mesogens. RSC Adv. **6**, 82890–82899 (2016)
65. Sasaki, H., Takanishi, Y., Yamamoto, J., Yoshizawa, A.: Achiral flexible liquid crystal trimers exhibiting chiral conglomerates. Soft Matter **12**, 3331–3339 (2016)
66. Sasaki, H., Takanishi, Y., Yamamoto, J., Yoshizawa, A.: Achiral flexible liquid crystal trimers exhibiting gyroid-like surfaces in chiral conglomerate phases. Soft Matter **13**, 6521–6528 (2017)
67. Seddon, J.M., Templer, R.H.: Polymorphism of lipid-water systems. In: Lipowsky, R., Sackmann, E. (eds.) Handbook of Biological Physics, vol. 1, 1st edn, pp. 97–160. Elsevier, London, UK (1995)
68. Szepke, D., Zarzeczny, M., Pawlak, M., Jarmuła, P., Yoshizawa, A., Pociecha, D., Lewandowski, W.: Disentangling optical effects in 3D spiral-like, chiral plasmonic assemblies templated by a dark conglomerate liquid crystal. J. Chem. Phys. **160**, 074201 (2024)
69. Oikawa, R., Sasaki, H., Takanishi, Y., Yamamoto, J., Yoshizawa, A.: Linear sym- metric liquid crystal trimers exhibiting supramolecular chiral architectures. Soft Matter **15**, 3179–3187 (2019)
70. Tazuke, S., Kurihara, S., Ikeda, T.: Amplified image recording in liquid crystal media by means of photochemically triggered phase transition. Chem. Lett. **16**, 911–914 (1987)
71. Ikeda, T., Tsutsumi, O.: Optical switching and image storage by means of azobenzene liquid-crystl films. Science **268**, 1873–1875 (1995)

72. Ichimura, K., Suzuki, Y., Seki, T., Hosoki, A., Aoki, K.: Reversible change in alignment mode of nematic liquid crystals regulated photochemically by command surfaces modified with an azobenzene monolayer. Langmuir **4**, 1214–1216 (1988)
73. Ichimura, K.: Photoalignment of liquid-crystal systems. Chem. Rev. **10**, 1847–1874 (2000)
74. Ikeda, T.: Photomodulation of liquid crystal orientations for photonic applications. J. Mater. Chem. **13**, 2037–2057 (2003)
75. Chigrinov, V.G., Kozenkov, V.M., Kwok, H.-S.: Photoalignment of Liquid Crystalline Materials: physics and Applications, pp. 1–248. Wiley, Hoboken, NJ, USA (2008)
76. Chigrinov, V.G., Kudreyko, A.A., Kozenkov, V.M.: Kinetics of photoinduced phase retardation in azo dye layer. Liq. Cryst. **49**, 1376–1383 (2022)
77. Chigrinov, V.G.: Liquid crystal applications in photonics. SID **47**, 927–936 (2016)
78. Yu, Y., Nakano, M., Ikeda, T.: Directed bending of a polymer film by light. Nature **425**, 145 (2003)
79. Yamada, M., Kondo, M., Mamiya, J., Yu, Y., Kinoshita, M., Barrett, C.J., Ikeda, T.: Photomobile polymer materials: towards light-driven plastic motors. Angew. Chem. Int. Ed. **47**, 4986–4988 (2008)
80. Sang, Y., Han, J., Zhao, T., Duan, P., Liu, M.: Circularly polarized luminescence in nanoassemblies: generation, amplification, and application. Adv. Mater. **32**, 1900110 (2020)
81. Nikolova, L., Todorov, T., Ivanov, M., Andruzzi, F., Hvilsted, S., Ramanujam, P.S.: Photoinduced circular anisotropy in side-chain azobenzene polyesters. Opt. Mater. **8**, 255–258 (1997)
82. Iftime, G., Labarthet, F.L., Natansohn, A., Rochon, P.: Control of chirality of an azobenzene liquid crystalline polymer with circularly polarized light. J. Am. Chem. Soc. **122**, 12646–12650 (2000)
83. Vera, F., Tejedor, R.M., Romero, P., Barberá, J., Ros, M.B., Serrano, J.L., Sierra, T.: Light-driven supramolecular chirality in propeller-like hydrogen-bonded complexes that show columnar mesomorphism. Angew. Chem. Int. Ed. **46**, 1873–1877 (2007)
84. Cheng, X., Miao, T., Qian, Y., Zhang, Z., Zhang, W., Zhu, X.: Supramolecular chirality in azobenzene-containg polymer system: traditional photopolymerization self-assembly versus in situ supramolecular self-assembly strategy. Int. J. Mol. Sci. **21**, 6186 (2020)
85. Zep, K., Sitkowska, K., Pociecha, D., Górecka, E.: Photoresponsive helical nanofilaments of B4 phase. J. Mater. Chem. C **2**, 2323–2327 (2014)
86. Paterson, D.A., Xiang, J., Singh, G., Walker, R., Agra-Kooijman, D.M., Martínez-Felipe, A., Gao, M., Storey, J.M.D., Kumar, S., Lavrentovich, O.D., et al.: Reversible isothermal twist–bend nematic–nematic phase transition driven by the photoisomerization of an azobenzene-based nonsymmetric liquid crystal dimer. J. Am. Chem. Soc. **138**, 5283–5289 (2016)
87. Alaasar, M., Poppe, S., Dong, Q., Liu, F., Tschierske, C.: Isothermal chirality switching in liquid-crystalline azobenzene compounds with non-polarized light. Angew. Chem. Int. Ed. **56**, 10801–10805 (2017)
88. Sasaki, H., Takanishi, Y., Yamamoto, J., Yoshizawa, A.: Photo-driven chirality switching in a dark conglomerate phase of an achiral liquid crystal trimer. ChemistrySelect **3**, 3278–3283 (2018)
89. Briquet, L., Vercauteren, D., Perpete, E., Jacquemin, D.: Is solvated trans-azobenzene twisted or planar? Chem. Phys. Lett. **417**, 190–195 (2006)
90. Szyrszyng, M., Nowak, E., Gdaniec, M., Milewska, M.J., Polonski, T.: Optical activity of the guest azobenzene molecule generated by inclusion complexation with steroidal bile acids. Tetrahedron Asymmetry **15**, 3257 (2004)
91. Sasaki, H., Abe, K., Sagisaka, M., Yoshizawa, A.: Photo-induced guest–host interactions produce chiral conglomerates accompanying grain boundaries in a smectic phase. J. Mater. Chem. C **9**, 12928–12937 (2021)
92. Lehmann, A., Alaasar, M., Poppe, M., Poppe, S., Prehm, M., Nagaraj, M., Sreenilayam, S.P., Panarin, Y.P., Vij, J.K., Tschierske, C.: Stereochemical rules govern the soft Self-Assembly of achiral compounds: understanding the heliconical liquid-crystalline phases of bent-core mesogens. Chem. Europ. J. **26**, 4714–4733 (2020)

93. Sayama, S., Yoshizawa, A.: Achiral H-shaped liquid crystals exhibiting an electric-field-induced chiral nematic phase. J. Mater. Chem. C **7**, 6905–6913 (2019)
94. Chen, W., Li, B., Gao, G., Sun, T.: Chiral supramolecular nanomaterials: From chirality transfer and amplification to regulation and applications. Interdiscip. Mater. **29**, 689–713 (2023)
95. Peluso, P., Chankvetadze, B.: Recognition in the domain of molecular chirality: from noncovalent interactions to separation of enantiomers. Chem. Rev. **122**, 13235–13400 (2022)
96. Haq, S., Humblot, V., Jansen, A.P., Raval, R.: Drastic symmetry breaking in supramolecular organization of enantiomerically unbalanced monolayers at surfaces. Nat. Chem. **1**, 409–414 (2009)
97. Das, A., Ghosh, S., George, S.J.: Chiroptical amplification of induced circularly polarized luminescence in nucleotide-templated supramolecular polymer. Angew. Chem. Int. ed. **62**, e2023088281 (2023)
98. Castles, F., Day, F.V., Morris, S.M., Ko, D.-H., Gardiner, D.J., Qasim, M.M., Nosheen, S., Hands, P.J.W., Choi, S.S., Friend, R.H., et al.: Blue-phase templated fabrication of three-dimensional nanostructures for photonic applications. Nat. Mater. **11**, 599–603 (2012)
99. Yoshizawa, A., Kagami, T., Kato, K., Abe, K., Hamada, K., Uemura, T., Yamaguchi, M., Sagisaka, M.: Helical polymerization using a polymer network derived from a blue phase as a template for chirality transfer. Liq. Cryst. **50**, 2268–2279 (2023)
100. Lee, J.-J., Kim, B.-C., Choi, H.-J., Bae, S., Araoka, F., Choi, S.-K.: Inverse helical nanofilament networks serving as a chiral nanotemplate. ACS Nano **14**, 5243–5250 (2020)
101. Kurata, M., Yoshizawa, A.: The formation of a chiral supramolecular structure acting as a template for chirality transfer. Chem. Commun. **56**, 8289–8292 (2020)
102. Yoshizawa, A., Kurata, M.: Porous surface of an achiral trimer in the chiral conglomerate phase catalyzes a direct aldol reaction. New J. Chem. **43**, 8865–8868 (2019)
103. Pavlov, V.A., Klabunovskii, E.I.: The origin of homochirality in nature: a possible version. Russ. Chem. Rev. **84**, 121–133 (2015)
104. Pavlov, V., Pavlova, T.: Paradoxes of symmetry: homochirality; cryptochiral reactions; chiral field, memory, and induction; chiral and racemic environment. Curr. Org. Chem. **21**, 872–888 (2017)
105. Pavlov, V.A., Zlotin, S.G.: Homochirality, stochastic chiral reactions, spontaneous chiral ordering of achiral molecules, and similar chiral effects. Is there a physical basis for these mirror symmetry breaking phenomena? Curr. Org. Chem. **22**, 2029–2054 (2018)
106. He, Y.J., Qi, F., Qi, S.C.: Earth's orbital chirality and driving force of biomolecular evolution. Med. Hypotheses **56**, 493–496 (2001)
107. Pavlov, V.A., Shushenachev, Y.V., Zlotin, S.G.: Chiral and racemic fields concept for understanding of the homochirality origin, asymmetric catalysis, chiral superstructure formation from achiral molecules, and B-Z DNA conformational transition. Symmetry **11**, 649 (2019)
108. Naraok, H., Takano, Y., Dworkin, J., Oba, Y., Hamase, K., Furusho, A., Ogawa, N.O., Hashiguchi, M., Fukushima, K., Aoki, D., et al.: Soluble organic molecules in samples of the carbonaceous asteroid (162173) Ryugu. Science **379**, eabn9033 (2023)
109. Glavin, D.P., et al.: Abundant ammonia and nitrogen-rich soluble organic matter in samples from asteroid (101955) Bennu. Nat. Astronomy **9**, 199–210 (2025)

Chapter 6
Biological Activities of Liquid Crystals

Abstract Liquid crystal states have been identified in numerous biological tissues within the body, including cell membranes. The fluidity and orientation order of liquid crystals are implicated in cell membrane functions, such as the maintenance of homeostasis within cells. The correlation between liquid crystals and life phenomena has been identified since the discovery of liquid crystals in the late nineteenth century, and in the twenty-first century, the utilization of liquid crystals in the medical field has become active. This chapter delineates sensors based on the visualization of interactions with biomolecules, drug delivery systems that utilize cell membrane permeability, and the antitumor effects of thermotropic liquid crystals on tumors. It is hypothesized that liquid crystal molecules with active sites will penetrate the cell membranes of solid tumors due to their structural similarity and act as drugs that inhibit the proliferation of cancer cells.

Keywords Liquid crystals · Sensors · Drug delivery systems · Antitumor medicine

6.1 Introduction

Liquid-crystalline materials play critical roles in biology and are increasingly studied for bioinspired technologies and medical applications. Examples of liquid crystals are abundant in nature, and their unique physical properties play a fundamental role in many biological processes that sustain life on Earth. Deoxyribonucleric acid (DNA) molecules at high concentrations can align into cholesteric liquid-crystalline phases, forming helical structures [1, 2]. Lipid bilayers in cell membranes exhibit liquid-crystalline behavior, enabling membrane fluidity and functionality [3]. Myosin and action, the proteins that make up muscles, exhibit an orderly arrangement similar to liquid crystals under specific conditions [4]. It is believed that this liquid-crystalline property enables muscles to contract and relax efficiency. Silk exhibits liquid-crystalline behavior, particularly, during its natural spinning process in organisms like silkworms and spiders [5].

A. Yoshizawa, *Liquid-Crystalline Supermolecules*, The Materials Research Society Series, https://doi.org/10.1007/978-981-92-2524-8_6

Since their discovery, liquid crystals have been known to be related to life phenomena [6]. The smallest unit of life is the cell. Lehmann was the first to point out the relationship between cell structure and liquid crystal structure [7]. The liquid crystal state is capable of meeting the requirements of biological systems, which must maintain a high degree of orientation over a wide range while retaining softness and responding to external stimuli. The fluid mosaic model of cell membrane, proposed by Singer and Nicholson, is generally accepted [3]. In the cell membrane, phospholipids form a single bilayer structure, into which molecules such as proteins, glycolipids, and cholesterol are inserted as shown in Fig. 6.1. These lipids form a lyotropic lamellar liquid-crystalline phase. Figure 6.1 also shows the molecular structure of phosphatidylcoline as a typical example of phospholipids. The phospholipids form a liquid-crystalline phase, allowing the membrane proteins to move fluidity. The exchange of substances between the inside and outside of the cell membrane occurs through these membrane proteins. Phospholipids are amphiphilic molecules consisting of a hydrophilic group and hydrophobic alkyl chains, and form a single bilayer structure through micro-segregation. The maintenance of body temperature in thermoregulating creatures such as humans is closely related to the lipid composition of the cell membrane. The cell membrane has a temperature adaptation mechanism that changes its lipid composition in response to changes in body temperature to maintain a constant fluidity [1]. Changes in the liquid-crystalline properties of cell membranes in response to stimuli are promoting the development of liquid crystal-based biosensors [1, 8, 9]. One of the processes by which substances pass through the cell membrane is uptake accompanied by changes in cell membrane morphology. A drug delivery system containing phospholipids in the carrier has been developed [10], and its membrane permeability mechanism has been reported to be based on this mechanism [11].

Some drugs exhibit liquid-crystalline properties. The liquid-crystalline properties of drugs are thought to be related to drug functions such as stability, solubility,

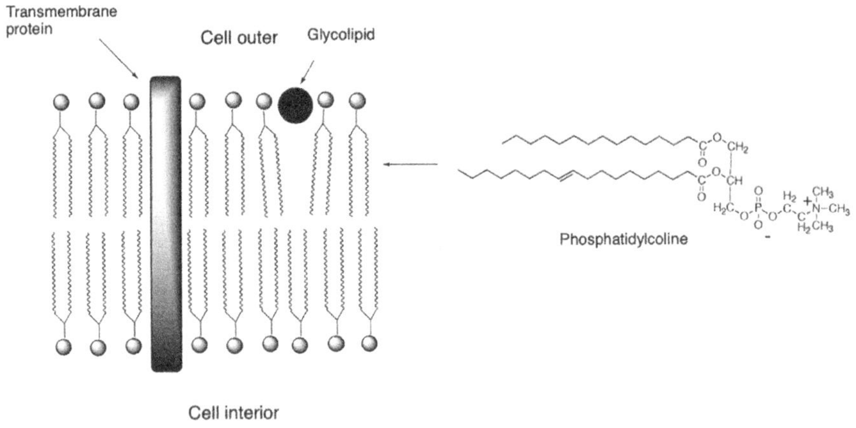

Fig. 6.1 Schematic image of the cell membrane

dissolution rate, and metabolism in the body. Stevenson et al. have investigated and summarized the thermotropic and lyotropic liquid-crystalline properties of existing drugs [12]. Furthermore, drugs characterized by liquid-crystalline properties are also being designed.

This chapter discusses liquid crystal-based sensors, drug delivery systems, and pharmaceutical liquid crystals that exhibit antitumor effects.

6.2 Liquid Crystals for Biomedical Applications

A lot of liquid crystal systems has been applied in biomedical applications such as biosensing, bioimaging, drug delivery, and artificial implants [13, 14]. We focus on the biosensing and drug delivery systems.

6.2.1 Liquid Crystal-Based Sensors

Abbott et al. reported an attempt to visualize the interaction between receptors and ligands using liquid crystals [15]. By converting microscopic intermolecular interactions into macroscopic changes in liquid crystal orientation and amplifying them, it becomes possible to observe them with the naked eye. Brake et al. demonstrated that intermolecular interactions between nematic liquid crystal [4-cyano-4′-pentylbiphenyl (**5CB**)] and phospholipid [L-α-dilauroyl phosphatidylcholine (**L-DLPC**)] induced the reorientation of the liquid crystal alignment (Fig. 6.2) [16]. Films of **5CB** were deposited into the pores of gold grids supported on octadecyltrichlorosilane (**OTS**)–coated glass substrates (Fig. 6.2a). The liquid-crystalline molecules in contact with the substrate are oriented perpendicular to the substrate. On top of that, an aqueous solution containing dispersed vesicles composed of **L-DLPC** was immersed. The texture appeared bright under crossed polarizers (Fig. 6.2b). The liquid-crystalline molecules in contact with the aqueous phase are oriented parallel to the substrate due to their hydrophobicity. Within 10 min of immersion, black domains were observed to nucleate and grow in the LC (Fig. 6.2c). These black domains correspond to regions where the liquid-crystalline molecules are oriented homeotropically. This indicates that the alignment of **5CB** molecules from parallel to homeotropic due to the interaction between the hydrophobic phospholipid tails and the LC molecules at the interface. After 2 h of immersion, the optical appearance of the LC was uniformly dark (Fig. 6.2d). Abbott et al. demonstrated that biological lipids can self-assemble at the aqueous interface of LCs, resulting in the formation of a phospholipid-decorated LC interface whose mobility is comparable to that of biological membranes [16–18].

This amplification mechanism offers new approaches for the detection of protein-binding event and enzymatic activity. Go et al. reported a molecular dynamics (MD)

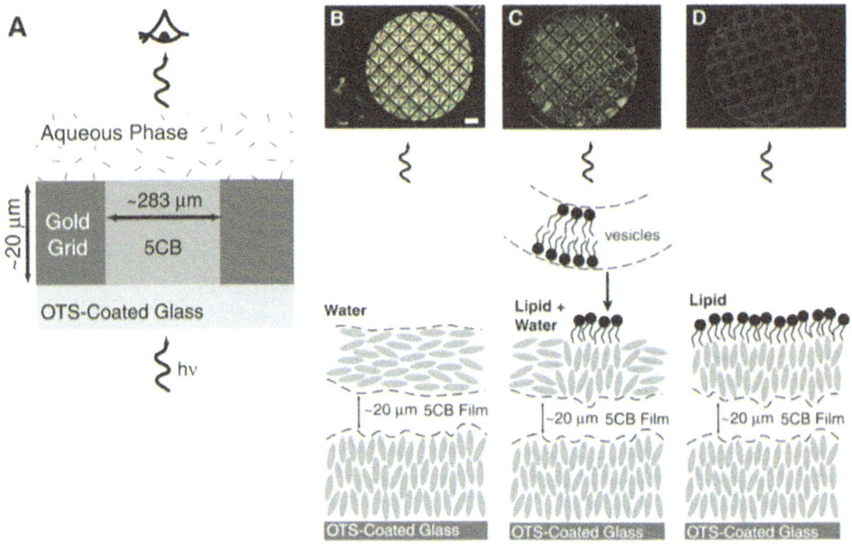

Fig. 6.2 **a** Schematic illustration of the experimental system for detecting intermolecular interactions between **5CB** and **L-DLPC**. **b–d** POM textures and molecular alignment models of anchoring **5CB** by contacting with or without the aqueous vesicle dispersion of **L-DLPC**. Reprinted with permission from [16]. Copyright 2003 American Association for the Advancement of Science

simulation of protein binding at the aqueous–liquid crystal interfaces of bioconjugated mesogenic molecules [19]. They used monolayers composed biotin-conjugated or biotin-free amphiphilic mesogenic molecules and streptavidin in water. The molecular structures of the amphiphilic LCs are shown in Fig. 6.3. The all-atom MD simulations reveal that the binding of streptavidin to the biotin mesogenic monolayer is significantly stronger than that to biotin-free mesogenic monolayers. Figure 6.4 shows binding approach of streptavidin to the monolayer composed of the biotin-conjugated mesogenic molecules.

The relative distance in Fig. 6.4 is expressed below

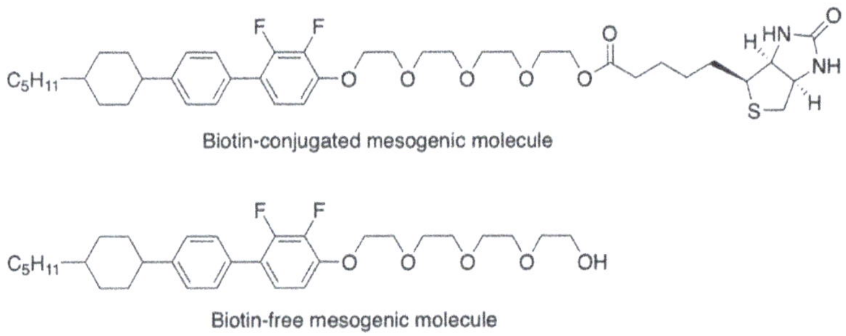

Fig. 6.3 Molecular structures of the biotin-conjugated LC and the biotin-free LC

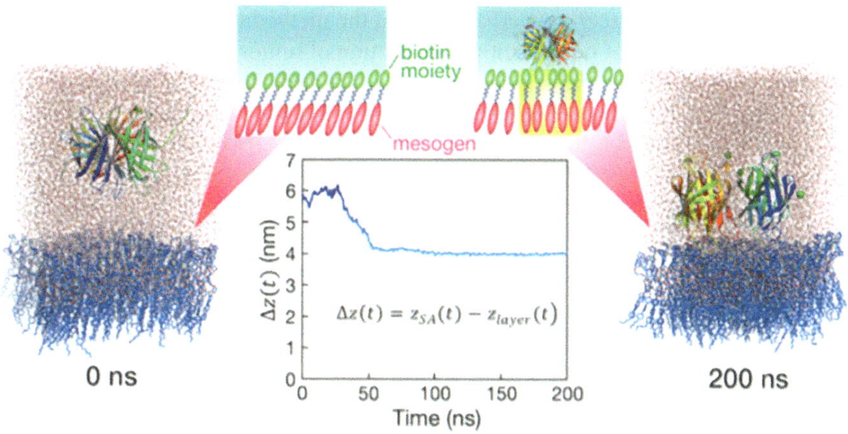

Fig. 6.4 Snapshots of MD simulation for streptavidin binding at the aqueous–LC interfaces of the biotin-conjugated mesogenic molecules at 0 ns and 200 ns. Time profile of the relative distance along the z-direction between the center of the mass of streptavidin and the monolayer is displayed. Reprinted with permission from [19]. Copyright 2020 American Chemical Society

$$\Delta z(t) = z_{SA}(t) - z_{layer}(t). \tag{6.1}$$

where $z_{SA}(t)$ and $z_{layer}(t)$ represent the z-coordinates of the position of center of the mass of streptavidin and the monolayer, respectively. As shown in Fig. 6.4, streptavidin was stably bound at the interface of monolayer composed of the biotin-conjugated mesogenic molecules. Significant changes were observed in the tilt of biotin-conjugated mesogenic molecules of the monolayer near the bound protein. Uchida and Kato et al. proposed a molecular design of bioconjugated mesogens to couple biomolecular interactions at the aqueous interfaces with the ordering transition of the liquid crystals [20]. Liquid crystals have been used to construct functional interfaces for detecting environmental pollutants, toxic metal ions, and pathogens that cause disease [21–24].

Biomarkers are physiological indicators or substances in the body that serve as indicators of the presence or absence of a disease. The detection of tumor biomarkers such as proteins, DNA, and mRNA is of fundamental importance in cancer treatment. They are sometimes used to investigate the cancer in advance, predict its effectiveness, and determine the cause of treatment. Chen et al. proposed an LC-based system for detecting CA125, one of the most widely studied biomarkers for epithelial ovarian cancer [25, 26]. The LC-based sensor significantly lowered the detection limit of CA125 to 0.01–0.1 μg/mL from that with the traditional fluorescence immunoassay (4–5 μg/mL) [27]. Furthermore, Qi et al. demonstrated a strategy for simultaneous detection of multiple tumor markers such as carcinoembryonic antigen (CEA), alpha-fetoprotein (AFP), and prostate specific antigen (PSA) in blood by functional LC-based sensors assisted with target-induced dissociation of aptamer [28]. The excellent performance of the method can meet the demand for practical applications of rapid multiplex detection of tumor markers in blood.

An LC-based immunosensor for diagnosis of tuberculosis was developed [29]. Tuberculosis (TB) caused by bacteria remains one of the major health problems in the world. Many methods have been developed to detect TB effectively. For example, nucleic acid amplification with polymerase chain reaction (PCR) has high specificity, but it requires 1–2 days for diagnosis of TB. Kim et al. reported an immunosensor to diagnose TB sensitively in clinical serum specimens with a rapid optical response to the orientational change of LCs occurring on the nanostructured surface (Fig. 6.5) [29]. The specific binding of anti-TB antibodies from TB patient clinical samples to the TB antigen-immobilized surface disrupted the anisotropic surface topology, inducing the orientational transition of nematic LCs (**5CB**). This resulted in change of the optical texture of the nematic phase from uniform to schlieren. On the other hand, in the case of clinical samples from healthy people, the TB antigen–antibody reaction did not occur. Therefore, the orientation of LCs was retained, resulting a uniform texture of the N phase. The sensitivity and specificity of the LC-based sensor for TB diagnosis were 76% and 91%, respectively. The authors stated as follows. The LC-based immunosensor is a promising platform, particularly in clinical TB diagnostics, which does not require complicated preparation of clinical specimens or complex instrumentation.

6.2.2 Liquid Crystal-Based Drug Delivery Systems

cells that formA drug delivery system (DDS) is a technology designed to deliver drugs to the necessary amounts and for the necessary amount of time. This maximizes the effectiveness of the drug and minimizes side effects. Development of DDS utilizing liquid-crystalline properties has been actively pursued [30–34]. One of the cell membrane transport mechanisms is caveolae-mediated endocytosis [35] (Fig. 6.6). Caveolae are small flask-shaped invaginations with a diameter of 50–100 nm found in cell membranes [36]. After the drug interacts with the caveolae region on the cell membrane surface, it is encapsulated by morphological changes in the cell membrane and taken up into the cell. The liquid-crystalline nature of the cell membrane plays an important role in endocytosis. Sahay et al. reported a cancer cell-selective DDS that utilizes different endocytosis pathways in normal cells and tumor cells [11]. They prepared polymeric micelles with cross-linked ionic core of poly(methacrylic acid) and nonionic shell of poly(ethylene oxide) (*cl*-micelles). The *cl*-micelles internalized in epithelial cancer cells proceeded through caveolae-mediated endocytosis. On the other hand, this endocytosis toute was absent in normal epithelial cells that form tight junctions. They designed *cl*-micelles bound to cytotoxic drug, doxorubicin, linked via pH-sensitive hydrazone bonds. The bonds were broken in the acidic environment of the lysosome, and the drug accumulated in the nucleus. Such pH-sensitive *cl*-micelles showed selective toxicity against cancer cells but were non-toxic to normal epithelial cells.

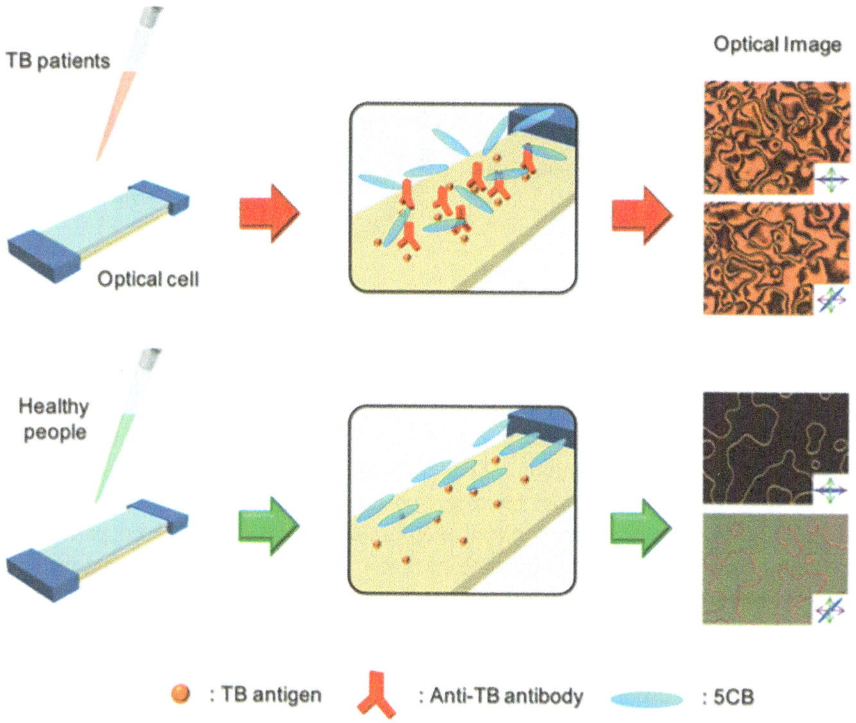

Fig. 6.5 Schematic illustration of TB diagnosis in clinical specimens, using a LC-based immunosensor. Reprinted with permission from [29]. Copyright 2017 American Chemical Society

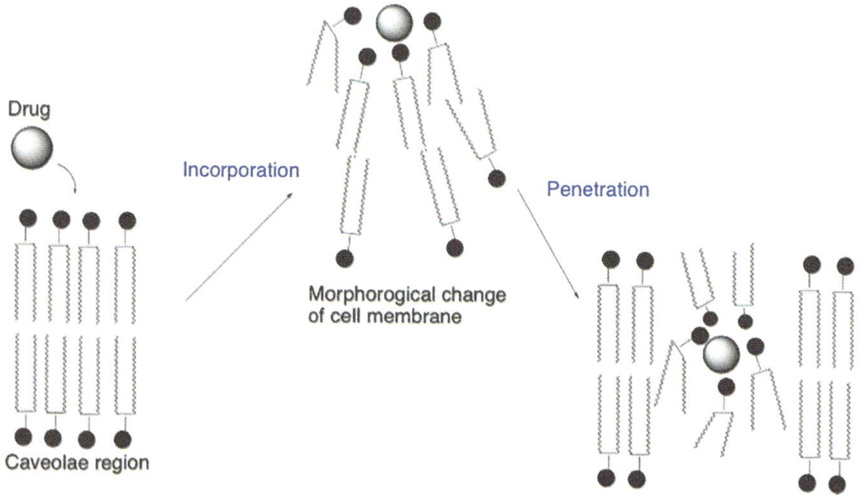

Fig. 6.6 Schematic illustration of caveolae-mediated endocytosis

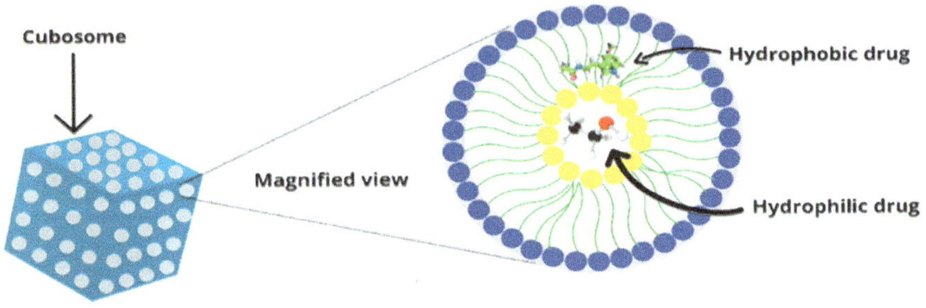

Fig. 6.7 Schematic illustration of a cubosome. Reproduced form [43]

Amphiphilic lipids such as glyceryl monooleate (GMO) can form various lyotropic liquid crystal phases, i.e. lamellar, bicontinuous cubic, and hexagonal phases, depending on the amount of water added. Particularly, the bicontiuous cubic phase is of great interest and has been observed in many biological specimens, including nuclear and mitochondrial membranes [37, 38]. Dispersed particles of biscontinuous cubic phase, cubosomes, are self-assembled nanostructured particles that can be formed in aqueous lipid and surfactant systems [39–41]. They are characterized by a double structure with two continuous but non-intersecting water channels separated by a lipid layer. This structure gives them a large surface area and the ability to encapsulate hydrophilic and hydrophobic drugs as shown in Fig. 6.7. Gustafsson et al. demonstrated practical examples of forming colloidally stable cubosomes [42]. Cubosomes are categorized as a gyroid ($Ia3d$), primitive ($Im3m$), or diamond ($Pn3m$), and they maintain cubic symmetry like their bulk parent phase [41, 43]. The important advantages of cubosomes are their biocompatibility and the ability to be loaded with numerous medications.

Research is being conducted on controlling the release rate of encapsulated drugs by changing the structure of lyotropic liquid crystal phases in response to stimuli such as temperature and pH. The release of drug from lipid-based liquid crystal systems is diffusion controlled, with generally larger aqueous channels of the bicontinuous cubic phase providing a faster release rate than the inverse hexagonal phase [44]. Fong et al. designed temperature-responsive liquid crystals providing 'on-demand' drug delivery [45]. Negrini and Mezzenga proposed a DDS that utilizes the control of liquid crystal structures caused by pH changes dependent on differential biological sites, such as the intestine and stomach [46]. Monolinolein, linoleic acid as a pH sensitive molecule, and phloroglucinol as a model hydrophilic drug were used for the DDS. Their molecular structures are shown in Fig. 6.8. The linoleic acid (intrinsic pKa $\approx$ 5) can be in either deprotonated or protonated state when changing the pH from neutral to acidic conditions. Therefore, the system self-assembles into a reverse bicontinuous cubic phase of Im3m symmetry at pH 7 and transforms into a reverse columnar hexagonal phase at pH 2. The cubic phase has the 3D organization of the water channels, therefore, diffusion of the model drug at pH 7 occurs 4 times faster than at pH 2. The pH-responsive nanocarrier can be efficiently used to spontaneously trigger the release of drugs in the gastrointestinal tract (pH 7), while preventing premature release in the stomach environment (pH < 2) as shown in Fig. 6.9.

The temperature-driven phase transition of thermotropic liquid crystals can also function as "on-demand" drug release. With respect to skin drug delivery, the largest barrier to drug penetration is the stratum corneum (SC) forming a

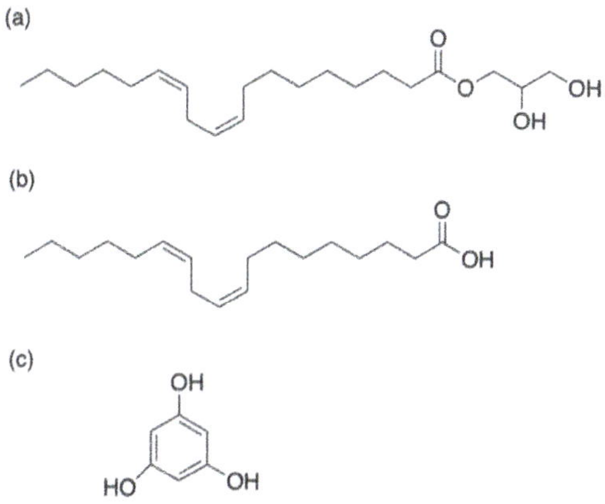

Fig. 6.8 Molecular structures of **a** momolinolein, **b** linoleic acid, and **c** phloroglucinol

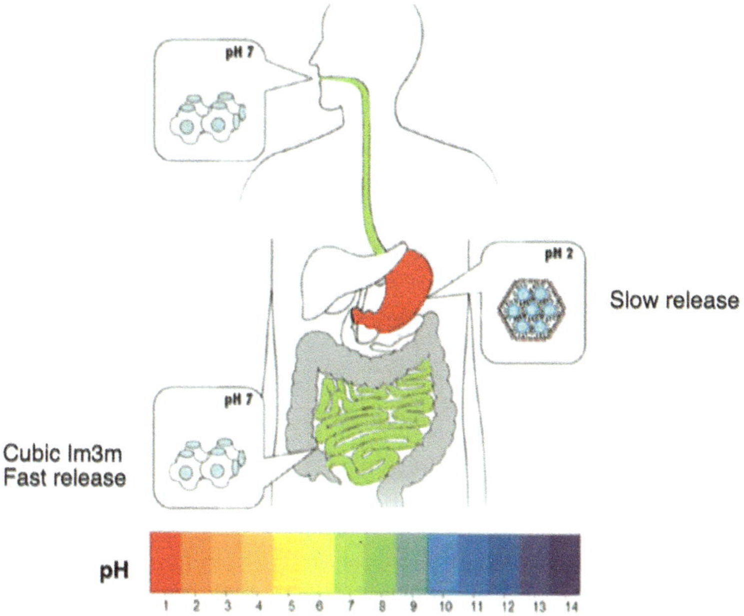

Fig. 6.9 pH-responsive lyotropic liquid crystal-based drug delivery across the gastrointestinal tract. Reprinted with permission from [46]. Copyright 2011 American Chemical Society

lamellar liquid-crystalline structure consisting of the intercellular lipids. Liquid-crystalline molecules are expected to be compatible the SC due to their structural similarity to the lipid molecules. Nesterkina et al. reported thermoresponsive cholesteric liquid–crystal systems doped with terpenoids [47]. These systems switch to the liquid-crystalline state in response to variations in skin temperature enabling temperature-dependent drug release control.

Miyata et al. reported reversibly temperature-responsive micelles as smart DDS carriers, formed from amphiphilic side-chain liquid crystal polymers (LCPs) having grafted oligo(ethylene glycol) OEG (**LCP-*g*-OEGs**) [48]. Figure 6.10a presents the conceptional illustration of the DDS system. Figure 6.10b shows the molecular structure of **LCP-*g*-OEGs**. They focused on **LCP-*g*-OEG$_{360}$** with a mesogen content 84.7 mol% and an OEG content of 5.4 mol%. The amphiphilic **LCP-*g*-OEG$_{360}$** was prepared by hydrosilylation of hydrophobic mesogens and hydrophilic OEG monomethacrylate with an average molecular weight of 360 onto poly(methylsiloxane). Micelles of the amphiphilic **LCP-*g*-OEG$_{360}$**, which has a T_{NI} of 36.2 °C, were formed in water. They formed spherical micelles with a diameter of approximately 130 nm and a polydispersity index (PDI) of 0.073 (Fig. 6.11).

Fluorescein was loaded as a model drug into **LCP-*g*-OEG$_{360}$** micelles at a fluorescein/**LCP-*g*-OEG$_{360}$** ratio of 0.5/5.0 (mg/mg). The diameter and PDI of the fluorescein-loaded **LCP-*g*-OEG$_{360}$** micelles were 146 nm and 0.052. The micelles

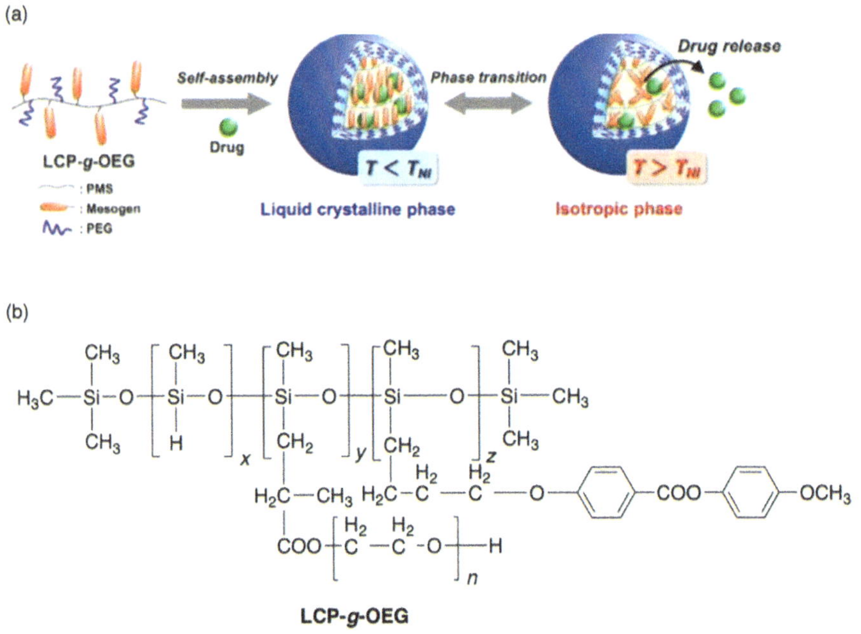

Fig. 6.10 **a** Preparation of LCP-g-OEG micelles and drug release behavior caused by the nematic–isotropic phase transition. Reprinted with permission from [48]. Copyright 2011 American Chemical Society. **b** The molecular structure of **LCP-*g*-OEG**

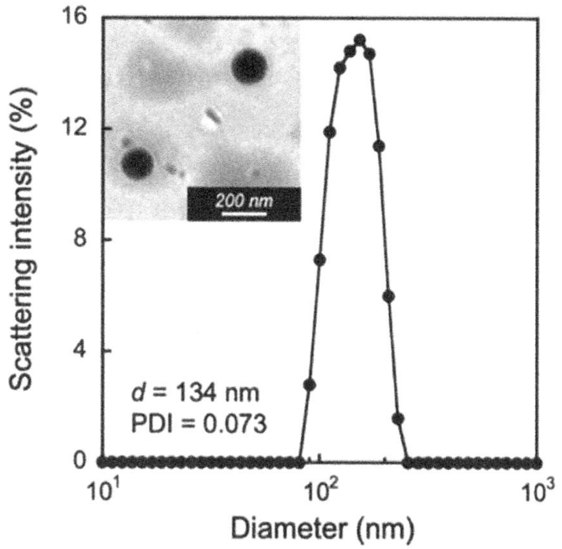

Fig. 6.11 Size distribution and TEM image of the **LCP-*g*-OEG$_{360}$** micelles. Reprinted with permission from [48]. Copyright 2011 American Chemical Society

have a suitable size for enhanced permeability and retention effect (see Sect. 6.3.1). The micelles remain stably dispersed in aqueous media over a wide temperature range. The release behavior from **LCP-*g*-OEG$_{360}$** micelles changes dramatically at T_{NI}. The release rate remains almost constant at temperatures lower than T_{NI}, but increases monotonically with the increasing temperature at temperatures higher than T_{NI}. The authors investigated the reversible ON–OFF regulation of fluorescein release from the micelles in response to stepwise changes in temperature between 25 and 50 °C. The fluorescein release remained completely stopped at 25 °C but was promoted at 50 °C. The cellular uptake of the fluorescein-loaded **LCP-*g*-OEG$_{360}$** micelles into normal and tumor cells was investigated at 30 °C. The fluorescein functionalized as a fluorescent probe. When L929 as a normal cell was incubated with fluorescein-loaded **LCP-*g*-OEG$_{360}$** micelles, the green fluorescence of the fluorescein was not observed inside the cell. By contrast, green fluorescence was clearly observed inside HepG2 tumor cell (Fig. 6.12). The authors discuss the cellular uptake behavior as follows. The micelles are transported into the tumor cell via endocytosis. The PEG content in the micelle plays a crucial role in the tumor-selective cellular uptake. The fluorescence intensity from the fluorescein-loaded **LCP-*g*-OEG$_{360}$** micelles within HepG2 disappeared when the temperature was increased to 37 °C, indicating that fluorescein is releases from the **LCP-*g*-OEG$_{360}$** micelles within HepG2. The drug-loaded **LCP-*g*-OEG$_{360}$** micelles are selectively taken up by tumor cells via endocytosis and regulated drug release reversibly in response to changes in temperature.

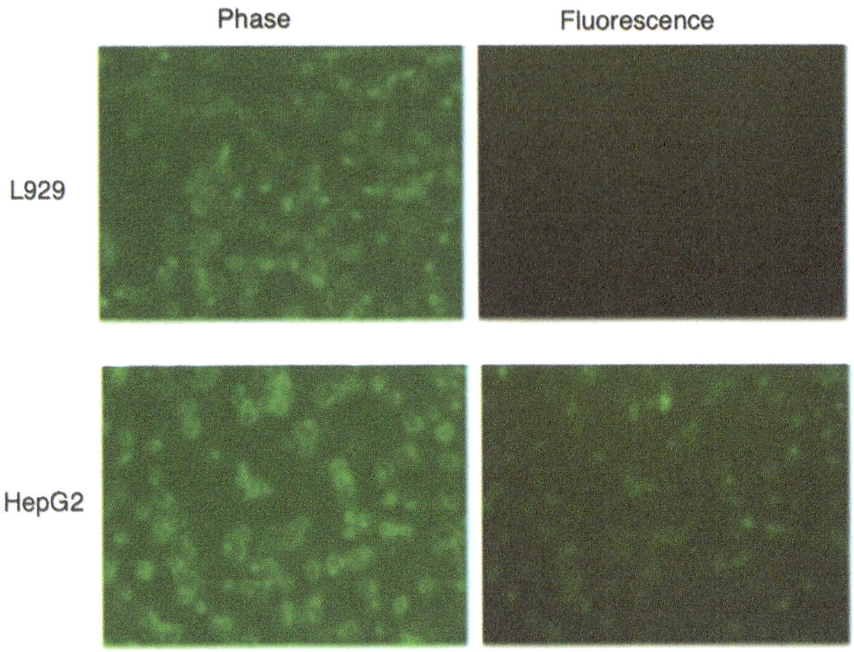

Fig. 6.12 Phase and fluorescence images of L929 and HepG2 cells incubated with the fluorescein-loaded micelles after exposure at 30 °C for 12 h. The **LCP-*g*-OEG$_{360}$** micelles concentration was 0.09 mg mL^{-1}. Reprinted with permission from [48]. Copyright 2011 American Chemical Society

6.3 Pharmaceutical Liquid Crystals

6.3.1 Chemotherapy Medecines

Antitumor medicines, also called as anticancer drugs, are medications used to treat tumors by suppressing the proliferation of cancer cells, shrinking tumors, or preventing their spread. The mechanism of action varies depending on type of drug and the type of cancer. Anticancer medicines fall mainly into four groups: chemotherapy, targeted therapy, immunotherapy, and hormonal therapy. The chemotherapy drugs kill or inhibit the growth of rapidly dividing cell. They interfere with DNA synthesis, mitosis, or other steps of the cell cycle. Cell division occurs actively in cancer cells which lead to the tumor formation. The purpose of chemotherapy is to regulate the uncontrolled proliferation of cancer cells. Most chemotherapeutic drugs are classified as alkylating agents, monoclonal antibodies, anti-metabolites, anthracyclines, plant alkaloids, topoisomerase inhibitors, or other anticancer agents [49–55]. The cell cycle is usually divided into two major phases: Interphase and Mitotic (M) phase. Interphase consists of three sub-phases: G1 phase, S phase, and G2 phase. G1 phase is marked by synthesis of enzymes needed for DNA replication. During S phase, DNA is replicated. In G2 phase, cell prepared for division and checks DNA for errors. M phase consists of mitosis and cytokinesis. Mitosis is the division of the nucleus. After mitosis, the cytoplasm divides in a process called cytokinesis, producing two identical daughter cells. The cell cycle is carefully regulated by checkpoints and proteins such as cyclins and cycline-dependent kinases (CDKs). Many of chemotherapy drugs target the S phase of cell cycle. For example, nitrogen mustard is intercalated to DNA double helix via the alkylation of nucleic-acid base, which disturbs the DNA replication as shown in the Fig. 6.13 [56]. Because they also affect normal fast-growing cells, they often cause side effects like hair loss. Furthermore, in order to develop the imperative therapeutic agents against solid cancer cells, they have to penetrate the layers of cell membrane. Atmosphere within the membrane is hydrophobic, thus a hydrophilic drug has difficulty to penetrate the cell membrane [57].

Due to not only reducing the side effects but also increasing the membrane penetration, drug delivery systems have been investigated extensively [58–60]. A lot of nanocarriers entrapping cytostatic agents have been designed [61–67]. For example, doxorubicin's nanocarrier which is based on polyethylene glycol-phosphatidylethanolamine is shown in Fig. 6.14 [61]. If nanocarriers have a suitable size for enhanced permeability and retention (EPR) effect [65–67], they are expected to show selectivity. Compared to normal blood vessels, the gaps in tumor tissue blood vessels are larger. Therefore, particles of a certain size become more permeable to tumor tissue blood vessels and accumulate selectively in tumor tissue (Fig. 6.15). The nanocarriers can be taken up by the cells via endocytosis followed

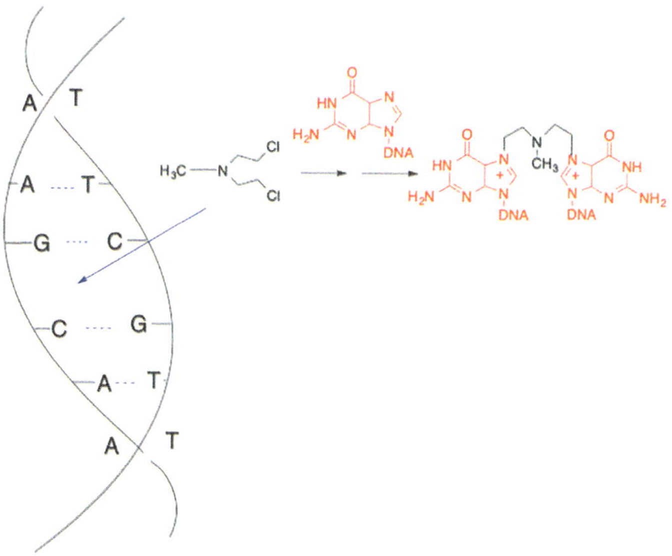

Fig. 6.13 The intercalation to DNA double helix of nitrogen mustard via the alkylation of guanine. Reprinted with permission from [58]. Copyright [2022] Springer Nature

by releasing the agent within a cell. In addition to these carriers, prodrug systems have been developed for increasing the anticancer selectivity [68, 69]. Non-toxic prodrugs can be activated by an enzyme that is naturally over-expressed in the tumor microenvironment.

In recent years, supramolecular nanostructures have attracted significant attention in the development of effective anticancer drugs [58, 70, 71]. Research has been conducted on drug molecules that promote self-assemble in to micellar structures [72–75]. Cheetham et al., reported a supramolecular strategy for assembling the

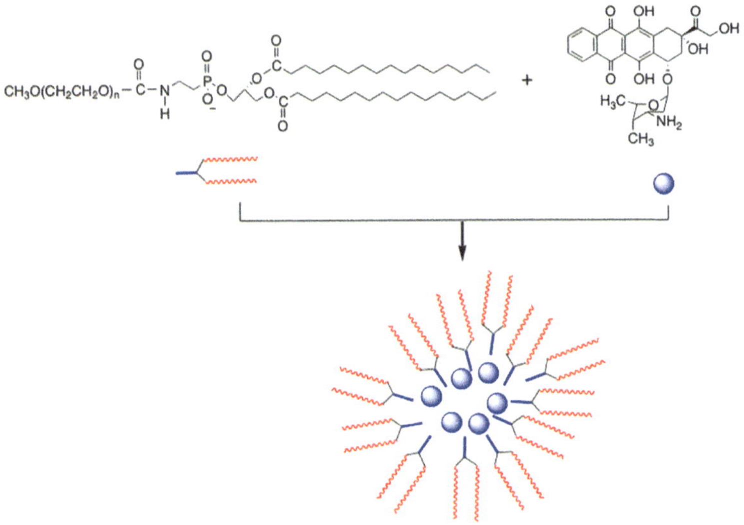

Fig. 6.14 Polyethylene glycol-phosphatidylethanolamine-based nanocarrier of doxorubicin [61]. Reprinted with permission from [58]. Copyright [2022] Springer Nature

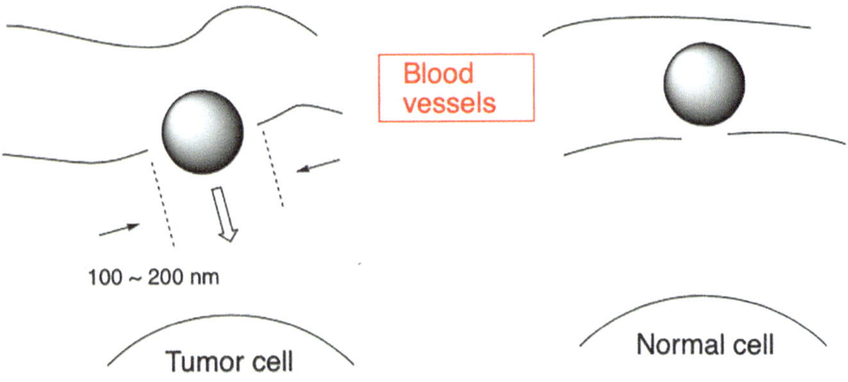

Fig. 6.15 Schech of the EPR effect

hydrophobic anticancer drug camptothecin (CPT) into stable filamentous nanostructures with high concentrations and consistent drug loading [76]. These drug nanostructures are capable of releasing the bioactive form of CPT and exhibit significant cytotoxicity against cancer cells (Fig. 6.16). Supramolecular drug delivery systems based on linear-dendritic copolymers have also been investigated [77]. These delivery technologies can improve water solubility and enable targeted delivery of anticancer drugs. Antitumor activity has also reported for supramolecular metal complexes. The highly charged $[Pt_6L_4]^{12+}$ supramolecular complex exhibited cytotoxicity against human cancer cells equivalent to that of cisplatin, while showing only one-seventh the toxicity of cisplatin against normal lung cells [78]. Pöthig and Casini reviewed the imaging and anticancer role of supramolecular architectures based on metals [79]. Supramolecular nanostructures based on cyclodextrin-derived polyrotaxanes have been developed. Yu et al., prepared therapeutic and diagnostic shell-crossinked nanoparticles (SCNPs) using β-cyclodextrin-based polyrhotaxane. This approach prevents premature drug leakage and enables contolled, targeted release [80]. Drug-loaded SCNPs exhibited the significant anti-tumor activity against the orthotopic breast cancer.

Cell membranes contain phospholipids and form a layered bilayer mesophase that exhibits liquid-crystalline properties. In the medical field, there are several applications of liquid crystals, such as cancer biomarkers and drug delivery systems [12, 81]. Recently, characterizations of pharmaceutical compounds by their liquid-crystalline states have been increased (Fig. 6.17) [82]. Yoshizawa et al., reported the influence of liquid-crystalline properties on the cell proliferation-inhibiting activity against A549 human lung cancer cells [83]. Due to the structural similarity between liquid crystal molecules and cell membranes, liquid crystal-based pharmaceuticals will improve

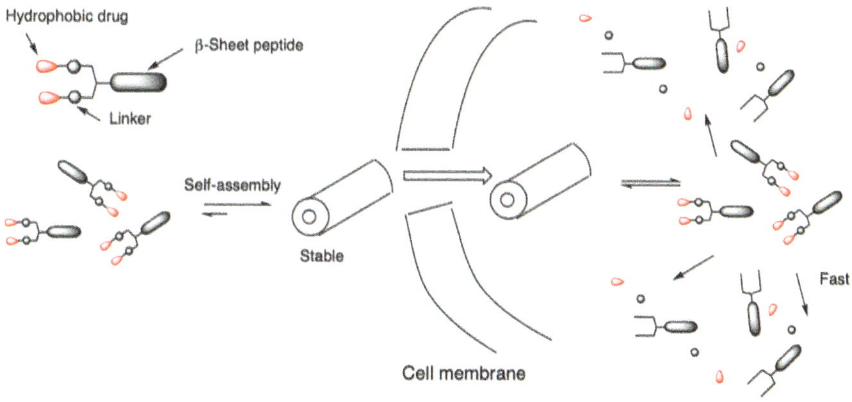

Fig. 6.16 Schematic illustration of the cytostatic mechanism of the self-assembled nanostructures containing hydrophobic drugs [76]. Reprinted with permission from [58]. Copyright [2022] Springer Nature

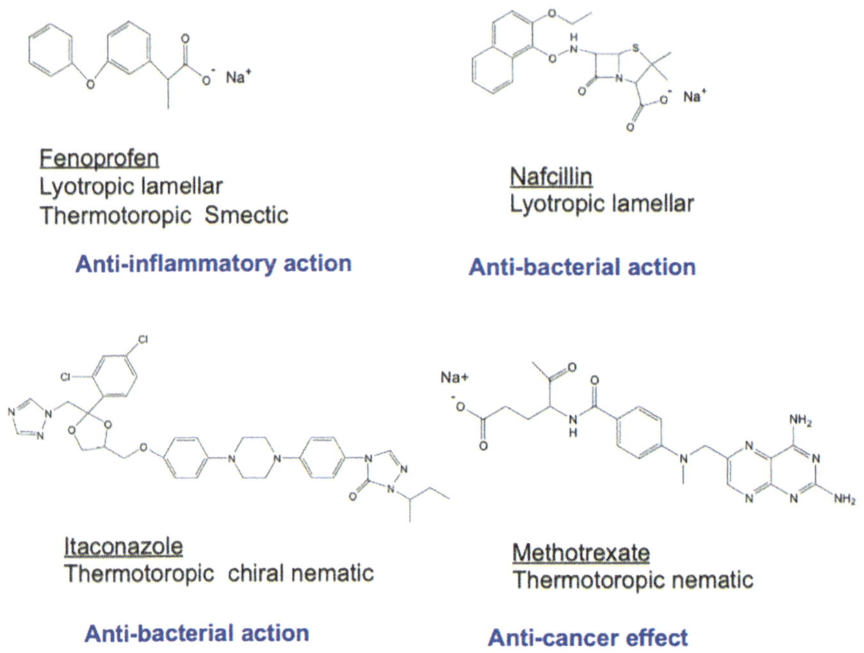

Fig. 6.17 Drugs exhibiting liquid crystallinity

the cell membrane permeability of drugs, which is currently a challenge in the treatment of solid tumors. Therefore, supramolecular assemblies of mesogenic molecules containing active sites are expected to serve as a new treatment for solid tumors.

6.3.2 Anti-Proliferative Potential of Thermotropic Liquid Crystals Against Solid Cancers

Many amphiphilic polyhydroxy derivatives have been synthesized, and their physical properties have been investigated [84, 85]. Yoshizawa et al., designed an amphiphilic compound **PPY-G** in which phenylpyrimidine and *D*-glucamine are connected through a flexible spacer [83]. Figure 6.18 shows their molecular structures and phase transition temperatures. The **PPY-G** exhibited two liquid crystalline phases, i.e., uniaxial smectic A phase and unidentified tilted smectic X phase. In contrast, no liquid-crystalline properties were observed in its precursors, **PPY** and *D*-Glucamine. X-ray diffractions measurements revealed that both smectic phases possess an interdigitated structure with different layer spacings [83]. The molecular length is about 37 Å. Figure 6.19 shows molecular arrangement models for the uniaxial smectic A (SmA) and the tilted smectic X (SmX) phase. In the high-temperature SmA phase, **PPY-G** molecules form a lamellar bilayer structure in which the hydrophilic portions face outward (Fig. 6.19a), whereas in the low-temperature SmX phase, they form a structure in which the hydrophobic portions face outward (Fig. 6.19b). Therefore, mesogenic amphiphilic molecules can reorganize their molecular assemblies depending on the external circumstance.

We investigated the cytotoxicity of **PPY-G** against human lung cancer cells (A549). A549 cells (4 × 10^3 cells/mL) were seeded at 1 mL per well in a 24-well plate and cultured at 37 °C in a humidified atmosphere containing 5% CO_2 for 24 h. The cells were then exposed to various concentrations of **PPY-G** at 37 °C for 96 h. Cells adherent to the plates were detached by trypsinization and counted using a Coulter counter. Cell viability is expressed as a percentage relative to the results of control groups treated with the solvent (DMSO). We also examined the cytotoxicity of **PPY**, *D*-glucamine, and their equimolar mixture. The results are shown in

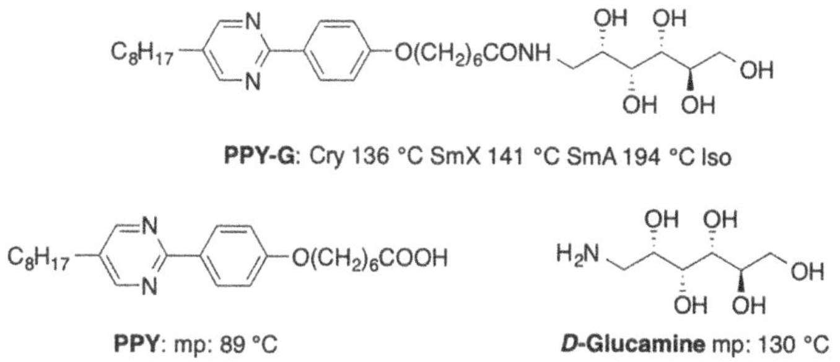

Fig. 6.18 Molecular structures of **PPY-G**, **PPY** and *D*-glucamine with their phase transition temperatures [83]

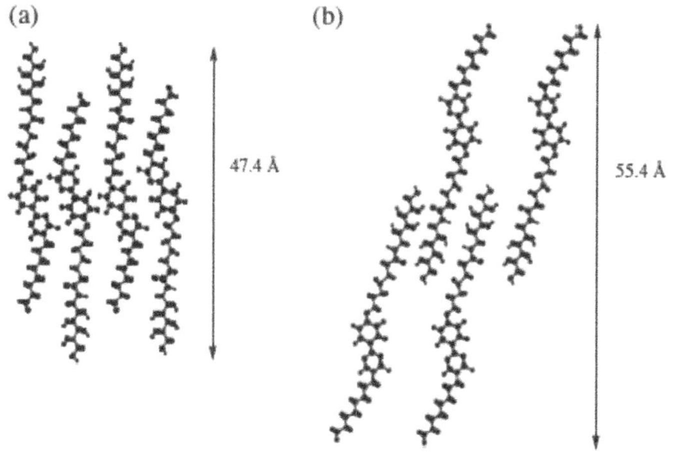

Fig. 6.19 Molecular organization model for the smectic A phase (**a**) and for the smectic X phase (**b**). Reprinted with permission from [83]. Copyright 2009 The Chemical Society of Japan

Table 6.1. **PPY-G** exhibited a significant inhibition of cell proliferation at concentrations exceeding 10 μM, whereas no such marked inhibition was observed for **PPY**, *D*-glucamine, or the mixture. The liquid crystal properties play a key role in suppressing the proliferation of the human lung cancer cells. We surmise the penetration mechanism as follows. The **PPY-G** molecules form nano-assemblies in which the hydrophilic glucamine units face outward in water (Fig. 6.19a). That nanostructure transforms into another nanoaggregate within the cell membrane, with its hydrophobic tails facing outward (Fig. 6.19b). Unfortunately, it showed suppressive effects on the growth of WI-38 normal fibroblast cells [86].

Table 6.2 shows the structure–property correlations in the antiproliferative activity of thermotropic liquid-crystalline compounds with representative core structures [86]. All compounds were screened for their effects on the proliferation of A549 lung cancer cells at a concentration of 10 μM concentration. 4-Cyano-4′-(ω-hydroxyhexyloxy)biphenyl and 4-butylphenyl 4-(6-hydroxyhexyloxy)benzoate exhibited a significant inhibitory effect on the proliferation of A549 lung cancer cells.

Table 6.1 Cell viability of A549 lung cancer cells expressed as a percentage (%) relative to solvent (DMSO)-treated control incubations [83]

Conc./μM	PPY-G	PPY	*D*-Glucamine	Mixture[a]
1	102 ± 4	105 ± 4	110 ± 3	103 ± 5
10	66 ± 0	91 ± 5	93 ± 1	86 ± 6
20	31 ± 10	89 ± 2	101 ± 3	80 ± 5

[a] An equimolar mixture of **PPY** and *D*-glucamine

Table 6.2 Effects of liquid-crystalline compounds on the growth of the A549 human lung cancer cell [86]. Cell proliferation is expressed as relative value against control. The culture was carried out in liquid media containing 10 μM of each compound for 96 h

Structure	Cell proliferation	Structure	Cell proliferation
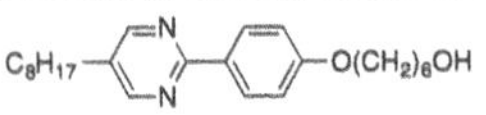	0.79	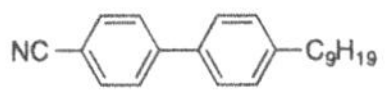	0.9
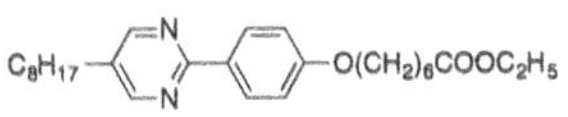	0.82	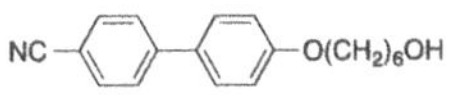	0.25
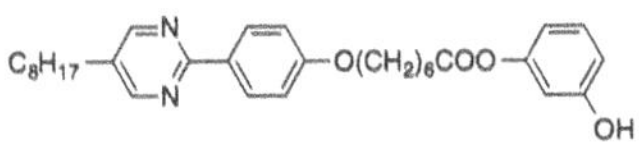	0.67	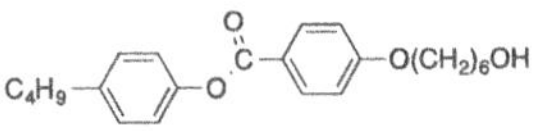	< 0.01
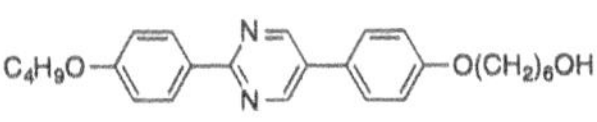	0.81	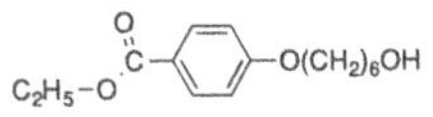	1.1
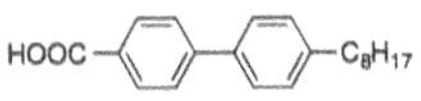	0.83	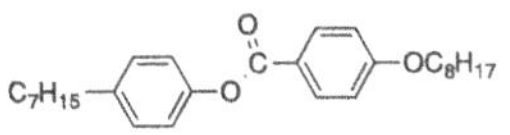	0.81

Fig. 6.20 Molecular structures of cyanobiphenyl derivatives **I-*n*** and **II-4**

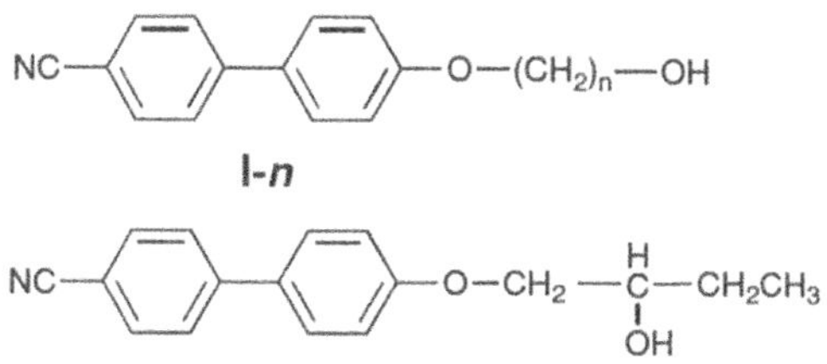

Therefore, we investigated effects of a homologues series of conventional nematic (N) liquid crystals, 4-cyano-4′-(ω-hydroxyalkyloxy)biphenyl (**I-*n***), against human lung cancer A549 cell growth [87]. For comparison, we investigated the cytotoxicity of compound **II-4**, which contains a racemic secondary alcohol. Figure 6.20 shows their molecular structures, and Table 6.3 provides their phase transition temperatures.

Figure 6.21 presents effects of compounds **I-*n*** on human lung cancer A549 cell growth. Compound **I-7**, which has the widest nematic phase temperature rage, exhibited the strongest inhibitory effect on cell proliferation at concentrations exceeding 5 μM. With the exception of **I-10** and **I-12**, the other compounds exhibited noticeable inhibitory effects at a concentration of 10 μM. Cell viability by compound **II-4** at a concentration of 10 μM was 0.56. No significant difference was observed between compounds **I-4** and **II-4** in terms of their inhibitory effects on cell proliferation. To investigate the effects of these cyanobiphenyl derivatives on the proliferation of normal cells, an experiment was conducted on WI-38 fibroblast cells under identical conditions for 168 h. Compound **I-*n*** ($n = 4$–7) did not show any inhibition of cell proliferation. However, compound **II-4**, which contains a secondary alcohol, inhibited cell proliferation not only in the cancer cells but also in the normal cells

Table 6.3 Phase transition temperatures (°C) of compounds **I-*n*** and **II-4**. Square brackets denote a monotropic transition [87]

Compound	Cry	N	Iso	T_N[a]/K
I-4	• 132	• [120]	•	0
I-5	• 90	• 104	•	14
I-6	• 97	• 109	•	12
I-7	• 78	• 100	•	22
I-8	• 89	• 102	•	13
I-9	• 85	• 96	•	11
I-10	• 96	• 97	•	1
I-12	• 101		•	0
II-4	• 85	• [68]	•	0

[a] Temperature range of the enantiotropic nematic phase

as shown in Table 6.4. The cyanobiphenyl derivatives possessing a primary alcohol may distinguish between the cancer cells and the normal cells.

Compound **I-7**, which exhibited the strongest anticancer activity, caused the most pronounced cell cycle arrest in the G1 phase (59.8% for control; 78.6% for compound **I-7**), indicating that compound **I-7** slows cell progression from the G1 into the S phase. However, no significant difference in apoptosis-induced cell death was observed between the control group and compound **I-7** (1.9% for control; 2.2% for compound **I-7**). Figure 6.22 shows the western blot profiles of cell cycle related enzymes in A549 lung cancer cells. Compound **I-7** suppressed the expression of cyclin-dependent kinase 2 (CDK2). CDK2 is a cell cycle-regulating protein in the G1 phase; it binds to cyclin E via one or more hydrogen bonds. The binding complex acts an enzyme that enables cells to transition from the G1 phase to the S phase. Compound

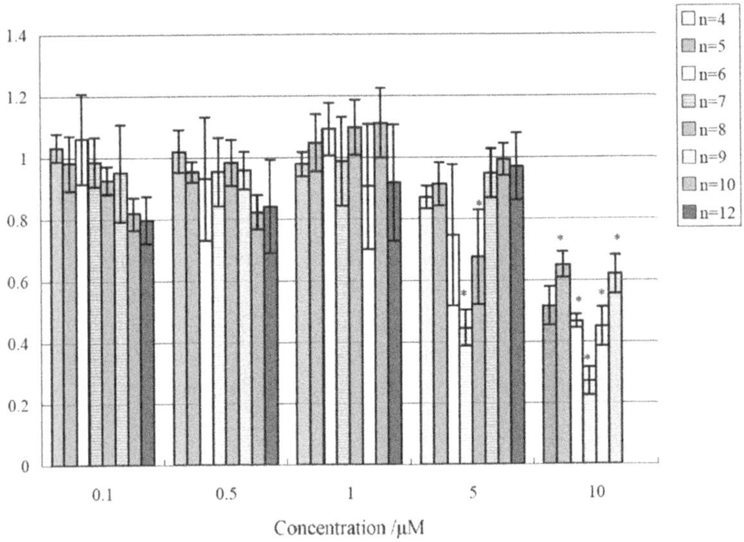

Fig. 6.21 Inhibition of cell proliferation by compound **I-*n*** at concentrations of 0.1–10 μM in A549 human lung cancer cell line for 96 h [87]. The survival ratio to solvent (DMSO)-treated control culture is shown. *$p < 0.05$ by student's *t*-test. Compounds **I-10** and **I-12** were crystallized at a concentration of 10 μM. Reprinted with permission from the reference. Copyright 2009 The Chemical Society of Japan

Table 6.4 Inhibition of cell proliferation by compounds **I-4** and **II-4** at 10 μM in A549 human lung cancer cells for 96 h or in WI-38 fibroblast cells for 168 h. The survival ratio as a percentage (%) to solvent (DMSO)-treated control culture is shown [87]

Compound	A549	WI-38
I-4	56 ± 9	140 ± 32
II-4	55 ± 5	47 ± 20

Reprinted with permission from the reference. Copyright 2009. The Chemical Society of Japan

I-7 interacts with CDK2 via hydrogen bonding, thereby preventing cyclin E from interacting with CDK2 (Fig. 6.23). Compound **I-7** inhibits the synthesis of enzymes necessary for DNA replication in A549 lung cancer cells [86]. Compound **PPY-G** also suppressed CDK2 expression [86]. Both compounds inhibit the proliferation of A549 lung cancer cells via the same mechanism. However, compound **PPY-G** possessing both primary and secondary alcohols showed cytotoxicity against WI-38 normal cells.

Takahashi et al. reported the correlation between liquid crystallinity and anti-tumor activity in supramolecular assembly of different mesogenic compounds possessing a ω-hydroxyl unit [89]. The mesogenic compounds are 4-cyano-4′-(hydroxyhexyloxy)biphenyl (**I-CN**), 4-methyloxy-4′-(hydroxyhexyloxy)biphenyl (**I-OMe**), and 4-fluoro-4′-(hydroxyhexyloxy)biphenyl (**I-F**) (Fig. 6.24). These compounds have a terminal hydroxy group. While compounds **I-CN** and **I-F** contain electron-withdrawing groups, compound **I-OMe** contains an electron-donating group.

Figure 6.25 shows the effects of these mesogenic compounds and their equimolar binary mixtures at a concentration of 5 μM on A549 lung cancer cells. No significant

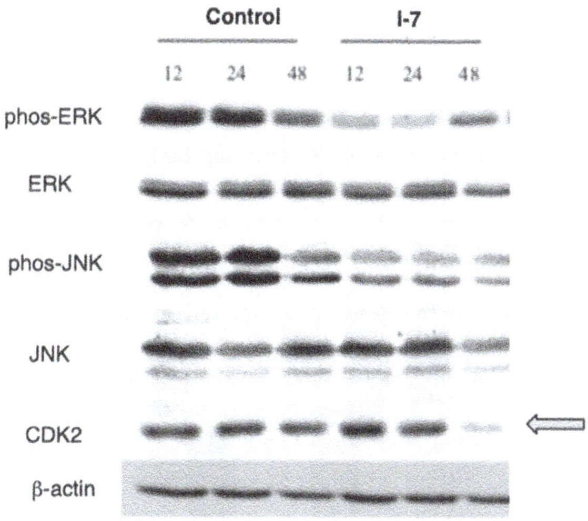

Fig. 6.22 The profiles of western blotting for MAPK and cell cycle related molecules in A549 cells. A549 cells were treated with 10 μM compound **I-7** for 12–48 h. Reprinted with permission from [86]. Copyright [2011] Springer Nature

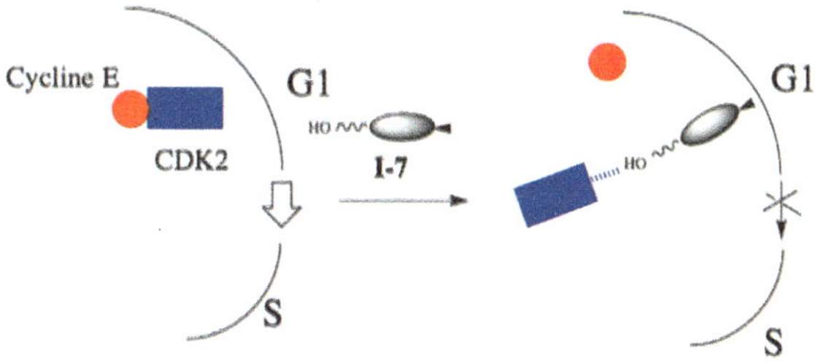

Fig. 6.23 Mechanism for the G1 arrest. Reprinted with permission from [88]. Copyright 2019 Taylor & Francis

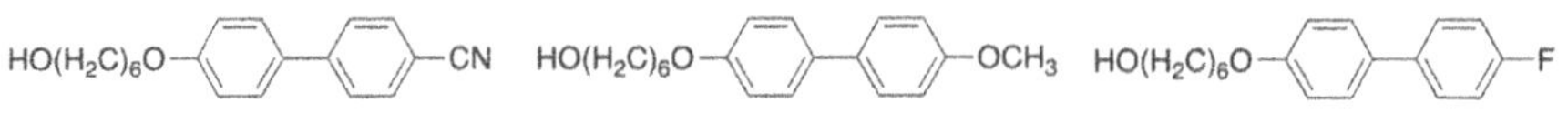

I-CN I-OMe I-F

Fig. 6.24 Molecular structures of the mesogenic compounds

difference in the inhibitory effects on the cell proliferation was observed among **I-CN**, **I-OMe**, **I-F**, and a binary mixture of **I-CN** and **I-F**. Interestingly, a binary mixture of **I-CN** and **I-OMe** shows marked inhibition of cell proliferation.

Figure 6.26 presents how the mixing ratio of **I-OMe** and **I-CN** in the binary system affects the inhibition of cell proliferation of A549 lung cancer cells. A mixture of **I-CN** and **I-OMe** in a 1:1 molar ratio exhibited the strongest inhibitory effect against the cancer cells, suggesting that the one-to-one molecular complex of **I-CN** and **I-OMe** is an effective species for inhibiting the proliferation of the cancer cells.

The equimolar binary mixture of **I-CN** and **I-OMe** inhibited the proliferation of WI-38 fibroblast cells, whereas compound **I-7** did not such an effect (Fig. 6.27). The mechanism of the antitumor effect may differ between that mixture and **I-7** [89].

We will discuss the self-assemble properties of these mesogenic materials. Compound **I-CN** exhibited a nematic phase; its isotropic-to-nematic transition temperature T_{IN} was 109 °C. In contrast, neither of the compounds **I-OMe** and **I-F** showed a liquid-crystalline behavior; their isotropic-to-crystal temperatures were 144 °C and 93 °C, respectively. In constats, the equimolar mixture of **I-CN** and **I-OMe** exhibited a nematic phase, and its T_{IN} was 16 K higher than that of **I-CN**. Figure 6.28 shows a photomicrograph of the mixture (500 μM) in DMSO–H_2O (1:19) solution at 37 °C. Although some crystalline textures were seen, numerous

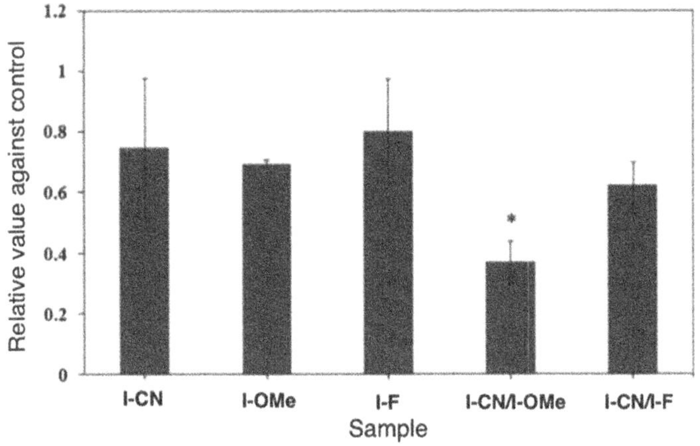

Fig. 6.25 Effects of mesogenic materials on the growth of the A549 human lung cancer cell line. The culture was conducted in liquid media containing 5 μM of each sample for 96 h. The binary mixtures contained 2.5 μM of each component. Reproduced from [89] with permission from the Royal Society of Chemistry

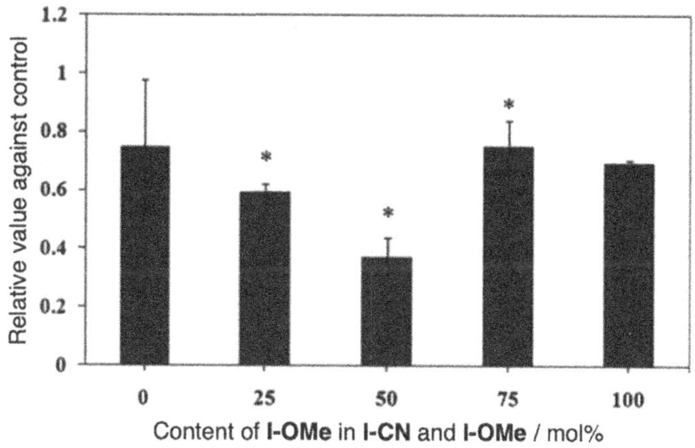

Fig. 6.26 Effects of concentration of **I-OMe** in the **I-CN** and **I-OMe** binary system on the growth of the A549 human lung cancer cell line. The culture was conducted in liquid media containing 5 μM of each sample for 96 h. Reproduced from [89] with permission from the Royal Society of Chemistry

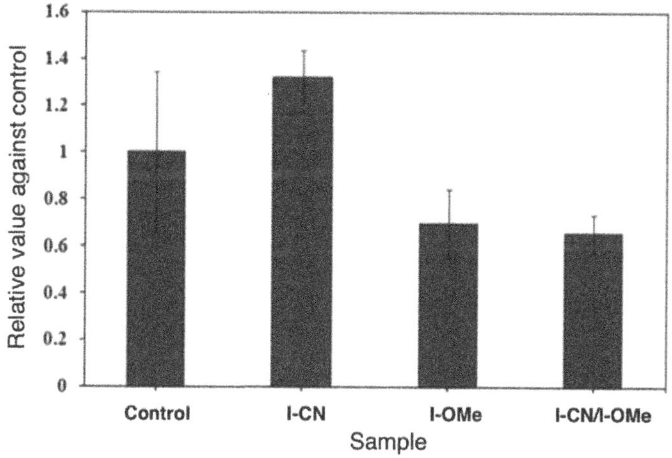

Fig. 6.27 Effects of the mesogenic materials on the growth of WI-38 fibroblast cells. The culture was conducted for 168 h in liquid media containing 5 μM of each sample. Reproduced from [89] with permission from the Royal Society of Chemistry

droplet textures were observed, suggesting the presence of lamellar and/or micellar liquid-crystalline phases. Table 6.5 shows critical aggregated concentration (CAC) values of the mesogenic materials. CAC is the minimum concentration of amphiphilic molecules in which they start to from aggregates in solution. The equimolar mixture of **I-CN** and **I-OMe** shows the lowest CAC among them, suggesting that it has

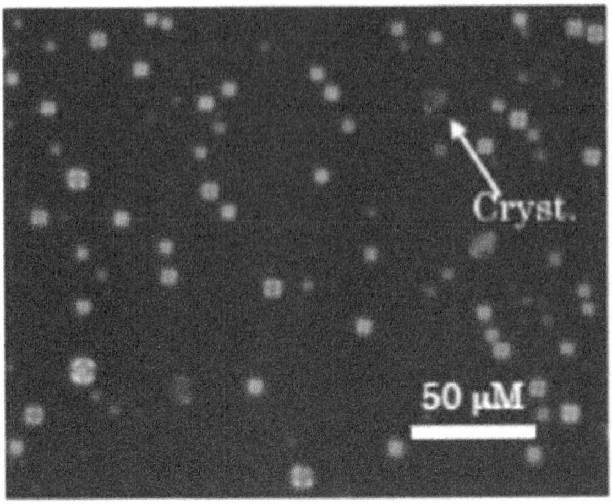

Fig. 6.28 Photomicrograph of a sample of the equimolar mixture of compounds I-CN and I-OMe at 500 μM in DMSO–H_2O (1:19) solution at 37 °C. Reproduced from [89] with permission from the Royal Society of Chemistry

Table 6.5 Critical aggregation concentration (CAC) of each sample. The turbidity of each sample at various concentrations was detected by observing the absorbance at 500 nm. The CAC is defined as a concentration at which the absorbance begins to increase

Sample	CAC/μM
I-CN	30 ± 8
I-OMe	30 ± 11
I-F	57 ± 38
Equimolar mixture of **I-CN** and **I-OMe**	8 ± 5
Equimolar mixture of **I-CN** and **I-F**	44 ± 10

Reproduced from [89] with permission from the Royal Society of Chemistry

the highest aggregation ability. The supramolecular assemblies of these different mesogenic molecules exhibit antitumor activity.

Next, we will discuss mesogenic compounds containing phenyl benzoate. As shown in Fig. 6.29, the phenyl benzoate derivative possessing a terminal hydroxy group (denoted as compound **III**) showed the much stronger cytotoxicity against A549 lung cancer cells than the other benzoic acid esters, i.e., ethyl 4-(6-hydroxyhexyloxy)benzoate (**IV**) and 4-heptylphenyl 4-octyloxybenzoate (**V**) [90]. Compound **IV** does not exhibit liquid-crystalline properties, whereas compound **V** does not contain a hydroxyl group. With respect to compound **III**, it is believed that the phenyl benzoate unit contributes to the formation of the N phase, while the terminal hydroxyl group serves as the active site. According to the dose-dependent cytotoxicity of compound **III** on A549 cells as shown in Fig. 6.30, the 50% inhibitory concentration (IC_{50}) for 96 h is calculated as 4.7 μM. Figure 6.31 shows the inhibitory effect of compound **III** on the cell cycle profile in A549 cells. Treatment with 20 μM of compound **III** for 24 and 48 h, respectively, resulted in sub-G1 cell cycle arrest rates of 33 and 61% in A549 cells, compared to 1% in the control group. The marked

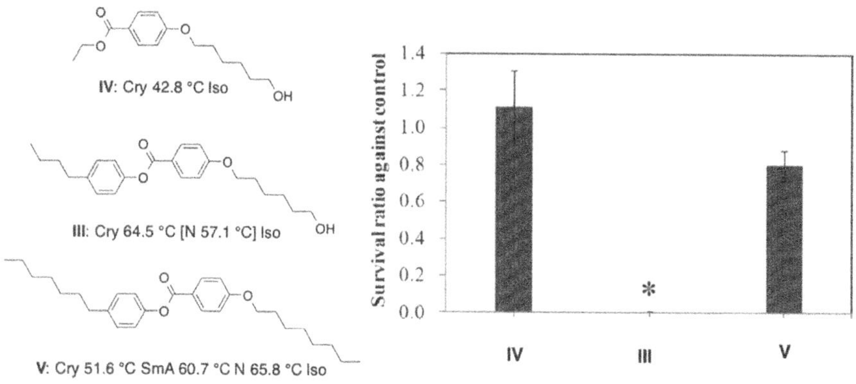

Fig. 6.29 Effects of compounds **III–V** on human lung cancer A549 cell growth at a concentration of 10 μM for 96 h. Reproduced from [90] with permission from the Royal Society of Chemistry

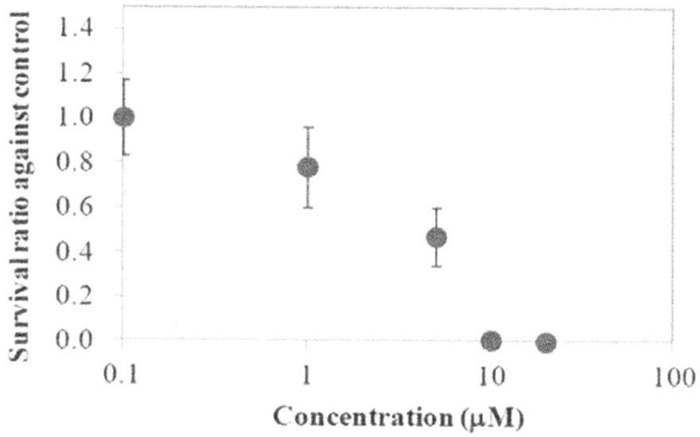

Fig. 6.30 Dose-dependent cytotoxicity of compound **III** on A549 cells for 96 h. Reproduced from [90] with permission from the Royal Society of Chemistry

accumulation in sub-G1 phase suggests that the phenyl benzoate derivative **III**, unlike the cyanobiphenyl derivatives **I-*n***, induces apoptosis.

The cytotoxic effects of compound **III** on WI-38 normal fibroblasts, and certain cancerous cell lines, such as THP1 leukemic cells and SW480 colon cancer cells, were also investigated. As shown in Fig. 6.32, compound **III** significantly inhibits cell proliferation in these solid tumors compared to the hematological cancerous cells. This supports the hypothesis that liquid-crystalline molecules have high cell membrane permeability. Unfortunately, compound **III** exhibits significant cytotoxicity even toward the normal cells. The ester group may be causing cytotoxicity in the normal cells.

To elucidate the correlation between structure and properties, we synthesized two liquid-crystalline supermolecules–namely 1,2-; and 1,3-benzene derivatives–as shown in Fig. 6.33, and investigated their cell proliferation-inhibiting effects. properties. Neither of them exhibited liquid-crystalline. Figure 6.34 shows the concentration-dependent cytotoxicity of compound **VI** (1,2-derivative) and compound **VII** (1,3-derivative) against A549 cells. As the concentration of compound

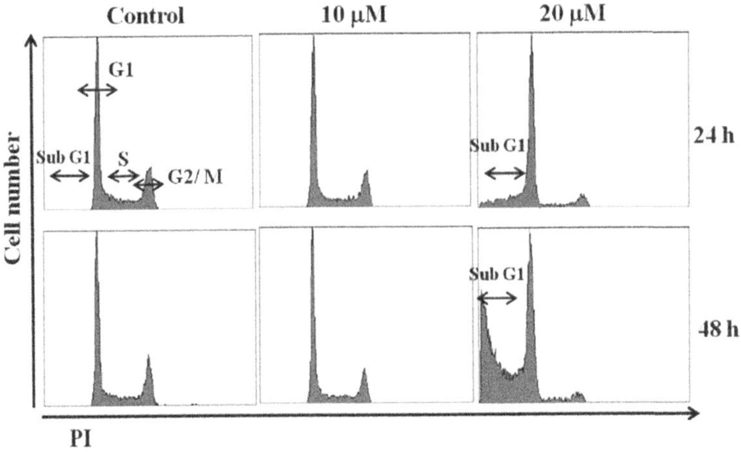

Fig. 6.31 Dose-dependent effects of compound **III** on the cell cycle distribution in A549 lung cancer cells. Reproduced from [90] with permission from the Royal Society of Chemistry

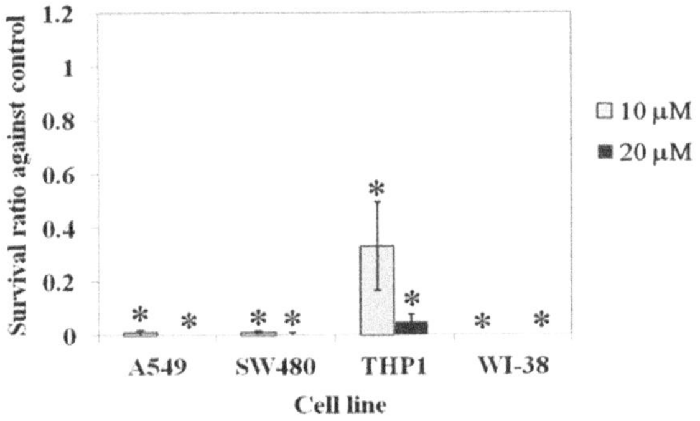

Fig. 6.32 Effects of compound **III** on cell proliferation of A549 lung cancer cells, SW480 colon cancer cells, THP1 leukemic cells and WI-38 normal fibroblast cells at 10 and 20 μM. Reproduced from [90] with permission from the Royal Society of Chemistry. Those cells cultured in the presence of compound **III** were harvested (A549: 96 h, SW480: 96 h, THP1: 72 h, WI-38: 168 h). Survival cells were not detected in A549 cells in the presence of compound **III** at 20 μM. Those were not detected in WI-38 cells in the presence of compound **III** at 10 and 20 μM

VII increases from 5 to 10 μM, its cytotoxicity rises sharply. The IC_{50} for compounds **VI** and **VII** were calculated as 5.8 μM and 8.9 μM, respectively, for 96 h.

Table 6.6 shows the cell accumulation in sub-G1 phase after compounds **VI** and **VII** treatments. The cell proliferation inhibition rates of both compounds relative to the control are shown in parentheses. For compound **VI**, the cell accumulation is minimal and does not correlate with the inhibition rate. In contrast, compound **VII** exhibits significant cell accumulation corresponding to its inhibition rate. Therefore, compound **VII** may cause the DNA fragmentation which leads toward the cell death. However, compound **VI** showed the inhibition of cell growth but does not induce

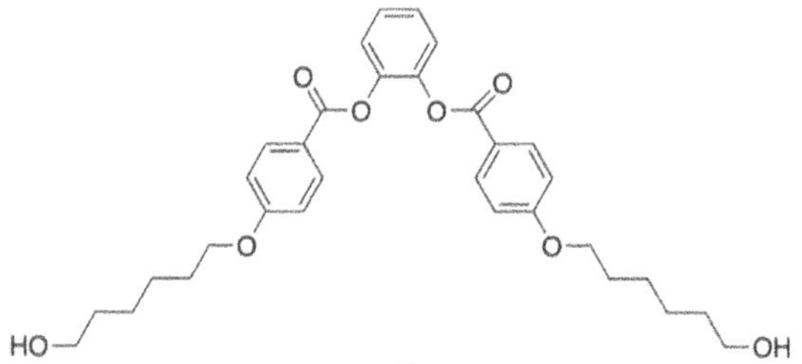

VI

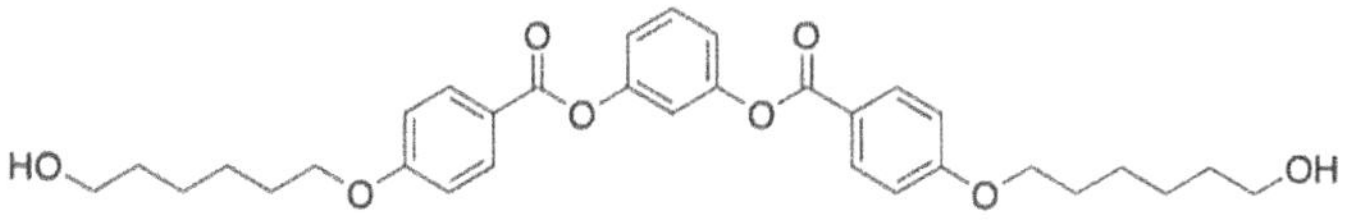

VII

Fig. 6.33 Molecular structures of compounds **VI** and **VII**

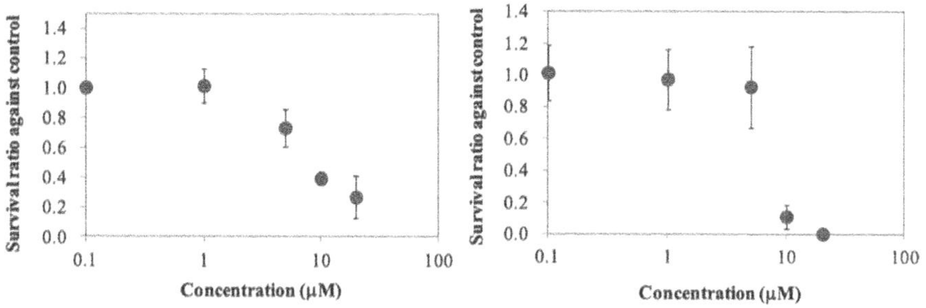

Fig. 6.34 Dose-dependent cytotoxicity of compound **VI** and that of compound **VII** on A549 cells for 96 h. Reproduced from [90] with permission from the Royal Society of Chemistry

their death. Figure 6.35 represents the growth-inhibitory effects of compounds **VI** and **VII** on WI-38 normal cells and various cancer cells. Compound **VII**, like compound **III**, exhibits significant cytotoxicity not only against solid cancer cells (A549 and SW480) but also against WI-38 normal cells. Interestingly, compound **VI** suppressed the cell proliferation of A549 lung cells, whereas it did not affect that of the normal cells at 10 μM as same as compound **I-*n***.

Figure 6.36 shows MOPAC models of compounds **VI** and **VII**. Compound **VII** might form a pseudo-linear dimer to interact with A549 lung cancer cells via the hydroxy and ester groups, inducing apoptosis. As described above, the cyanobiphenyl derivative **II-4** possessing a secondary alcohol shows the cytotoxicity against WI-38 normal fibroblast cells. Therefore, it is believed that these ester groups are involved in the cytotoxicity toward the normal cells. On the other hand, compound **VI** forms a V-shaped structure and it is likely that the ester groups do not contribute to the pharmacological activity.

Table 6.6 Cell accumulation (%) in the sub-G1 phase and the inhibition ratio (%) against control in brackets for compounds **VI** and **VII** for 96 h

	VI		VII	
	10 μM	20 μM	10 μM	20 μM
24 h	0.6 (21)	1.5 (41)	24.1 (52)	46.4 (56)
48 h	0.8 (34)	0.8 (60)	20.7 (81)	63.8 (93)

Reproduced from [90] with permission from the Royal Society of Chemistry

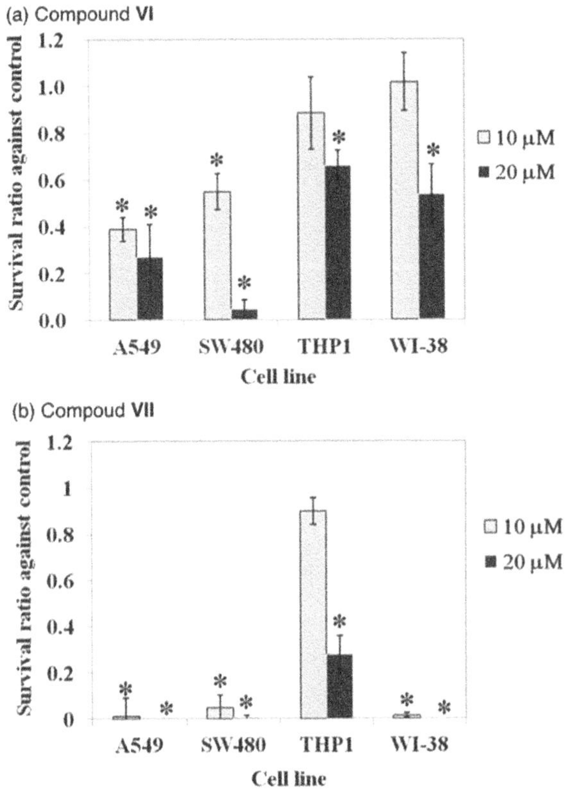

Fig. 6.35 Effects of compounds **VI** and **VII** on cell proliferation of A549 lung cancer cells, SW480 colon cancer cells, THP1 leukemic cells and WI-38 normal fibroblast cells at 10 and 20 μM. Reproduced from [90] with permission from the Royal Society of Chemistry. Those cells cultured in the presence of each compound were harvested (A549: 96 h, SW480: 96 h, THP1: 72 h, WI-38: 168 h). Survival cells were not detected in A549 and WI-38 cells in the presence of compound **VII** at a concentration of 20 μM

6.3.3 Effects of Liquid Crystallinity on the Anti-Proliferative Activity

We discuss the effect of liquid crystallinity on antitumor activity in some benzoate derivatives (Fig. 6.37) [91]. Compounds **III** and **VIII** showed a nematic phase, whereas compound **IX** showed no liquid-crystalline phase. The topological polar area (tPSA) and octanol–water partition coefficient (Log *P*) of a compound are useful parameters for evaluating cell membrane permeability. According to Egan

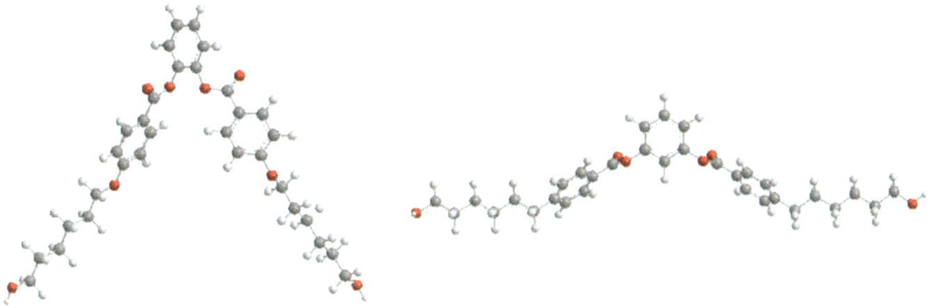

VI VII

Fig. 6.36 MOPAC models of compounds **VI** and **VII**. Their molecular structures are obtained by semi-empirical calculations using MOPAC-6/PM3. Reproduced from [90] with permission from the Royal Society of Chemistry

et al., suitable ranges of those parameters for high penetration ability are $-1 < \text{Log}\ P < 5.9$ and $0 < \text{tPSA} < 132\ \text{Å}^2$ [92]. Table 6.7 portrays the cell proliferation of A549 lung cancer cells, Log *P* and tPSA of compounds **III**, **VIII** and **IX**. All the compounds have suitable Log *P* and and tPSA values for permeability through the cell membrane. Particularly, they have almost the same Log *P*. The anti-proliferative activity of compound **IX**, which does not liquid-crystalline properties, is significantly weaker than that of compounds **III** and **VIII**, which exhibit a nematic phase.

Compound **III** showed a Maltase cross texture at 100 μM in a DMSO-H_2O (1: 99) solution at 37 °C under crossed polarizers (Fig. 6.38). The maltase cross texture is a typical signature of a radical nematic or spherulitic liquid crystal phase. It was also observed for compound **VIII**. Both compounds **III** and **VIII** exhibit a thermotropic nematic phase, therefore, they may form a nematic-like spherical

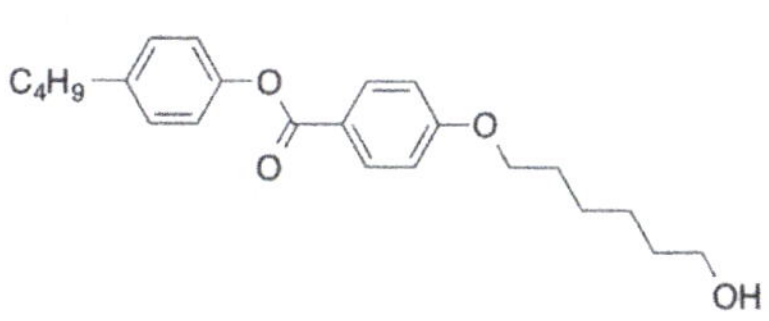

III: Cry 64.5 [N 57.1 (0.74)] Iso

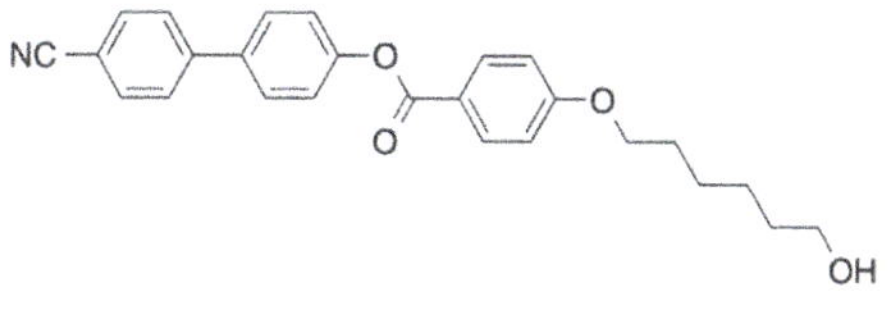

VIII: Cry 120 N 252 (1.0) Iso

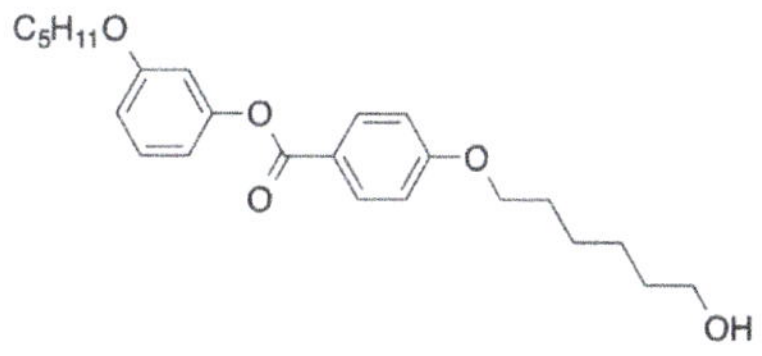

IX: Cry 67.8 Iso

Fig. 6.37 Molecular structures, phase transition temperatures (°C), and enthalpies (kJ mol^{-1}, in parentheses) of transitions for compounds **III**, **VIII** and **IX**. A square bracket denotes a monotropic transition

Table 6.7 Cell proliferation of A549 human lung cancer cells in the presence of **III**, **VIII** or **IX** at a concentration of 10 μM for 96 h [91]. Calculated Log *P* and tPSA values of compounds **III**, **VIII** and **IX**

	III	**VIII**	**IX**
Cell proliferation	0.01	0.20	0.75
Log *P*	6.0	6.0	5.8
tPSA	55.8	79.6	68.0

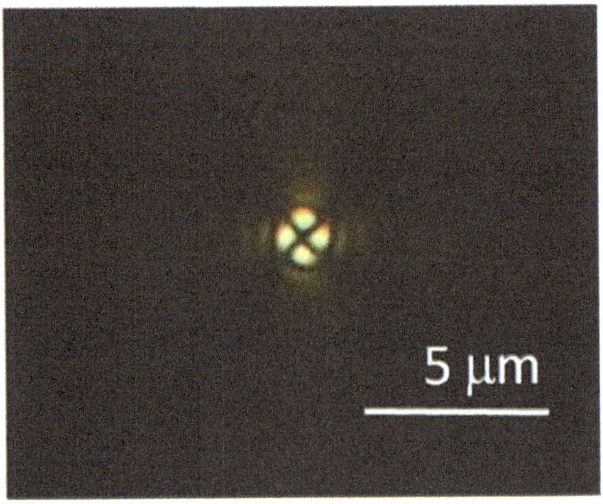

Fig. 6.38 Photomicrograph of compound **III** at 100 μM in DMSO-H_2O (1: 99) solution at 37 °C under crossed polarizers. Reproduced from [90] with permission from the Royal Society of Chemistry

aggregation in a DMSO-H_2O solution. We then used dynamic light scattering to detect the formation of particles by each compound at a concentration of 10 μM in a solution of DMSO and H_2O (1:999). Average diameters of compounds **III**, **VIII** and **IX** were 143,157, and 167 nm, respectively. There is no significant difference in the formation of particles among them. Compound **IX** exhibited neither the maltase cross texture nor any thermotropic liquid-crystalline phase. Thermotropic liquid-crystalline ability to form a spherical molecular aggregation plays an important role in the anti-proliferative activity.

6.3.4 *Selective Anti-Proliferative Effect*

We will now discuss the selectivity exhibited by the liquid-crystalline compounds described above toward cancer cells and normal cells. Table 6.8 summarizes their cytotoxic effects on A549 lung cancer cells and WI-38 normal cells. While compounds **I-4** and **I-7**, which contain primary alcohols, exhibit cytotoxicity only against the cancer cells, compound **II-4**, which contains a secondary alcohol, exhibits

Table 6.8 Cell proliferation of A549 human lung cancer cells for 96 h and that of WI-38 normal fibroblast cells for 168 h in the presence of each mesogenic compound at a concentration of 10 μM. The survival ratio to solvent (DMSO)-treated control culture is shown

Compound	A549 cancer cells	WI-38 normal cells
PPY-G	0.66	0.8
I-7	0.25	1.2
I-4	0.56	1.4
II-4	0.55	0.47
III	<0.01	Not detected
VI	0.40	1.0
VII	<0.1	<0.1

the cytotoxicity against not only the cancer cells but also the normal cells. **PPY-G** possessing both primary and secondary alcohols shows the cytotoxicity against the normal cells. On the other hand, compound **III** forms molecular aggregates of a size suitable for the EPR effect, but exhibits strong cytotoxicity not only against the cancer cells but also against the normal cells. It is believed that the ester group of compound **III** causes strong, non-selective cytotoxicity. Although linear-shaped compound **VII** shows the strong cytotoxicity against the normal cells as same as compound **III**, V-shaped compound **VI** does not. It is thought that this is because compound **VI** adopts the V-shaped conformation, causing its ester groups to remain distant from the interaction site. Directional hydrogen bonding from the primary alcohol group plats a crucial role in selectivity.

6.3.5 Pharmaceutical Liquid Crystals Other Than Anticancer Medicines

Reports on liquid crystal pharmaceuticals are scare. Shibuya and Ohta reported thermotropic liquid-crystalline properties of immunosuppressant 2-amino-2-[2-(4-alkylphenyl)ethyl]propane-1,3-diol hydrochloride (**C_n-C_2-APD**) [93]. Immunosuppressants have important clinical roles in organ transplantation and the treatment of autoimmune diseases. Fujita et al. isolated a unique immunosuppressant, **ISP-I**, from the fermentation broth of *Isaria sinclairii*, and the structure was determined as shown in Fig. 6.39 [94]. They et al. modified the structure of **ISP-I** in order to reduce its toxicity, to improve its physicochemical properties, and to identify the structure essential for immunosuppressive activity. Adachi et al. discovered the novel immunosuppressant **FTY720** (Fig. 6.33) [95]. Kikuchi et al. synthesized various derivatives of **FTY720** and investigated their immunosuppressive effects on rat skin Allograft of compounds [96].

Shibuya and Ohta resynthesized **FTY720** and its derivatives with extended side chains, and investigated their liquid crystallinity. All the compounds **C_n-C_2-APD**

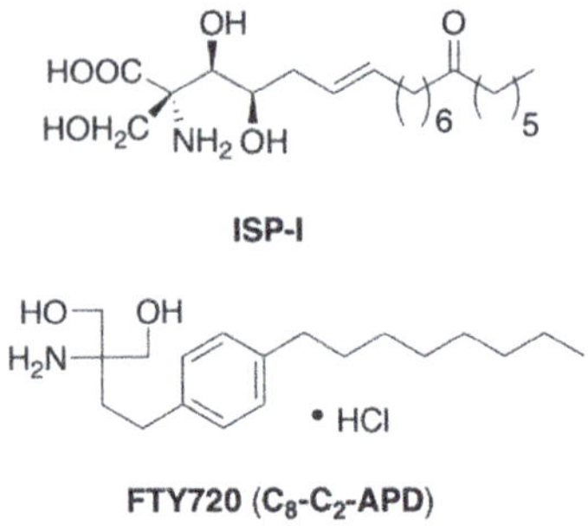

Fig. 6.39 Molecular structures of **ISP-I** and **FTY720**

(n = 8 ~ 14) exhibited an enantiotropic smectic A phase. Their SmA–Iso transition was not observed due to thermal decomposition. XRD measurements reveal that the SmA phases have a bilayer structure. Figure 6.40 shows a phase structure model of the SmA phase of **C_8-C_2-APD**.

According to Kikuchi et al., compounds **C_n-C_2-APD** (n = 8 ~ 12) exhibited the immunosuppressive activity, however, no dependence on side-chain length was observed for the activity. It has been demonstrated that compounds with immunosuppressive effects exhibit liquid-crystalline properties; however, the correlation between liquid-crystallinity and immunosuppressive activity remains unclear [93].

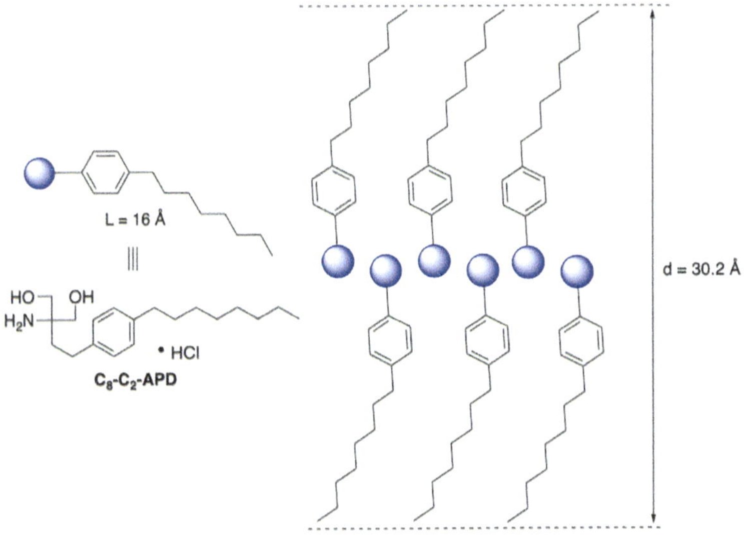

Fig. 6.40 Phase structure model of the SmA phase of **C_8-C_2-APD**

6.4 Concluding Remarks

This chapter describes the anti-proliferative effects of compounds containing typical mesogenic units, such as phenylpyrimidine, cyanobiphenyl, or phenyl benzoate, on the A549 human lung cancer cells. Based on the correlation between the structures and physical properties of these compounds, it has become clear that their thermotropic liquid-crystallinity plays a key role in their anti-proliferative effects. Interestingly, cyanobiphenyl derivatives **I-*n*** bearing a hydroxyl group at the terminal end of the side chain exhibited a marked inhibitory effect on the proliferation of A549 cells without inhibiting WI-38 normal fibroblast cells. Figure 6.41 shows a schematic diagram of the anti-proliferative mechanism exhibited by the liquid-crystalline compound **I-*n*** against A549 cells. Supramolecular assemblies composed of mesogenic molecules with active sites penetrate the cell membranes of the solid tumors, after which the mesogenic molecules rearrange to exert cytotoxic effects. No carrier is needed to transport the medication. We surmise the mechanism as follows. That mesogenic molecules form a spherical molecular aggregate in which the hydroxy units face outward in the hydrophilic environment outside the cell membrane. Within the cell membrane, these molecules form an aggregate in which the cyanobiphenyl units face outward. After penetrating the cell membrane, the molecules reorganize and then interact with CDK2 via hydrogen bonding. As a result, cyclin E can no longer interact with CDK2, inhibiting the synthesis of the enzyme responsible for advancing the cell cycle of cancer cells from G1 to S phase. Although the molecules shown here are low-molecular-weight compounds with simple structures, their pharmacological activity is manifested when they aggregate. Not only the membrane permeability but also the cytotoxicity against solid cancer cells can be easily adjusted by a modification of a mesogenic molecule. Further systematic research is needed to elucidate the antiproliferative effects of liquid-crystalline compounds. Although the IC_{50} values of liquid-crystalline compounds introduced in this chapter are in the μM range, they need to be reduced to the nM range. Furthermore, it is necessary to investigate their cytotoxicity and selectivity in vivo. Although there are still challenges to be addressed, thermotropic liquid-crystalline systems will play a key role in the design of cancer-specific agents.

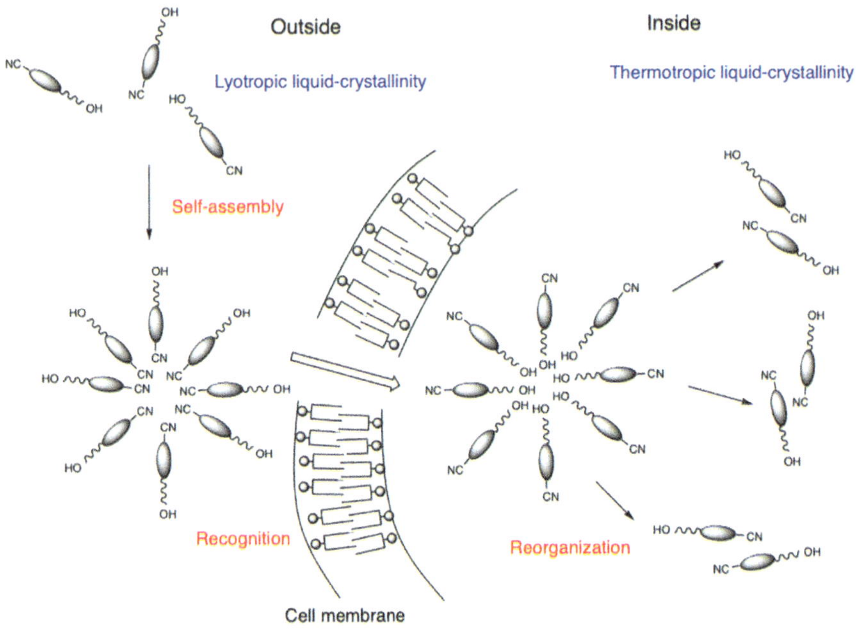

Fig. 6.41 Schematic illustration of the proposed anti-proliferative mechanism of liquid-crystalline compound **I-*n*** against A549 lung cancer cell lines. Reprinted with permission from [58]. Copyright [2022] Springer Nature

References

1. Jewell, S.A.: Liquid crystals in living systems and liquid crystals in the development of life. In: Goodby, J.W., Collings, P.J., Kato, T., Tschierske, C., Gleeson, H.F., Raynes, P. (eds.) Handbook of Liquid Crystals, 2nd edn, vol. 7, pp. 3–27. Wiley-VCH Verlag GmbH, Weinheim, Germany (2014)
2. Strelecka, T.E., Davidson, M.W., Rill, R.L.: Multiple liquid crystal phases of DNS at high concentrations. Nature **331**, 457–460 (1988)
3. Slinger, S.J., Nicolson, G.L.: Fluid mosaic model of structure cell-membranes. Science **175**, 720–631 (1972)
4. Furukawa, R., Kundra, R., Fechheimer, M.: Formation of liquid crystals from actin filaments. Biochemistry **32**, 12346–12352 (1993)
5. Knight, D.P., Vollrath, F.: Liquid crystals and flow elongation in a spider's silk production line. Proc. R. Soc. Lomdon Ser. B: Biol. Sci. **266**, 519–523 (1999)
6. Dunmur, D., Sluckin, T.: Soap, Science, and Flat-Screen Tvs: a History of Liquid Crystals. Oxford University Press, Oxford, UK (2011)
7. Lehmann, O.: Über flussende Kristalle, *Zeitschrift für Physikallische Chemie* **4**, 462–472 (1889)
8. Woltman, S.J., Jay, G.D., Crawford, G.P.: Liquid-crystal materials find a new order in biomedical applications. Nat. Mater. **6**, 929–938 (2007)
9. Cronin, T.: Biosenaors and liquid crystals. In: Goodby, J.W., Collings, P.J., Kato, T., Tschierske, C., Gleeson, H.F., Raynes, P., (eds.) Handbook of Liquid Crystals, 2nd edn, vol. 8, pp. 909–929 Wiley-VCH Verlag GmbH, Weinheim, Germany (2014)
10. Gabizon, A., Catane, R., Uziely, B., Kaufman, B., Safra, T., Cohen, R., Martin, F., Huang, A., Barenholz, Y.: Prolonged circulation time and enhanced accumulation in malignant exudates of

doxorubicin encapsulated in polyethylene-glycol coated liposomes. Cancer Res. **54**, 987–992 (1994)
11. Sahay, G., Kim, J.O., Kabanov, A.V., Bronich, T.K.: The exploitation of differential endocytic pathways in normal and tumor cells in the selective targeting of nanoparticulate chemotherapeutic agents. Biomaterials **31**, 923–933 (2010)
12. Stevenson, C.L., Bennett, D.B., Lechuga-Ballesteros, D.: Pharmaceutical liquid crystals: The relevance of partially ordered systems. J. Pharm. Sci. **94**, 1861–1880 (2005)
13. Zhang, Z., Yang, X., Zhao, Y., Ye, F., Shang, L.: Liquid crystal materials for biomedical applications. Adv. Mater. **35**, 2300220 (2023)
14. Blanco-Fernandez, G., Blanco-Fernandes, B., Fernandez-Ferreiro, A., Otero-Espinar, F.J.: Lipids lyotropic liquid crystals: insights on biomedical applications. Adv. Colloid Interface Sci. **313**, 102867 (2023)
15. Gupta, V.K., Skaife, J.J., Dubrovsky, T.B., Abbott, N.L.: Optical amplification of ligand-receptor binding using liquid crystals. Science **279**, 2077–2080 (1998)
16. Brake, J.M., Daschner, M.K., Luk, Y.-Y., Abbott, N.L.: Biomolecular interactions at phospholipid-decorated surfaces of liquid crystals. Science **302**, 2094–2097 (2003)
17. Brake, J.M., Abbott, N.L.: An experimental system for imaging the reversible adsorption of amphiphiles at aqueous-liquid crystal interfaces. Langmuir **18**, 6101–6109 (2002)
18. Brake, J.M., Daschner, M.K., Abbott, N.L.: Formation of characterization of phospholipid monolayers spontaneously assembled at interfaces between aqueous phases and thermotropic liquid crystals. Langmuir **21**, 2218–2228 (2005)
19. Watanabe, G., Eimura, H., Abbott, N.L., Kato, T.: Biomolecular binding at aqueous interfaces of Langmuir monolayers of bioconjugated amphiphilic mesogenic molecules: a molecular dynamics study. Langmuir **36**, 12281–12287 (2020)
20. Uchida, J., Niwa, A., Hasome, M., Makiura, R., Abobott, N.L., Kato, T.: Self-assembly of peptide-conjugated forklike mesogens at aqueous/liquid crystalline interfaces: Molecular design for ordering transition induced by specific binding of biomolecules. ACS Appl. Mater. Interfaces **15**, 36657–36666 (2023)
21. Wang, Z., Xu, T., Noel, A., Chen, Y.-C., Liu, T.: Applications of liquid crystals in biosensing. Soft Matter **17**, 4674–4702 (2021)
22. Qu, R., Li, G.: Overview of liquid crystal biosensors: from basic theory to advanced applications. Biosensors **12**, 205 (2022)
23. Wang, H., Xu, T., Fu, Y., Wang, Z., Lesson, M.S., Jiang, J., Liu, T.: Liquid crystal biosensors: Principles, structure and applications. Biosensors **12**, 639 (2022)
24. Pani, I., Sil, S., Pal, K.: Liquid crystal biosensors: a new therapeutic window to point-of-care diagnostics. Langmuir **39**, 908–917 (2023). References cited therein
25. Scholler, N., Urban, N.: CA125 in ovarian cancer. Biomark. Med. **1**, 513–523 (2007)
26. Sun, S.-H., Lee, M.-J., Lee, Y.-H., Lee, W., Song, X., Chen, C.-Y.: Immunoassays for the cancer biomarker CA125 based on a large-birefringence nematic liquid-crystal mixture. Biomed. Opt. Express **6**, 245–256 (2015)
27. Su, H.-W., Lee, Y.-H., Lee, M.-J., Hsu, Y.-C., Lee, W.: Label-free immunodetection of the cancer biomarker CA125 using high-Δn liquid crystals. J. Biomed. Opt. **19**, 077006 (2014)
28. Qi, L., Liu, S., Jiang, Y., Lin, J.-M., Yu, L., Hu, Q.: Simultaneous detection of multiple tumor markets in blood by functional liquid crystal sensors assisted with target-induced dissociation of aptamer. Anal. Chem. **92**, 3867–3873 (2020)
29. Kim, H.J., Rim, J.J., C.-H.: Liquid-crystal-based immunosensor for diagnosis of tuberculosis in clinical specimens. ACS Appl. Mater. Interfaces **9**, 21209–21215 (2017)
30. Guo, C., Wang, J., Cao, F., Lee, R.J., Zhai, G.: Lyotropic liquid crystal systems in drug delivery. Drug Discov. Today **15**, 1032–1040 (2010)
31. Chen, Y., Ma, P., Gui, S.: Cubic and hexagonal liquid crystals as drug delivery systems. Biomed. Res. Int. 815981 (2014)
32. Kim, D.-H., Jahn, A., Cho, S.-J., Kim, J.S., Ki, M.-H., Kim, D.-D.: Lyotropic liquid crystal systems in drug delivery: a review. J. Pharm. Inv. **45**, 1–11 (2015)

33. Chountoulesi, M., Pispas, S., Tseti, I., Demetzos, C.: Lyotropic liquid crystalline nanostructures as drug delivery systems and vaccine platforms. Pharmaceuticals **154**, 429 (2022)
34. Nesterkina, M., Kravchenko, I., Hirsch, A.K.H., Lehr, C.-M.: Thermotropic liquid crystals in drug delivery: a versatile carrier for controlled release. Eur. J. Phar. Biopharm. **200**, 114343 (2024)
35. Kiss, A.L., Botos, E.: Endocytosis via caveolae: alternative pathway with distinct cellular compartments to avoid lysosomal degradation? J. Cell. Mol. Med. **13**, 1228–1237 (2009)
36. Palade, G.E.: Fine structure of blood capillaries. J. Appl. Phys. **24**, 1424–1432 (1953)
37. Hyde, S.T.: Swelling and structure. Analysis of the topology and geometry of lamellar and sponge lyotropic mesophases. Langmuir **13**, 842–851 (1997)
38. Hyde, S.T.: Identification of lyotropic liquid crystalline mesophases. In: Holmberg, K. (ed.) Handbook of Applied Surface and Colloid Chemistry, pp. 299–332. Wiley (2011)
39. Larsson, K.: Cubic lipid-water phases: structures and biomembrane aspects. J. Phys. Chem. **93**, 7304–7314 (1989)
40. Spicer, P.T.: Progress in liquid crystalline dispersions: cubosomes. Curr. Opin. Colloid Interface Sci. **10**, 274–279 (2005)
41. Sivadasan, D., Sultan, M.H., Alqahtani, S.S., Javed, S.: Cubosomes in frug delivery-a comprehensive review on its structural compounds, preparation techniques and therapeutic applications. Biomedicines **11**, 1114 (2023)
42. Gustafsson, J., Ljusberg-Wahren, H., Almgren, M., Larsson, K.: Submicron particles of reversed lipid phases in water stabilized by a nonionic amphiphilic polymer. Langmuir **13**, 6964–6971 (1997)
43. Rizwan, S.B., Boyd, B.J.: Cubosomes: structure, preparation and use as an antigen delivery system. In: Foged, C., Rades, T., Perrie, Y., Hook, S. (eds.) Subunit Vaccine Delivery, pp. 125–140. Springer, New York, NY (2015)
44. Boyd, B.J., Whittaker, D.V., Khoo, S.-M., Davey, G.: Lyotropic liquid crystalline phases formed from glycerate surfactants as sustained release drug delivery systems. Int. J. Pharm. **309**, 218–226 (2006)
45. Fong, W.-K., Hanley, T., Boyd, B.J.: Stimuli responsive liquid crystals provide 'on-demand' drug delivery in vitro and in vivo. J. Control. Release **135**, 218–226 (2009)
46. Negrini, R., Mezzenga, R.: PH-Responsive lyotropic liquid crystals for controlled drug delivery. Langmuir **27**, 5296–5303 (2011)
47. Nesterkina, M., Vashchenko, O., Vashchenko, P., Lisetski, L., Kravchenko, I., Hirsch, A.K.H., Lehr, C.-M.: Thermotropic cholesteric liquid-crystal systems doped with terpenoids as drug delivery systems for skin applications. Eur. J. Phar. Biopharm. **191**, 139–149 (2023)
48. Inoue, Y., Takada, K., Kawamura, A., Miyata, T.: Amphiphilic liquid crystalline polymer micelles that exhibit a phase transition at body temperature. ACS Appl. Mater. Interfaces **14**, 31513–31524 (2022)
49. Kuo, P.L., Hsu, Y.L., Lin, T.-C., et al.: Induction of cell cycle arrest and apoptosis in human non-small cell lung cancer A549 cells by casuarinin from the bark of Terminalia arjuna Linn. Anticancer Drugs **16**, 409–415 (2005)
50. Abe, A., Yamane, M., Tomoda, A.: Prevention of growth of human lung carcinoma cells and induction of apoptosis by a novel phenoxazinone, 2-amino-4,4alpha-dihydro-4alpha, 7-dimethyl-3H-phenoxazine-3-one. Anticancer Drugs **12**, 377–382 (2001)
51. Skobridis, K., Kinigopoulou, M., Theodorou, V., et al.: Novel Imatinib derivatives with altered specificity between Bcr-Abl and FMS, KIT, and PDGF receptors. ChemMedChem **5**, 130–139 (2010)
52. Cheng, J.-H., Hung, C.-F., Yang, S.-C., et al.: Synthesis and cytotoxic, anti-inflammatory, and anti-oxidant activities of 2′,5′-dialkoxylchalcones as cancer chemopreventive agents. Bioorg. Med. Chem. **16**, 7270–7276 (2008)
53. Goel, A., Prasad, A.K., Parmar, V.S., et al.: 7,8-Dihydroxy-4-methylcoumarin induces apoptosis of human lung adenocarcinoma cells by ROS-independent mitochondrial pathway through partial inhibition of ERK/MAPK signaling. FEBS Lett. **581**, 2447–2454 (2007)

54. Chou, T.-C., Zhang, X., Zhong, Z.-Y., et al.: Therapeutic effect against human xenograft tumors in nude mice by the third-generation microtubule stabilizing epothilones. Proc. Natl. Acad. Sci. U.S.A. **35**, 13157–13162 (2008)
55. Zhang, J.-H., Fan, C.D., Zhao, B.-X., et al.: Synthesis and preliminary biological evaluation of novel pyrazolo[1,5-*a*]pyrazin-4(5*H*)-one derivatives as potential agents against A549 lung cancer cells. Bioorg. Med. Chem. **16**, 10165–10171 (2008)
56. Papirmeister, B., Gross, C.L., Meier, H.L., Petrali, J.P., Johnson, J.B.: Molecular basis of mustard-induced vesication. Fundam. Appl. Toxicol. **5**, S134–S149 (1985)
57. Primeau, A.J., Rendon, A., Hedley, D., et al.: The distribution of the anticancer drug doxorubicin in relation to blood vessels in solid tumors. Clin. Cancer Res. **11**, 8782–8789 (2005)
58. Yoshizawa, A.: Research and development of liquid-crystalline supramolecular assemblies as anticancer drugs. In: Goel, N., Kumar, N. (eds.) Pharmaceutical Applications of Supramolecules, pp. 89–112. Springer, Cham (2022)
59. Zhang, Y., Chan, H.-F., Leong, K.W.: Advanced materials and processing for drug delivery: the past and the future. Adv. Drug Deliv. Rev. **65**, 104–120 (2013)
60. Vargason, A.M., Anselmo, A.C., Mitragotri, S.: The evolution of commercial drug delivery technologies. Nat. Biomed. Eng. **5**, 951–967 (2021)
61. Tang, N., Du, G., Wang, N., et al.: Improving penetration in tumors with nanoassemblies of phospholipids and doxorubicin. J. Natl. Cancer Inst. **99**, 1004–1015 (2007)
62. Ding, J., Shi, F., Xiao, C., et al.: One-step preparation of reduction-responsive poly(ethylene glycol)-poly(amino acid)s nanogels as efficient intracellular drug delivery platforms. Polym. Chem. **2**, 2857–2864 (2011)
63. Wang, J., Wang, Y., Liang, W.: Delivery of drugs to cell membranes by encapsulation in PEG-PE micelles. J. Control. Release **160**, 637–651 (2012)
64. You, J., Zhang, R., Xiong, C., et al.: Effective photothermal chemotherapy using doxorubicin-loaded gold nanospheres that target EphB4 receptors in tumors. Can. Res. **72**, 4777–4786 (2012)
65. Maruyama, Y., Maeda, H.: A new concept for macromolecular therapeutics in cancer chemotherapy: mechanism of tumoritropic accumulation of proteins agent smancs. Can. Res. **46**, 6387–6392 (1986)
66. Maruyama, K.: Intracellular targeting delivery of liposomal drugs to solid tumors based on EPR effects. Adv. Drug Deliv. Rev. **63**, 161–169 (2011)
67. Wu, J.: The enhanced permeability and retention (EPR) effect: the significance of the concept and methods to enhance its application. J. Pers. Med. **11**, 771 (2021)
68. Rautio, J., Meanwell, N.A., Di, L., Hageman, M.J.: The expanding role of prodrugs in contemporary drug design and development. Nat. Rev. Drug Discov. **17**, 559–587 (2018)
69. Boreski, D., Schmid, V.F., Bosquesi, P.L., et al.: Current trends in clinical trials of prodrugs. Pharmaceuticals **18**, 210 (2025)
70. Li, S., Wang, Y., Li, C., et al.: Supramolecular nanomedicine in the intelligent cancer therapy: recent advances and future. Front. Pharmacol. 1490139 (2024)
71. Gao, F., Baddi, S., Qiu, X., Feng, C.: Translating supramolecular design principles to camptothecin for precision and targeted cancer therapy. Acc. Mater. Res. **7**, 368–381 (2026)
72. Mackay, J.A., Chen, M., McDaniel, J.R., et al.: Self-assembling chimeric polypeptide–doxorubicin conjugate nanoparticles that abolish tumors after a single injection. Nat. Mater. **8**, 993–999 (2009)
73. Shen, Y., Jin, E., Zhang, J.-B., et al.: Prodrugs forming high drug loading multifunctional nanocapsules for intracellular cancer drug delivery. J. Am. Chem. Soc. **132**, 4259–4265 (2010)
74. Li, X.-Q., Wen, H.-Y., Dong, H.-Q., et al.: Self-assembling nanomicelles of a novel camptothecin prodrug engineered with a redox-responsive release mechanism. Chem. Comm. **47**, 8647–8649 (2011)
75. Gao, Y., Kuang, Y., Guo, Z.F.: Enzyme-instructed molecular self-assembly confers nanofibers and a supramolecular hydrogel of taxol derivative. J. Am. Chem. Soc. **131**, 13576–13577 (2009)
76. Cheetham, A.G., Zhang, P., Lin, Y., et al.: Supramolecular nanostructures formed by anticancer drug assembly. J. Am. Chem. Soc. **135**, 2907–2910 (2013)

77. Gheybi, H., Adeli, M.: Supramolecular anticancer drug delivery systems based on linear–dendritic copolymers. Poly. Chem. **6**, 2580–2615 (2015)
78. Zheng, Y.-R., Suntharalingam, K., Bruno, P.M., et al.: Mechanistic studies of the anticancer activity of an octahedral hexanuclear Pt(II) cage. Inorg. Chim. Acta **452**, 125–129 (2016)
79. Pöthing, A., Casini, A.: Recent developments of supramolecular metal-based structures for applications in cancer therapy and imaging. Theranostics **9**, 3150–3169 (2019)
80. Yu, G., Yang, Z., Fu, X., et al.: Polyrotaxane-based supramolecular theranostics. Nat. Commun. **9**, 766–768 (2018)
81. Vallamkondu, J., Corgiat, E.B., Buchaiah, G., et al.: Liquid crystals: a novel approach for cancer detection and treatment. Cancers **10**, 462 (2018). https://doi.org/10.3390/cancers10110462
82. Bunjes, H., Rades, T.: Thermotropic liquid crystalline drugs. J. Pharma. Pharmacol. **57**, 807–816 (2005)
83. Yoshizawa, A., Takahashi, Y., Nishizawa, A., et al.: Biological activity of some cyanobiphenyl derivatives. Chem. Lett. **38**, 530–531 (2009)
84. Tschierske, C.: Molecular self-organization of amphotropic liquid crystals. Prog. Polym. Sci. **21**, 775–852 (1996)
85. Goodby, J.W., Görtz, V., Cowling, S.J., et al.: Thermotropic liquid crystalline glycolipids. Chem. Soc. Rev. **36**, 1971–2032 (2007)
86. Takahashi, Y., Hazawa, M., Takahashi, K., et al.: Suppressive effects of liquid crystal compounds on the growth of the A549 human lung cancer cell line. Invest. New Drugs **2**, 659–665 (2011)
87. Yoshizawa, A., Takahashi, Y., Terasawa, R., et al.: Liquid crystallinity and biological activity of a novel amphiphilic compound. Chem. Lett. **38**, 310–311 (2009)
88. Yoshizawa, A.: Nanostructured assemblies of liquid-crystalline supermolecules: from display to medicine. Liq. Cryst. **46**, 1950–1972 (2019)
89. Takahashi, Y., Hazawa, M., Takahashi, K., Sagisaka, M., Kashiwakura, I., Yoshizawa, A.: Supramolecular assembly composed of different mesogenic compounds possessing a ω-hydroxylalkyl unit exhibits suppressive effects on the A549 human lung cancer cell line. Med. Chem. Commun. **2**, 55–59 (2011)
90. Fukushi, Y., Yoshino, H., Ishikawa, J., et al.: Synthesis and anticancer properties of phenyl benzoate derivatives possessing a terminal hydroxyl group. J. Mater. Chem. B **2**, 1335–1343 (2014)
91. Fukushi, Y., Yoshino, H., Ishikawa, J., et al.: Effects of liquid crystallinity on anticancer activity of benzoate derivatives possessing a terminal hydroxyl group. Liq. Cryst. **43**, 1873–1878 (2014)
92. Egan, W.J., Merz, K.M., Baldwin, J.J.: Prediction of drug absorption using multivariate statistics. J. Med. Chem. **43**, 3867–3877 (2000)
93. Shibuya, T., Ohta, K.: Thermotropic liquid crystalline medicine: establishment of liquid crystalline properties and the X-ray phase structure analysis of immunosuppressant 2-amino-2-[2-(4-alkylphenyl)ethyl]propane-1,3-diol hydrochloride. Liq. Cryst. **50**, 1007–1013 (2023)
94. Fujita, T., Inoue, K., et al.: Fungal metabolites. Part 11. A potent immunosuppressive activity found in *Isaria sinclairii* metabolite. J. Antibiot. **47**, 208–215 (1994)
95. Adachi, K., Kohara, T., Nakao, N., Arita, M., Chiba, K., Mishina, T., Sasaki, S., Fujita, T.: Design, synthesis, and structure-activity relationships of 2-sustituted-2-amino-1,3-propanediols: discovery of a novel immunosuppressant, FTY720. Bioorg. Med. Chem. Lett. **5**, 853–856 (1995)
96. Kikuchi, M., Adachi, K., et al.: Synthesis and immunosuppressive activity of 2-substitued 2-aminopropane-1,3-diols and 2-aminoethanols. J. Med. Chem. **43**, 2946–2961 (2000)

The manufacturer's authorised representative in the EU is Springer Nature Customer Service Centre GmbH, Europaplatz 3, 69115 Heidelberg, Germany. If you have any concerns regarding our products, please contact ProductSafety@springernature.com

Printed and bound by CPI Group (UK) Ltd, Croydon, CR0 4YY
26/10/2026
02239437-0002